Proceedings in Life Sciences

The Evolution of Insect Life Cycles

Edited by
Fritz Taylor and Richard Karban

With Contributions by

W.E. Bradshaw, V.K. Brown, H. Dingle,
C. Grüner, C.M. Holzapfel, R. Karban,
L.P. Lounibos, C.E. Machado-Allison, S. Masaki,
D. Neumann, D. Roff, K.P. Sauer, A.M. Shapiro,
J.B. Spalding, H. Spieth, F. Taylor,
E.B. Vinogradova, K.G. Wardhaugh, W. Wipking

With 90 Figures

Springer-Verlag
New York Berlin Heidelberg
London Paris Tokyo

Fritz Taylor
Department of Biology
University of New Mexico
Albuquerque, New Mexico 87131
U.S.A.

Richard Karban
Department of Entomology
University of California, Davis
Davis, California 95616
U.S.A.

On the front cover: An adult periodical cicada. Their emergence is synchronous, and densities are enormous. They occur in any locality only once every 13 or 17 years.

AGR
QL
495
.5
.E96
1986

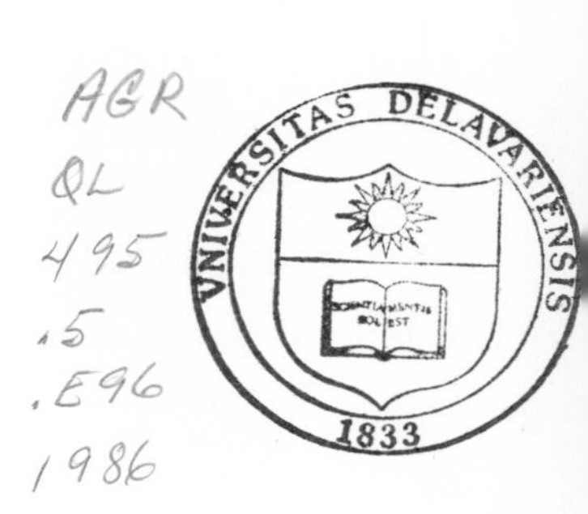

Library of Congress Cataloging in Publication Data
The Evolution of insect life cycles.
Papers from a symposium at the International Congress of Entomology held in Hamburg, Germany, on Aug. 21, 1984.
Includes bibliographical references and index.
1. Insects—Development—Congresses. 2. Insects—Evolution—Congresses. I. Taylor, Fritz. II. Karban, Richard. III. International Congress of Entomology (17th:1984:Hamburg, Germany)
QL495.5.E96 1986 595.7'03 86-13058

© 1986 by Springer-Verlag New York Inc.
All rights reserved. No part of this book may be translated or reproduced in any form without written permission from Springer-Verlag, 175 Fifth Avenue, New York, New York 10010, U.S.A. The use of general descriptive names, trade names, trademarks, etc. in this publication, even if the former are not especially identified, is not to be taken as a sign that such names, as understood by the Trade Marks and Merchandise Marks Act, may accordingly be used freely by anyone.
Permission to photocopy for internal or personal use, or the internal or personal use of specific clients, is granted by Springer-Verlag New York Inc. for libraries and other users registered with the Copyright Clearance Center (CCC), provided that the base fee of $0.00 per copy, thus $0.2 per page is paid directly to CCC, 21 Congress Street, Salem, MA 01970, U.S.A. Special requests should be addressed directly to Springer-Verlag, 175 Fifth Avenue, New York, NY 10010, U.S.A.
96349-9/1986 $0.00 + .20

Media conversion by David Seham Associates, Metuchen, New Jersey.
Printed and bound by Arcata Graphics/Halliday, West Hanover, Massachusetts.
Printed in the United States of America.

9 8 7 6 5 4 3 2 1

ISBN 0-387-96349-9 Springer-Verlag New York Berlin Heidelberg
ISBN 3-540-96349-9 Springer-Verlag Berlin Heidelberg New York

Preface

This book was developed out of a symposium at the XVII International Congress of Entomology held in Hamburg, Germany, on August 21, 1984. This symposium was organized by Drs. William Bradshaw and Hugh Dingle, who subsequently asked us to edit the proceedings. The chapters represent, for the most part, papers that were read in Hamburg but have been expanded and updated.

The goal of this volume is to provide a comprehensive view of current research on insect life cycles, including field and laboratory studies, broad comparisons among species or local populations, and intensive studies of single populations, as well as theoretical research. Of necessity, given the magnitude of research now being carried out on insects, some important research programs are not included, and therein lie the makings of future volumes.

This volume is divided into three parts. The first part, Geographical Patterns in Insect Life Cycles, explores various applications of a comparative method that has been valuable in investigating the potential for variability in life history parameters and the relation of these parameters to important variables in the environment.

The second part addresses the Diversity of Life Cycle Patterns. The hallmark of insect life cycle evolution is the remarkable diversity of adaptations that can be observed at all taxonomic levels. Thus, the chapers in the section demonstrate that it has proven fruitful, in terms of understanding life history traits, to compare closely related taxa in different environments, diverse taxa in the same location as the environment varies through time, and members of the same species in different environments. In the latter case, it is especially noteworthy that a bewildering array of life history options can be observed even within a population.

The chapters in the third section, Mechanisms of Insect Life Cycle Evolution, attempt more to study the processes of evolution as compared to the earlier chapters that tend to document the outcome or fact of evolution. A common thread throughout these papers is the use of demographic analyses to evaluate and compare various mechanisms.

The disparity between the complexity or diversity of life cycles that has frequently been documented and the relative simplicity of the theory presented here, as well as elsewhere in the literature, should provoke a healthy discomfort; for much remains to be done before we can truly say that we understand the evolution of insect life cycles. It should nonetheless be clear from the research presented in this volume that we already know a great deal about insect life cycles, and this fact suggests that the study of insects will continue to contribute in a major way to our understanding of the evolution of life cycles in general.

We would like to acknowledge the cooperation of the contributors to this volume and especially the expertise of Deborah Schippert, who entered the entire text onto a word processor for use by Springer-Verlag.

Fritz Taylor
Richard Karban

Contents

Contributors

WILLIAM E. BRADSHAW Department of Biology, University of Oregon, Eugene, Oregon 97403, U.S.A.

VALERIE K. BROWN Imperial College at Silwood Park, Ascot, Berks SL5 7PY, United Kingdom.

HUGH DINGLE Department of Entomology, University of California, Davis, Davis, California 95616, U.S.A.

CORNELIA GRÜNER Faculty of Biology, University of Bielefeld, 4800 Bielefeld 1, Federal Republic of Germany.

CHRISTINA M. HOLZAPFEL Department of Biology, University of Oregon, Eugene, Oregon 97403, U.S.A.

RICHARD KARBAN Department of Entomology, University of California, Davis, Davis, California 95616, U.S.A.

L.P. LOUNIBOS Florida Medical Entomology Laboratory, Vero Beach, Florida 32962, U.S.A.

C.E. MACHADO-ALLISON Institute of Tropical Zoology, Central University of Venezuela, Caracas 1041, Venezuela.

SINZO MASAKI Laboratory of Entomology, Faculty of Agriculture, Hirosaki University, Hirosaki 036, Japan.

DIETRICH NEUMANN Department of Zoology, University of Köln, D-5000 Köln 41, Federal Republic of Germany.

DEREK ROFF Department of Biology, McGill University, Montreal, Quebec H3A 1B1 Canada.

KLAUS PETER SAUER Faculty of Biology, University of Bielefeld, 4800 Bielefeld 1, Federal Republic of Germany.

ARTHUR M. SHAPIRO Department of Zoology, University of California, Davis, Davis, California 95616, U.S.A.

JOHN B. SPALDING Department of Biology, University of New Mexico, Albuquerque, New Mexico 87131, U.S.A.

HUBERT SPIETH Faculty of Biology, University of Bielefeld, 4800 Bielefeld 1, Federal Republic of Germany.

FRITZ TAYLOR Department of Biology, University of New Mexico, Albuquerque, New Mexico 87131, U.S.A.

E.B. VINOGRADOVA Department of Entomology, Leningrad State University and Zoological Institute, Leningrad, U.S.S.R.

K.G. WARDHAUGH Division of Entomology, CSIRO, Canberra City, A.C.T. 2601, Australia.

WOLFGANG WIPKING Department of Zoology, University of Köln, D-5000 Köln 41, Federal Republic of Germany.

Part I

Geographical Patterns in Insect Life Cycles

Chapter 1

Life Cycle Strategies of an Intertidal Midge Between Subtropic and Arctic Latitudes

DIETRICH NEUMANN

General Remarks

In most environments, organisms are confronted with seasonal fluctuations of physical and biotic factors. The amplitude of these fluctuations depends on geographical latitude, local climate, and local characteristics of habitats.Daily and, at the interface between land and sea, tidal fluctuations are superimposed on the seasonal environmental cycles. Consequently, to persist at a given locality, development, time of reproduction, and population dynamics of an animal species must be adapted specifically to the opportunities and disturbances created by these habitat fluctuations.

The adaptation of the life cycle of a species to the local environment involves a variety of physiological and behavioral features. To comprehend the coordination of these features, that is, to define the life cycle strategy of a species, many different attributes of life histories must be considered. These attributes include:

Growth rate. It should be noted that the rate of growth may differ greatly between species (cf. Table 1.1) independently of final body size and of temperature influences.

Number of generations per year. Among insects, this number ranges from several generations per year to only one per year, or, in extreme cases, one generation every 17 years as in the *Magicicada*.

Seasonality of dormancies. Diapause has served as the main focus for life cycle studies over the past 30 years, and studies concerning its onset, maintenance, and termination are still going on.

Adult lifetime. A huge variability exists ranging from about 1 hr, as in some intertidal chironomids, i.e., *Clunio,* to several months or even more than 1 year as in some carabid beetles.

Coincidence between the time of reproduction and adequate environmental conditions. This synchronization is of great importance for the reproductive success of a species in a fluctuating or unstable environment.

Fecundity. The reproductive effort of a female may reveal different strategies, such as brood care or high egg numbers, in correlation with a high risk of dispersal among young larvae or adults.

Correct physiological and behavioral adaptations. Adaptations such as cold resistance during diapause, reliable timing mechanisms for eclosion and reproduction, sexual pheromones, or orientational behavior for habitat selections are prerequisites for a high survival rate.

Population dynamics. The consequences over several years of all of the previous traits on population density, mortality, and gene flux between subpopulations should be considered.

Polymorphism. Polymorphic properties offer the possibility to examine the gene pool of populations and to discuss short-term evolutionary trends.

Many different methods have been used to examine these topics for life cycle studies. Phenological studies on populations in nature are fundamental for understanding the rate and timing of development and population growth in relation to climatic conditions. However, exact studies in insects are rare, even in most of those species well known from laboratory studies (Tauber and Tauber 1976). One can determine the influence of proximate environmental factors, mainly temperature and photoperiod, in laboratory experiments. In further experimental studies, one can elucidate the physiological processes that underlie specific traits, especially the timing of diapause, eclosion, and reproduction. One can determine gene-controlled components of life cycles, which are essential for successful adaptation to local conditions, by a comparison of traits that vary between geographical races or within polymorphic populations. As a result, one may finally be able to define different life cycle strategies that have been developed through evolution in different insect taxa.

What are the important principles in distinguishing different life cycle strategies in insects? From my point of view, which is that of a physiological ecologist, one of the decisive principles is the kind of developmental control that enables a species to correlate its ontogeny with fluctuating conditions during the course of a year. This correlation involves mainly the coordination of larval and pupal development with optimal periods for reproduction as well as the insertion of dormancies during unfavorable conditions for development. In this chapter, I shall describe the life history of an intertidal insect, the chironomid *Clunio*. This taxon occupies a marine environment, which is extremely unusual for an aquatic insect, with a benthic larva and a flying imago. Despite its unusual habitat, this midge offers a unique possibility to study the adaptation of an insect life cycle to seasonal, lunar-semimonthly, daily, and tidal fluctuations in its environment, and in different geographical populations. All the possible life cycle components of a chironomid insect are evident in this genus through consideration of (1) the temporal programming of its development by environmental time cues, (2) endogenous timing mechanisms, and (3) receptive physiological stages. Thus, one may infer, by the experimental study of *Clunio* populations, the physiological range of life cycle strategies found in the ontogeny of the chironomid family throughout their worldwide distribution.

Growth Rate

In studying the environmental influences on the growth rate of a species, one may consider the general temperature dependency of the net production, the Q^{10} values throughout the ecological range, the lower thermal threshold, and the effect of the temperature sum in fluctuating conditions. These different aspects have been analyzed in an exemplary way in an aquatic insect, the phantom midge (*Chaoborus crystallinus;* Ratte 1979, 1985). Nutrition and energy budgets belong to another complex of ecological problems in growth rate studies. In the context of life cycle strategies, I would like to focus attention on only two aspects: (1) the large differences in the size of eggs and adults between insect species and (2) the striking differences in the rate of development that occur independently of size between these species (Table 1.1).

The marine chironomid *Clunio* has relatively slow development rates in comparison with the common freshwater chironomid, *Chironomus thummi,* which grows three to seven times faster. For comparison the developmental rate of one of the big moths is added. As far as I know, any starting point for an experimental analysis of the endogenous conditions and factors controlling the specific growth rate of an insect is still lacking. Here, it should be mainly noted that the intertidal chironomid, *Clunio,* although living in an ecosystem of high primary production, shows a relatively slow growth rate.

The Clunio Life Cycle Between Subtropic and Arctic Latitudes

Clunio belongs to a group of about 70 known chironomid species that have successfully established themselves in the marine littoral zone. According to the comparative studies of Hashimoto (1962), three ecological types can be

Table1.1. Size of eggs and adults in some laboratory-bred insects, and the mean rate of development at temperatures between 20 and 26°C

	Fresh weight		Length of development (days)	Rate of development (% per day)
	Egg (μg)	Male pupa (mg)		
Manduca sexta, 24°C	1350	3720	21	4.7
Chaoborus crystallinus,[a] 26°C	1.9	0.4	22	4.6
		Adult		
Chironomus thummi,[b] 25°C	2.1	3.1	14	6.9
Clunio tsushimensis, 25°C	1.1	0.3	38–68[c]	2.6–1.4
Clunio marinus, 20°C	1.1	0.3	45–95[c]	2.2–1.0

[a]Values from Ratte 1979.

[b]From Kureck (unpublished).

[c]The duration of development between hatching of the larvae and eclosion of the adults characteristically varies over several weeks in *Clunio* species, even under optimal conditions, so that values of fast and slow development have been presented.

distinguished with regard to swarming behavior and swarming sites of the adults (Figure 1.1). The water surface-gliding type of the genera *Clunio* and *Pontomyia* represents the most successful marine chironomids whose larval habitat ranges far down into the sublittoral zone. Further characteristics of the adults of these genera are a small body size of only 2–3 mm, a striking dimorphism between the sexes with complete wing reduction in the female, and, finally, an extremely *short adult lifetime* of 30 min, to only a few hours so that these chironomids may be termed 1-hr midges. On the basis of laboratory experiments and field observations, some of my students and I have studied the development and reproduction of several populations of the European species *Clunio marinus* living between subtropic and arctic latitudes. This species inhabits rocky shores along the Atlantic and the North Sea coasts from northwest Spain to above the arctic circle in northern Norway (reviews: Neumann 1976, 1981a, b). The *generation time* of laboratory stocks was at least 6 weeks at high seawater temperatures of 20–25°C, but a high variation of up to 14 weeks always existed even among siblings under optimal conditions (cf Table 1.1 Neumann 1966). Similar values may be observed in nature during summer (Oka and Hashimoto 1959, Neumann 1966). In overwintering generations the larval period may exceed

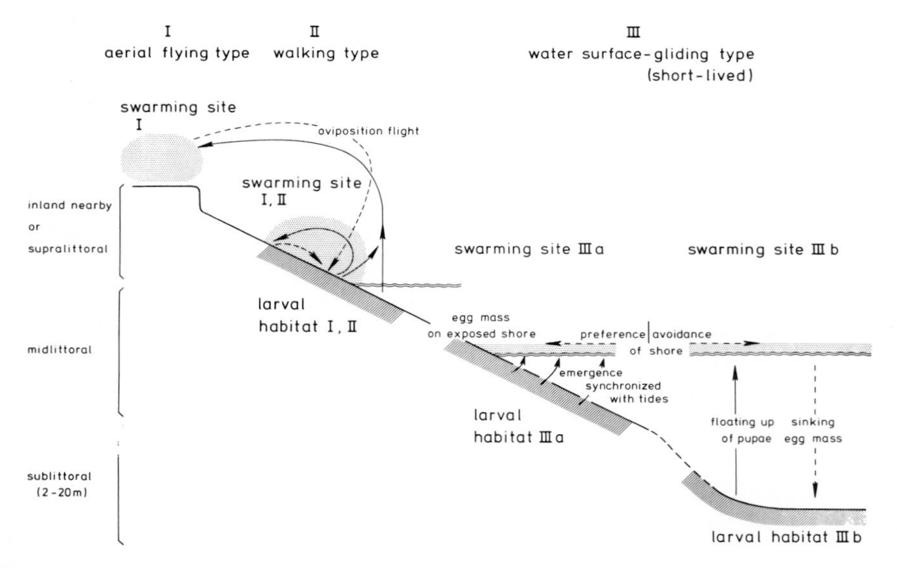

Figure 1.1. The range of larval habitats and swarming sites among intertidal chironomids. On the left is indicated zonation of the intertidal zone. Examples of three different types of locomotory behaviors are presented. Type I: swarming sites in the inland nearby (*Halocladius variabilis*) and swarming sites in sheltered places of a rocky shore (*Thalassosmittia thalassophila*). Type II: swarming and egg deposition on exposed rocky shore (*Thallassomyia frauenfeldi*). Type IIIa: emergence and swarming mostly about spring low water time and egg deposition on exposed substrates (*Clunio marinus, C. tsushimensis*). Type IIIb: emergence and egg deposition on the open sea of sheltered bays (*Clunio balticus*)(Neumann 1981a).

6–7 months, or even 12 months in the far north. However, in all locations the timing of metamorphosis and adult eclosion of intertidal populations is well synchronized with distinct and favorable tidal conditions for egg deposition, as will be explained later.

Figure 1.2 illustrates a geographical trend in the generation time. At the subtropical location (Shimoda), a strong lunar, semimonthly synchronization exists all year round. Emergence is correlated with the spring low water, occurring every 15 days at the same time of day. The generations overlap and therefore the approximate number per year is unknown, but in any case a multivoltine life cycle exists. At higher latitudes with seawater temperatures down to 1-4°C in winter, as at Helgoland, one overwintering and one or sometimes even two summer generations are present. A semilunar rhythm is manifested only during late summer when continuous temperatures between 15 and 20°C occur. In early summer the rhythm is disturbed, as a consequence of strong fluctuating temperatures (falling to about 5°C during high tide, and rising to as high as 15°C during exposure of the intertidal habitat at low tide) (Krüger and Neumann 1983). In the arctic (Tromsö), as well as in the northern Baltic Sea, only one generation per year is produced and is concentrated in one eclosion period.

In laboratory experiments with the Helgoland stock of *C. marinus* (Neumann and Krüger 1985), a *larval diapause* was induced by a combination of short

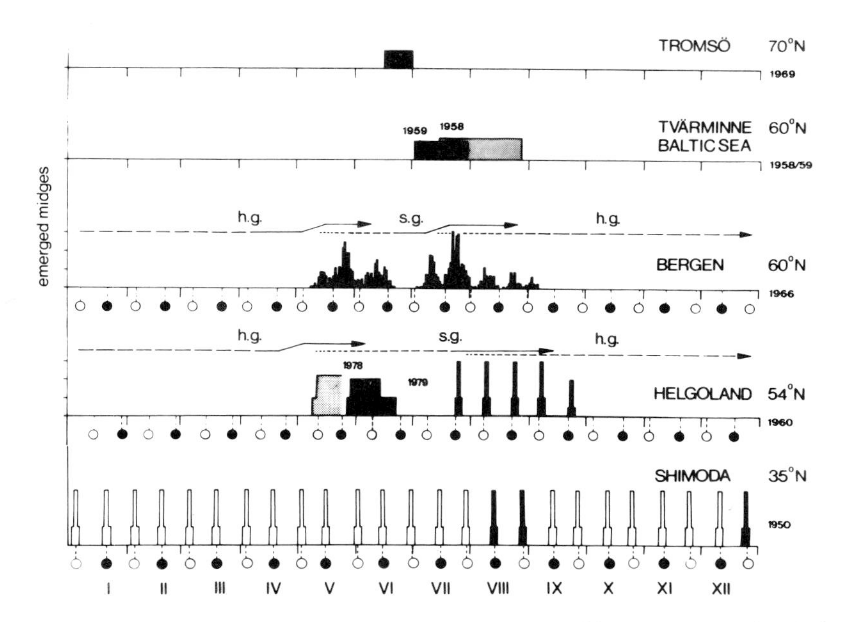

Figure 1.2. Seasonal life history of *Clunio* populations at different geographical latitudes. h. g., Hibernating generation, s. g., summer generation (expanded version of Figure 3 in Neumann 1976). Tromsö, Pflüger 1973; Tvärminne, Palmén and Lindeberg 1959; Bergen, Koskinen 1968; Helgoland, Krüger and Neumann 1983; Shimoda, Oka and Hashimoto 1959. Filled columns, published emergence data; open columns, emergence supposed). (○) Full moon; (●) new moon.

photoperiod and low temperatures (LD 8:16, 7° or 10°C). As in other aquatic chironomids, the individuals are still active during this diapause and they even feed and grow at these temperatures. Metamorphosis, starting with the imaginal disc formation during the last larval instar (see below), is suppressed. This dormancy can be terminated by the introduction of a long photoperiod (LD 16:8,7°C) or higher temperatures (LD 8:16, 18°C). The termination of the diapause at low temperatures (10°C) with a change from short to long photoperiods produces the first eclosion peak after 6 weeks. This type of diapause control may be classified as an oligopause (sensu Müller 1970).

In an intertidal environment, an organism such as *Clunio* is confronted not only with seasonal and daily fluctuations of the environmental conditions, but in addition with strong tidal influences. The short life span of the flying adult must be closely coordinated with the short intervals of advantageous conditions occurring around low tide when the larval habitat is exposed and when a direct egg deposition in this area is possible. Figure 1.3 illustrates the strong temporal correlation between the occurrence of spring tides around full and new moon, the exposure of a sublittoral *Clunio* habitat, and the times of eclosion and reproduction, as first described by the French entomologist Chevrel (1894). This *eclosion pattern* was named *lunar semimonthly* or *semilunar rhythm,* demonstrating an exact temporal control of metamorphosis in relation to the phases of the moon.

If one compares the intertidal range of the *Clunio* habitat between subtropical and arctic latitudes (Figure 1.4), one can observe that appropriate substrates for the settlement of the larval populations are displaced from lower to higher ranges of the midlittoral zone, respectively. In most locations between the subtropics and the northern fringe of the temperate zone (Shimoda and Helgoland), the preferred substrate is a thin, felt-like layer on rocks composed of filiform red algae and sand (Caspers 1951, Oka and Hashimoto 1959, for review see Neumann 1976). Under subtropical summer conditions these substrates survive only within the lower midlittoral and the upper sublittoral regions (unpublished observations at the Pacific coast of Honshu, Japan), i.e., the *Clunio* habitat is restricted to an intertidal region in which the upper parts are exposed only during spring low water occurring a few hours every 14–15 days. Thus, one may assume that the subtropical populations exist under strong selection for a synchronization between the egg deposition by the short-lived imagos and the lunar semimonthly time of exposure of the larval habitat. The same correlation should occur among the populations of the European *Clunio marinus* along the Atlantic coast of Spain and France (42–49°N). At higher European latitudes with temperate summer climates (e.g., Helgoland, Krüger and Neumann 1983) or with abnormal tides (e.g., Southern England, Heimbach 1976) parts of the *Clunio* substrates extend into the neap tide range which is exposed twice a day during each low water. Nevertheless, a clear-cut lunar semimonthly eclosion and reproduction rhythm with a peak around the days of spring tides is evident even at these northern locations (Figure 1.4, Helgoland), guaranteeing at least an advantageous mass concentration of the mating partners. By a comparison of these subtropical and more northern located *Clunio* populations one may

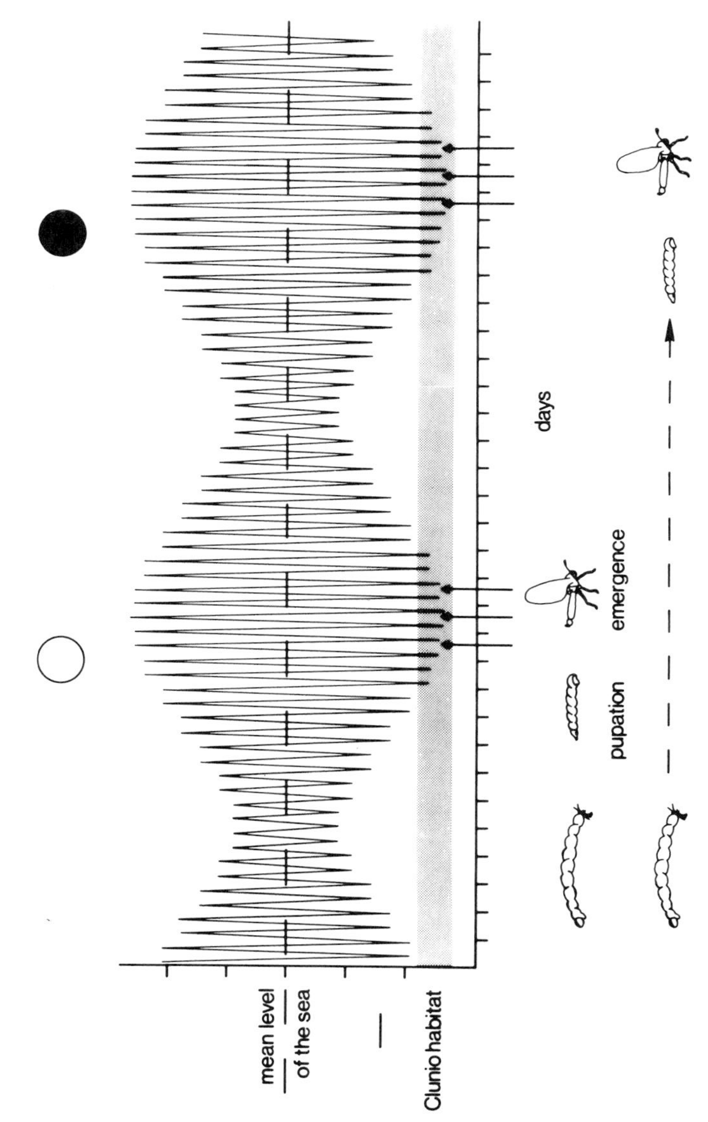

Figure 1.3. Schematic representation of the relationship between tides, range of the habitat of *Clunio marinus*, and the lunar semimonthly emergence of its adults, in this case just before the afternoon low water on days around spring tides (Neumann 1981a). (○), Full moon; (●), new moon.

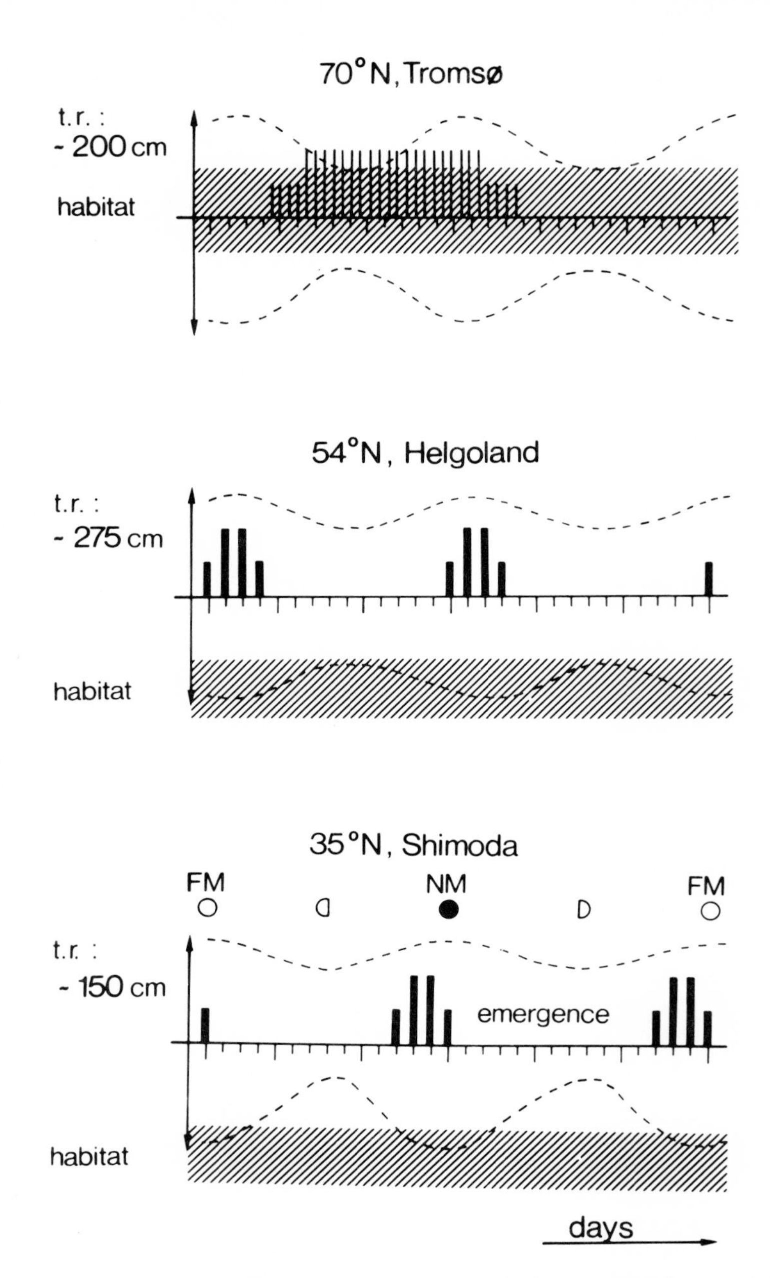

Figure 1.4. The correlation between emergence days of *Clunio* populations and the tidal amplitude at three different locations. Ordinate, local tidal range; dashed lines; height of the succeeding high or low water respectively; hatched area, range of the local larval habitat; abscissa, days; filled columns, days with emergence. t.r., Tidal range; FM, full moon, NM, new moon.

suppose that in this genus the evolution of lunar semimonthly reproduction rhythms had its primary cause and its origin in the more southern latitudes and from there populations moved north, invading northern Europe after the last glacial period.

The arctic population of *Clunio marinus* in northern Norway inhabits an area under the influence of the Gulf Stream and demonstrates an exceptional temporal adaptation (Figure 1.4, Tromso). At this location the larvae settle within the sandy mud flats of the middle and upper midlittoral zone which is warmed up in midsummer during each low tide for several hours. The univoltine life cycle of this population is characterized by a seasonal reproductive period of about 2 weeks and a synchronization of the adult's eclosion and egg deposition with each falling tide (Neumann and Honegger 1969; Pflüger 1973). Thus, the temporal programming of development and eclosion in the worldwide distributed intertidal *Clunio* populations corresponds very well with the tidal and climatic conditions of the local habitat of the larvae.

Endogenous Timing Mechanisms and External Time Cues

The timing of the lunar semimonthly emergence rhythms of *Clunio* has been the main topic of our laboratory studies for many years. By simulating the 24-hr light-dark cycle in combination with a monthly artificial moonlight program, or with artificial tidal conditions, it was possible to evoke, in laboratory cultures, a lunar semimonthly eclosion rhythm independent of the natural synodic month and its phases of the moon (Neumann 1965, 1966, 1968, 1978). Specifically, it was established that the development of each larva is controlled by two endogenous timing mechanisms of the physiological clock type, a circasemilunar oscillator for the onset of pupation and a circadian oscillator for the eclosion of the pharate imago a few days later (3–5 days at 20°C). The combination of these two physiological timing processes during metamorphosis allows the midge *Clunio* to synchronize its eclosion and reproduction with a specific tidal situation that recurs every 14–15 days at a distinct time of day, corresponding to the spring low water recurring at the times of full and new moon. This circasemilunar timing guarantees that pharate imagos within the pupal skin are present at exactly the days of spring tides or related optimal tidal amplitudes. The circadian timing adjusts the eclosion to the appropriate time of day when the optimal tidal phase generally occurs at a specific location.

Evidence for an endogenous timing mechanism was presented as free running circasemilunar eclosion rhythms resulting when synchronized cultures were exposed to conditions without external time cues (i.e., without moonlight or artificial tidal conditions). In recent experiments with a stock of the subtropical *Clunio tsushimensis*, the free-running semilunar rhythm did not disappear for 2–3 months when the offspring from the generation bred in the free-running conditions were emerging. By these persistent rhythms one may conclude that the rhythm of the laboratory population reflects a physiological long-term rhythm, i.e., an endogenous circasemilunar oscillator, which controls the occurrence of pupation in each individual. Corresponding evidence can easily be shown for the circadian mechanism with free-running rhythms of about 24 hr in constant dark or constant light (Neumann 1966), as it has been demonstrated in many insects and other organisms (Bünning 1973, Saunders 1976).

The *Clunio* species and their geographically separated populations possess different sensitivities to the external time cues that entrain the physiological timing mechanism and the developmental process under their control. Moonlight acts as a reliable time cue for lunar semimonthly rhythms only up to latitudes of 49°C (Neumann and Heimbach 1979). This limit was shown by experiments in which cultures were exposed to weak nocturnal dim light of 0.3 lux every

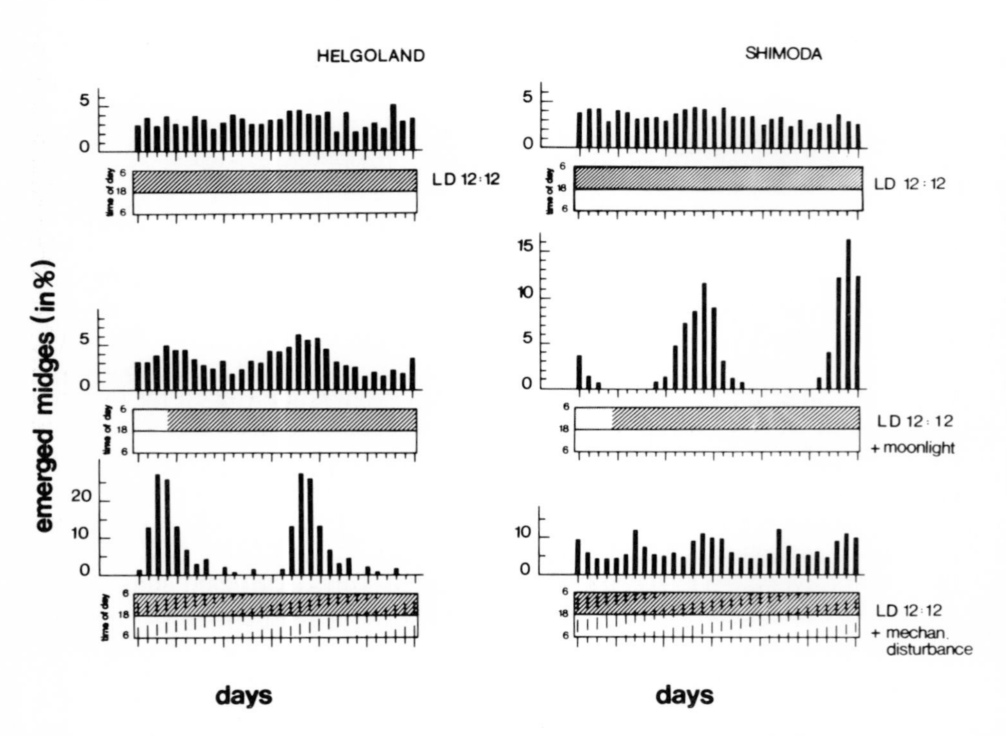

Figure 1.5. The emergence pattern of *Clunio* cultures under controlled laboratory conditions. On the left is shown a population of *Clunio marinus* (origin: Germany, 54°N, northern fringe of temperate latitudes); on the right, a subtropical population of *C. tsushimensis* (origin: Japan, 35°N). Above: control experiments in LD 12:12 day-night regimen, 18°C .Middle: experiments with simulated moonlight (four nights of 0.3 lux illumination every 30 days with LD 12:12). Below: experiments with a combination of LD 12:12 with a 12.4-hr tidal cycle of mechanical water disturbances (syn. turbulences) resulting in equal phase relationships with the 24-hr LD cycle every 15 days (Neumann 1985).

30 days during four succeeding nights (Figure 1.5, right side). Some European stocks from higher latitudes were only weakly entrained by moonlight but the imposition of an artificial tidal cycle of 12.4 hr and a 24-hr light-dark cycle resulted in a strong 15-day emergence rhythm (Figure 1.5, left side). The decisive tidal factor was the rising mechanical disturbance of the water during flood which could be easily simulated by the vibrations (50–200 cps, 6–10 hr per 12.4-hr cycle) of a synchronous electrical motor (Neumann 1968, 1978). Moonlight would be without doubt an unreliable time cue at higher latitudes during the summer season because the height of full moon is relatively low during these months and the moonlight additionally competes with the long duration of dusk and dawn as well as with the illuminated northern horizon. At sheltered seashores with low tidal range and few breaking waves during flood, e.g., within the fjords of western Norway, tidal temperature cycles act as reliable time cues whereas tidal turbulence has only a weak influence (Neumann and Heimbach 1984).

In summary, the length of the *Clunio* life cycle is controlled by larval diapause during winter, as observed in northern latitudes and, above all, by two physiological timing mechanisms, which control the eclosion time of the short-lived imagos to coincide with advantageous conditions in an extreme environment of high temporal complexity, as in the intertidal zone. Comparing populations between subtropical and temperate latitudes, one observes that there is successful adaptation by means of life cycle control mechanisms responding to a variety of reliable environmental factors. Photoperiod and temperature control diapause; moonlight or tidal factors in combination with the 24-hr light-dark cycle determine pupation; and, the light-dark cycle or even tidal temperature rises (arctic population, Figure 1.4 above) set the time of eclosion.

Coupling Between Physiological Timing Mechanisms and Developmental Events

For a more profound understanding of the temporal programming of the *Clunio* life cycle, one should not only examine the properties of the physiological clocks and their sensitivities to environmental time cues, but also should especially consider the coupling processes between the clock and the time of pupation and eclosion. Figure 1.6 illustrates, on the left side, the physiological components of a multioscillator timing mechanism as it has been established by a long series of experiments with the Helgoland stock of *Clunio marinus*. On the right side of the graph, the coupling between the timing mechanisms and the developmental events under temporal control are schematized together with a preliminary and hypothetical version of the hormonal control involved.

A more detailed analysis of the semilunar periodic metamorphosis showed that not only pupation but also a distinct step of the foregoing imaginal disc formation is under temporal control of the circasemilunar oscillator. This imaginal differentiation continues during the whole last larval instar (i.e., the fourth

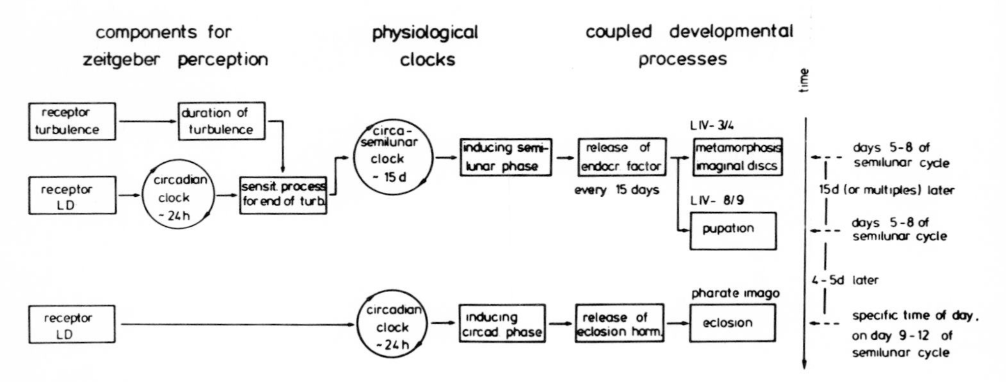

Figure 1.6. Schematic representation of the physiological components of the combined timing mechanisms controlling three physiological switches in the metamorphosis of *Clunio marinus* (type of Helgoland population, with lunar semimonthly pupation and daily emergence 4–5 days later).

in all chironomids). Wülker and Götz (1968) distinguished nine conspicuous stages of imaginal disc differentiation between the early fourth instar and prepupa in *Chironomus thummi*. This description was adopted to the imaginal disc formation in *Clunio,* as illustrated in Figure 1.7. An examination of the overwintering larvae, as well as of the lunar semimonthly synchronized larvae, re-

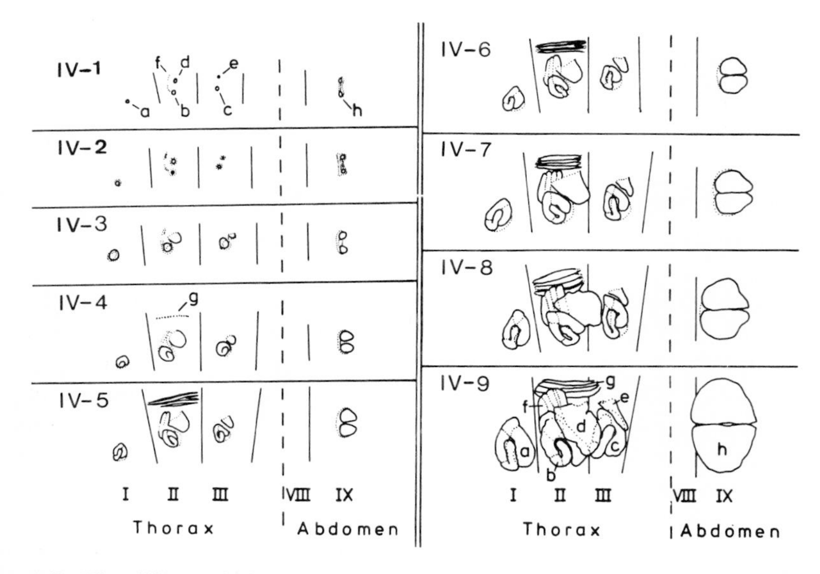

Figure 1.7. The differentiation of the imaginal discs during the last larval instar of the male of *Clunio marinus*. The numbers IV-1–IV-9 indicate nine characteristic stages within the development (Krüger and Neumann 1983).
a–g: Imaginal discs of the thorax (a–c, legs; d, wing; e, haltere; f, vertical; and g, longitudinal muscles of flight). h: Imaginal discs of hypopygium in the last abdominal segment.

vealed that the differentiation at about stage IV-4 represents an important physiological switching point in the temporal control of the *Clunio* life cycle. In diapausing larvae at Helgoland differentiation stops at this point, in nature at the latest in March, before the diapause is terminated by longer photoperiods and rising temperatures in April and May.

In synchronized populations under optimal conditions (LD 12:12 or even longer photoperiods and lunar semimonthly time cues), it is also about stage IV-4 where the imaginal differentiation is retarded while fat body deposition continues. Further imaginal disc differentiation is then induced only every 15 days depending on the semilunar zeitgeber cycle which is days 5–8 of the cycle at 20°C. Pupation is the second step which is under a strict temporal control in semilunar, periodically synchronized generations and is independent of the duration of fourth instar development. At all temperatures tested ranging between 8 and 24°C, pupation in the Helgoland stock occurred only on days 5–8 of the semilunar zeitgeber cycle (Figure 1.8).

In contrast, the developmental period of the pupa shows a normal Q^{10} temperature dependency with an obvious delay in the days of eclosion in synchronized populations of lower temperatures (Figure 1.8, right side). However, with regard to the summer swarming period in nature (it occurs at Helgoland during seawater temperatures above 15°C), the differences in pupation period and resulting day of eclosion are so small in the range of 15–23°C that here a

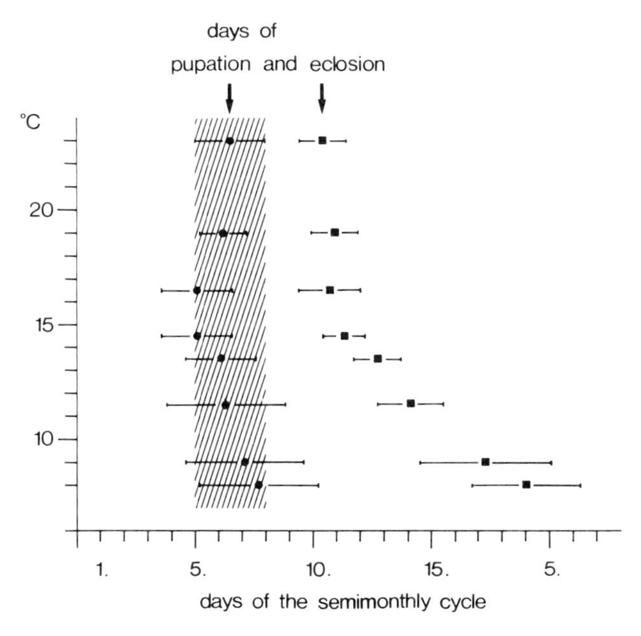

Figure 1.8. The phase relationship between the mean pupation day as well as the mean emergence day and the days of the lunar semimonthly zeitgeber cycle (exactly 15 days in the experiments) at different temperatures. (After Krüger and Neumann 1983.) Shown are mean values with standard deviations.

si;nificant temperature influence on the days of eclosion can be neglected, which is essential for an optimal semilunar periodic timing to distinct days of the semimonthly zeitgeber cycle and days of spring tides, respectively.

From experiments with free running, semilunar rhythms at different temperatures (15–25°C), one may conclude that the circasemilunar oscillator represents, as the circadian oscillator does, a temperature-compensated physiological timing mechanism (Neumann, unpublished results from a stock of*Clunio tsushimensis*). In connection with the previous experiments (Figure 1.8), it may be deduced that the timing of the semilunar periodic pupation is temperature-independent by a coupling with the temperature-compensated circasemilunar timing mechanism. It is worthy of emphasis that the initiation of imaginal disc formation at the switching point of stage IV-4 as well as pupation occurred at the same time within the semilunar zeitgeber cycle.

In summary, it may be concluded that in the chironomid *Clunio* at least three developmental switch points are involved in the temporal programming of its life cycle: imaginal disc formation at about stage IV-4, the occurrence of pupation, and the act of eclosion. The first step may be under seasonal (diapause) as well as under semilunar control, the second step (pupation) under the latter only, and the third under circadian control (Figure 1.6).

The simplest and most obvious hypothesis on the physiological coupling-between the circasemilunar oscillator and the reinitiation of imaginal disc differentiation or pupation might be that one and the same neuroendocrine hormone system triggers both. On the basis of observations in other insects (see Saunders 1976, for a review), prothoracicotropic hormone and ecdysone may provide the connecting links between the circasemilunar oscillator, probably located in the brain, and overt semilunar periodic events during development. To pursue this possibility, K. D. Spindler at the University of Düsseldorf and I have started making measurements of ecdysone in fourth instars, which, at 0.3mg, require a radioimmuno assay. Again by comparison with other insects (Truman et al. 1981), the last switch point, the circadian timing of eclosion, is probably mediated by the release of eclosion hormone from the brain.

Finally, it should be emphasized that the distinct phase relationship between the temporal occurrence of pupation or eclosion and the corresponding environmental cycles (semimonthly cycle of spring and neap tides, or 24-hr LD) are important and well-adapted features of the oscillatory timing mechanisms of each *Clunio* population. These phase relationships may differ between adjacent populations of a coast, in the case of the circadian eclosion time, in relation to the temporal displacements of the tides along a coastline. For instance, the time of day of spring low water differs along the French Atlantic coast by about 5 hr between the Normandy and the Basque coast. The daily eclosion times of the local populations differ correspondingly, even in the laboratory under a 24-hr light-dark regimen, which is the proximate factor for controlling the circadian timing. These different phase relationships of circadian eclosion are gene-controlled properties of the timing mechanism as shown by cross-breeding experiments (Neumann 1966). According to detailed experimental studies in *Drosophila pseudoobscura,* such phase-controlling properties may be involved inthe coupling processes between the circadian oscillator and the overt time of eclosion (Pittendrigh 1967, 1981).

Final Remarks

The aim of this chapter was to demonstrate the range of developmental control in the life cycle of a marine chironomid that is characterized by a slow growth rate, by a potentially multivoltine life cycle, by a larval diapause during the last instar; by an exact timing of metamorphosis at three physiological switching points in specific correlation to seasonal, tidal, and daily conditions of the local environment; and by an extremely short adult life with reproduction immediately after eclosion. The temporal control of the three succeeding developmental events (stage IV-4 of imaginal disc formation, pupation, eclosion) by oscillatory timing mechanisms offers the physiological prerequisites for the selection of advantageous phase relationships between the timing of reproduction (in relation to day and hour) and the short occurrence of favourable conditions at each appropriate intertidal location. During the geographical dispersion of the genus from subtropic to arctic latitudes, various species with geographical and physiological races have evolved, differing mainly with regard to the modus of diapause at higher latitudes, the sensitivity to reliable time cues for the physiological timing mechanisms (physiological clocks), and the correct phase relationship between time cue and the initiation of the relevant developmental event. In adapting to such an extreme environment as the intertidal zone, the genus *Clunio* demonstrates the principal capacities for the timing of a life cycle that can be realized in the aquatic insect family, Chironomidae. One may suppose that a similar kind of developmental life cycle strategy determining the time between the generations and time of eclosion may also be valid in other aquatic insect families with larval diapause.

Acknowledgments My thanks are due to the Deutsche Forschungsgemeinschaft for grants during many years, to Mrs. M. Winter-Bunnenberg for the careful preparation of the figures; and to Mrs. S. Meyen-Southard, Professor W. Bradshaw, and Professor F. Taylor for correcting the English text.

References

Bünning, E.: The Physiological Clock. Berlin Heidelberg New York: Springer-Verlag, 1973.

Caspers, H.: Rhythmische Erscheinungen in der Fortpflanzung von *Clunio marinus* (Dipt. Chiron.) und das Problem der lunaren Periodizität bei Organismen. Arch. Hydrobiol. Suppl. 18, 415–594 (1951).

Chevrel, R.: Sur un diptère marin du genre *Clunio* Haliday. Arch. Zool. Exp. Gén. Ser. III 2, 583–598 (1894).

Hashimoto, H.: Ecological significance of the sexual dimorphism in marine chironomids. Sci. Rep. Tokyo Kyaiku Daigaku (B) 157, 221–252 (1962).

Heimbach, F.: Semilunare und diurnale Schlüpfrhythmen südenglischer und norwegischer *Clunio*-Populationen (Diptera, Chironomidae). Diss. Univ. Köln, 121 pp. (1976).

Koskinen, R.: Seasonal emergence of *Clunio marinus* Haliday (Dipt., Chironomidae) in Western Norway. Ann. Zool. Fenn. 5, 71–75 (1968).

Krüger, M., Neumann, D.: Die Temperaturabhängigkeit semilunarer und diurnaler Schlüpfrhythmen bei der intertidalen Mücke *Clunio marinus* (Diptera, Chironomidae). Helgoländer Meeresunters. 36, 427–464 (1983).

Müller, H.J.: Formen der Dormanz bei Insekten. Nova Acta Leopoldina 35, 7–27(1970).

Neumann, D.: Photoperiodische Steuerung der 15-tägigen lunaren Metamorphose-Periodik von *Clunio*-Populationen (Diptera: Chironomidae). Z .Naturforsch. 206, 818–819 (1965).

Neumann, D.: Die lunare und tägliche Schlüpfperiodik der Mücke *Clunio*.Steuerung und Abstimmung auf die Gezeitenperiodik. Z. Vergl. Physiol. 53,1–61 (1966).

Neumann, D.: Die Steuerung einer semilunaren Schlüpfrhythmik mit Hilfe eines künstlichen Gezeitenzyklus. Z. Vergl. Physiol. 60, 63–78 (1968).

Neumann, D.: Adaptations of chironomids to intertidal environments. Annu. Rev. Entomol. 21, 387–414 (1976).

Neumann, D.: Entrainment of a semilunar rhythm by simulated tidal cycles of mechanical disturbance. J. Exp. Mar. Biol. Ecol. 35, 73–85 (1978).

Neumann, D.: Synchronization of reproduction in marine insects by tides. In: Advances in Invertebrate Reproduction. Clark Jr., W.H., Adams, T.S .(eds.). New York, Amsterdam, Oxford: Elsevier/North Holland, 1981a, pp. 21–35.

Neumann, D.: Tidal and lunar rhythms. In: Biological Rhythms. Handbook of Behavioral Neurobiology, Vol. 4, Aschoff, J. (ed.). New York and London: Plenum Press, 1981b, pp. 351–380.

Neumann, D.: Photoperiodic influences of the moon on behavioral and developmental performances of organisms. Proceedings 10th International Congress Biometeorology. Internat. J. Biometeorology 29, Suppl. 2, 165–177 (1985).

Neumann, D., Heimbach, F.: Time cues for semilunar reproduction in European populations of Clunio marinus. I. The influence of tidal cycles of mechanical disturbance. In: Cyclic phenomena in marine plants and animals, Naylor, E., Hartnoll, R.G. (eds.). Oxford and New York: Pergamon Press, 1979, pp. 423–433.

Neumann, D., Heimbach, F.: Time cues for semilunar reproduction rhythms in European populations of *Clunio marinus*. II. The influence of tidal temperature cycles. Biol. Bull. 166, 509–524 (1984).

Neumann, D., Honegger, H.W.: Adaptation of the intertidal midge *Clunio* to arctic conditions. Oecologia (Berl.) 3, 1–13 (1969).

Neumann, D., Krüger, M.: Combined effects of photoperiod and temperature on the diapause of an intertidal insect. Oecologia (Berl.) 67, 154–156 (1985).

Oka, H., Hashimoto, H.: Lunare Periodizität in der Fortpflanzung einer pazifischen Art von *Clunio* (Diptera, Chironomidae). Biol. Zbl. 78, 545–559 (1959).

Palmen, E., Lindeberg, B.: The marine midge, *Clunio marinus* Haliday (Diptera, Chironomidae) found in brackish water in the northern Baltic. Int. Revue Ges. Hydrobiol. Hydrogr. 44, 384–394 (1959).

Pflüger, W.: Die Sanduhrsteuerung der gezeitensynchronen Schlüpfrhythmik der Mücke *Clunio marinus* im arktischen Mittsommer. Oecologia (Berl.) 11,113–150 (1973).

Pittendrigh, C.S.: Circadian systems. I. The driving oscillation and its assay in *Drosophila pseudoobscura*. Proc. Natl. Acad. Sci. USA 58, 1762–1767(1967).

Pittendrigh, C.S.: Circadian systems: general perspective. In: Biological Rhythms, Handbook of Behavioral Neurobiology, Vol. 4, Aschoff, J. (ed.).
New York and London: Plenum Press, 1981, pp. 57–80.
Ratte, H.T.: Tagesperiodische Vertikalwanderung in thermisch geschichteten Gewässern: Einfluß von Temperatur- und Photoperiode-Zyklen auf *Chaoborus crystallinus* de Geer (Diptera: Chaoboridae). Arch. Hydrobiol. Suppl. 57, 1–37 (1979).
Ratte, H.T.: Temperature and insect development. In: Environmental Physiology and Biochemistry of Insects. Hoffmann, K. H. (ed.). Berlin, Heidelberg, New York, Tokyo: Springer-Verlag, 1985, pp. 33–66.
Saunders, D.S.: Insect Clocks. Oxford, New York: Pergamon Press, 1976.
Tauber, M.J., Tauber, C.A.: Insect seasonality, diapause maintenance, termination and postdiapause development. Annu. Rev. Entomol. 21, 81–107 (1976).
Truman, J.W., Taghert, P.H., Copenhaver, P.F., Tublitz, N.J., Schwartz, L.M.: Eclosion hormone may control all ecdyses in insects. Nature 291, 70–71 (1981).
Wülker, W., Götz, P.: Die Verwendung der Imaginalscheiben zur Bestimmung des Entwicklungszustandes von Chironomus-Larven (Dipt.) Z. Morph. Tiere 62, 363–388 (1968).

Chapter 2

Significance of Ovipositor Length in Life Cycle Adaptations of Crickets

SINZO MASAKI

Life cycle evolution is a holistic process (Roff 1981). It involves changes in many, if not most, functions and therefore structures of various body parts. Development time, one of the most crucial life cycle traits, often varies in association with adult size (Masaki 1967, 1973, 1978a,b Schoener and Janzen 1968, Roff 1978, 1981). Adult size in turn brings about not only specific consequences of ecological importance such as mating success, egg production, food utilization, competition, predation, etc. (Pianka 1974 and many others) but also general effects of allometric scaling on various body parts (Peters 1983, Schmidt-Nielsen 1984). The efficiency of an organ to perform its function may depend on its size relative to the whole or other coordinating parts of the body as well as on its absolute size.

The significance of a particular body part in the evolution of insect life cycles has not been discussed seriously. However, there might be various important interactions between the whole and parts, and also between different parts, of the body associated with life cycle evolution as a holistic process. Such interactions would proceed under the influence of various environmental stresses and cause adaptive variations in body shape. In this chapter I will explore how a particular body part, the ovipositor, is involved in the life cycle adaptations of crickets, especially to habitat, seasonal, and latitudinal conditions.

Selection Balance

The cricket's ovipositor is a prominent structure, protruding straight like a needle from the abdominal tip. It is used exclusively to insert eggs into the soil or plant stems. Together with other Ensifera, the crickets have inherited this remarkable organ from a Paleozoic ancestor. It has never been lost in many phyletic lines leading to the various modern groups (Alexander 1968). Laying eggs in the soil must have been a highly efficient means of adaptation for the crickets through their long history of evolution.

The soil protects eggs from heat, cold, drought, and terrestrial predators (Hubbell and Norton 1978), making it an almost ideal substratum in which to deposit eggs. Such protecting effects depend on the depth of the eggs in the soil, and the depth reached by the ovipositor is proportional to its length because the female inserts nearly the entire length of her ovipositor into the soil before releasing each egg. Obviously, a large cricket is able to have a long ovipositor and thereby lay eggs deeper than a small cricket. All other things being equal, a longer ovipositor would confer a larger benefit on the offspring than a short one. If, however, a female had a disproportionately long ovipositor, she could manipulate it only with difficulty, and egg laying would become an energy- and time-consuming task.

In spite of this inference, the cricket's ovipositor seems to be relatively free from allometric constraints compared with other protruding parts of the body such as appendages. There are indeed astonishing differences in its relative length between extreme cases. Burrowing species in general tend to have very short ovipositors. The giant cricket *Brachytrupes portentosus* is one extreme example from Southeast Asia, with an ovipositor of only one-tenth its body length (Sonan 1931). As their name implies, the short-tailed crickets of the genus *Anurogryllus* afford similar examples from North America and the Caribbean region (West and Alexander 1963, Walker 1973). The opposite extreme is represented by Australian species in the genera *Eurepa* and *Myara*. Some members of these genera look quite bizarre owing to their extraordinarily long ovipositors, two to three times as long as the whole body length (Otte and Alexander 1983).

The cricket's ovipositor is therefore an organ of great evolutionary plasticity and highly susceptible to selection. It might be a sensitive indicator of selection balance, leaning towards either the short or long side in response to the relative weights of the benefit gained by it and the cost paid for it (Figure 2.1). On the benefit side are general security and access to a water supply. Cricket eggs must absorb water through the chorion at a certain stage of embryogenesis (Browning 1953, McFarlane et al. 1959, Masaki 1960). They are in most cases highly susceptible to desiccation (Hogan 1967) so that the water supply alone may cause different survival rates of eggs laid at different depths in the soil.

On the other hand, there are several costs the crickets have to pay for their ovipositors. First, an ovipositor gives morphogenetic and metabolic loads. Second, in addition to the difficulty of using it, a very long ovipositor may hamper quick movement and be broken more easily than a short one. Third, the hatchlings from eggs buried deep by a long ovipositor would require a greater effort to emerge than those from eggs laid close to the soil surface. When the benefit is large, a high cost can be paid and natural selection favors a longer ovipositor. When the benefit is small and exceeded by the cost, the selection balance leans towards the opposite direction. In other words, there should be an optimal length of ovipositor that varies with the balance between cost and benefit. This model of selection balance sets the starting point for the following analyses of variations in ovipositor length.

Habitat Adaptations

The selection-balance model is tested first by comparing various species of Japanese ground crickets belonging to the Nemobiinae. They show an interesting example of adaptive radiation in habitat preference. Each species selects its own habitat of characteristic type from the following series: sandy beaches, pebbly river margins, sparsely weedy places, drained grassy situations, tall dense herb communities, forests, marshes, or paddy fields. This divergence is reflected in their cryptic coloration. For example, the sandy beach species *Pteronemobius csikii* is pale brown, sparsely mottled with dark spots, and its camouflage effect is perfect on beach sand, making a sharp contrast in color with the dark brown or black marsh species *P. ohmachii* and *P. nitidus* which are themselves cryptic on the wet dark soil.

This adaptive radiation is associated with interspecific variation in ovipositor length. Looking at this group of crickets, one may at once notice the long ovipositor of the sandy beach species and the short ovipositor of the marsh species although they are similar in body size. If the different species pay a similar cost for a unit length of ovipositor, the selection balance would establish such interspecific variation only when different amounts of benefit are gained in different species-specific habitats.

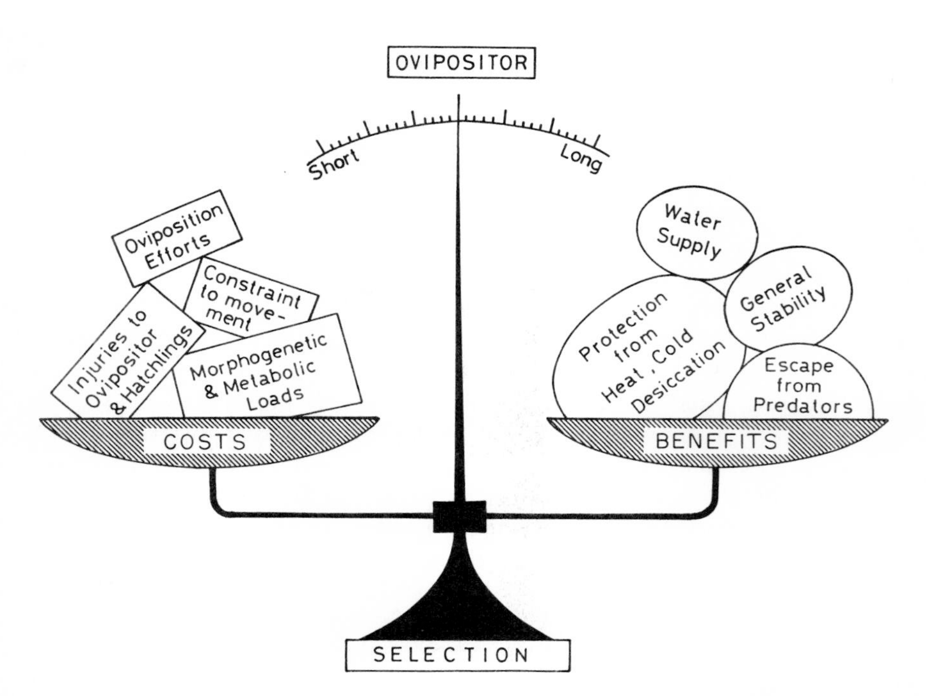

Figure 2.1. Selection balance weighs the benefit gained by lengthening the ovipositor and the cost paid for it to determine the ovipositor length.

The difference in benefit may be the result of the moisture conditions of the habitat soil—one of the most crucial factors for the survival of eggs. In Figure 2.2, the species are roughly arranged according to the tendency of the habitat soil to dry. The species mean of ovipositor length varies between 2.1 and 5.7 mm. This interspecific variation is to some extent a function of body size (Figure 2.3A), for the allometric regression of the ovipositor length on head width is significant ($t = 5.06$; df $= 20$; $p < 0.001$). It accounts for >50% of the total interspecific variance. A substantial proportion of the residual variance may be due to differential natural selection in the species-specific habitats. If a longer ovipositor is selected for, the ovipositor length (Y) should show a positive deviation from the allometric prediction ($\hat{Y}$). If a shorter one is selected for, a negative deviation would be expected.

To evaluate the effect of habitat type on the direction of selection, the deviation ($\Delta Y = Y - \hat{Y}$) of the mean ovipositor length from the allometric equation is plotted against the habitat types from dry (No. 1) to moist (Nos. 10, 11) (Figure 2.3B). Clearly, positive deviations tend to occur in dry habitats, and negative deviations in the moist habitats. In the habitats intermediate in moisture conditions, the deviations are not far from the zero level. According to the selection balance theory, this tendency means that the benefit derived from a longer ovipositor is larger in the harsh environment (loose, sandy soil) than in

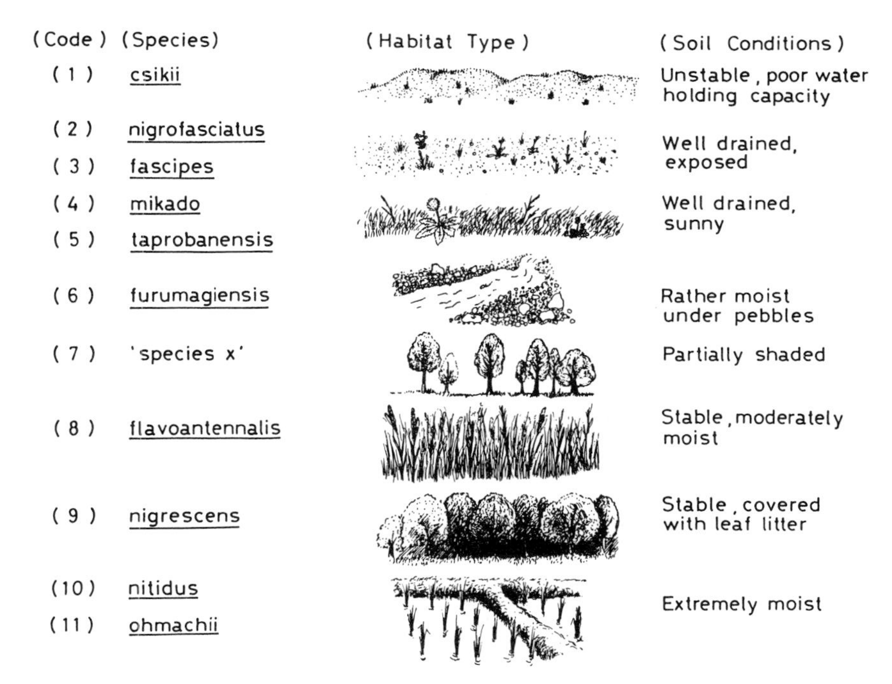

Figure 2.2. Japanese ground crickets in the genus *Pteronemobius* show adaptive radiation with respect to habitat preference. The species are roughly arranged in the order of the tendency of the habitat soil to become dry.

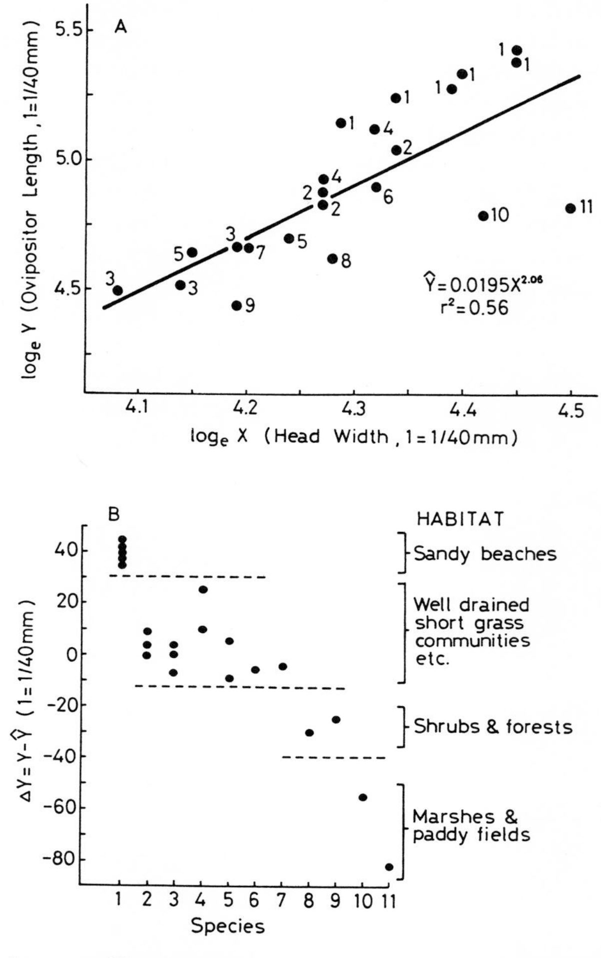

Figure 2.3. (A) Interspecific allometric regression of ovipositor length on head width in Japanese species of *Pteronemobius*. The data points represent means of 14–200 specimens. The species are identified by the code numbers given in Figure 2.2. (B) Deviation of the mean ovipositor length from the allometric equation is associated with the moisture conditions of the habitat soil.

the mild environment (stable, moist soil). This seems to be quite acceptable, because the benefit of having a longer ovipositor is obviously larger when the risk of death by desiccation is greater near the soil surface. Moreover, the easy insertion of the ovipositor into sandy soil would decrease the cost of using a longer ovipositor.

Several further examples illustrating this sort of selection balance are seen in other groups of crickets preferring different habitats. In Japan, *Teleogryllus infernalis* (referred to as *T. yezoemma* in previous papers) typically occurs in sandy beaches and has an ovipositor clearly longer than that of *T. emma* or *T. occipitalis* (referred to as *T. taiwanemma* in previous papers) inhabiting grassy places (Figure 2.4). In North America, three species of the *Allonemobius fasciatus* group show variation in ovipositor length corresponding to the moisture conditions of their segregated habitats (Alexander and Thomas 1959). *A. tinulus* occurs in xeric woodland and has the longest ovipositor among the three species. *A. fasciatus* has the shortest one and lives in marshes or other poorly drained grassy situations. *A. allardi* is found in lawns, pastures, fields, or roadsides, and has an ovipositor of intermediate length. This sort of adaptive response to the soil conditions can occur even in the same species, and both *Gryllus pennsylvanicus* and *G. veletis* have longer ovipositors in sandy areas than elsewhere (Alexander and Bigelow 1960). In spite of allometric constraints, the length of ovipositor thus varies in response to the selection balance in different environments.

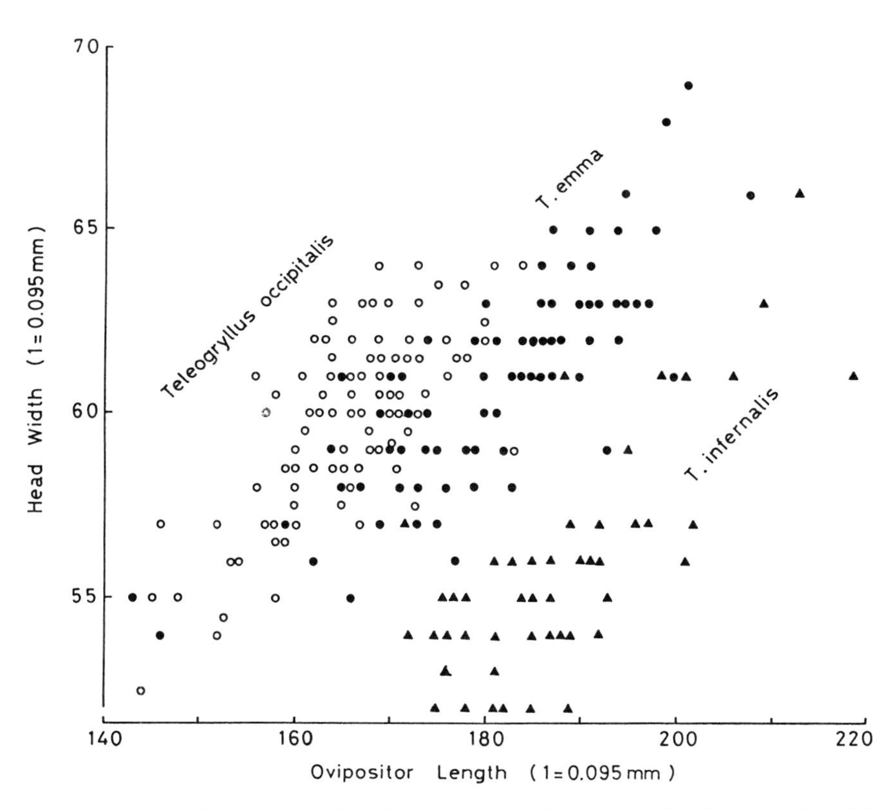

Figure 2.4. Scatter diagram showing the relationship between ovipositor length and head width in three closely related species of *Teleogryllus*. *T. infernalis* typically occurs in sandy beaches and the others in grassy situations. *T. occipitalis* hibernates as a nymph and the others as eggs.

Seasonal Adaptations

Figure 2.4 shows the difference in ovipositor length not only between the inhabitants of sandy beaches and grassland but also, though less conspicuous, between the two grassland species, *T. emma* and *T. occipitalis*. Physical conditions such as the water-holding capacity and the general stability of the soil are therefore not the sole factors selecting the ovipositor length. There should be another factor. Possibly, it is related to life cycle variation, because *T. emma* enters obligatory diapause and hibernates at the egg stage, whereas *T. occipitalis* has no egg diapause and hibernates at the nymphal stage (Masaki and Ohmachi 1967). In the field, the egg stage lasts for about 7–9months in *T. emma* but only for a few weeks in *T. occipitalis*. Obviously, the same degree of protection gives a larger benefit in the long egg stage than in the short because of the greater cumulative amount of time-proportional risks that can be avoided. The selection balance would therefore lean towards the long side.

If this reasoning is correct, there should be many other cases of interspecific variation corresponding to life cycle differences. Owing to differential selection of life cycles in different climatic regions, there are several pairs of closely related species of crickets, one with a stable egg diapause and the other without it (Masaki 1983). Under the name *Velarifictorus micado,* for example, two seasonally isolated species have been intermingled (Masaki 1961). The autumn form overwinters as an egg and matures in the autumn, whereas the summer form overwinters as a nymph and matures in late spring or early summer. Although they are to some extent interfertile in the laboratory, their seasonal cycles are based on genetic differences in their developmental physiology. At 25°C the mean incubation time is 24 days (*n* 2789) in the summer form, compared with 145 days (*n* 414) in the autumn form (M. Watanabe, unpublished master's dissertation). As predicted, the ovipositor is longer in the autumn form than in the summer form (Figure 2.5A). These two forms show different ranges of distribution in the Japanese Islands (Masaki 1983). However, the difference in ovipositor length persists in the zone of overlapping distributions as indicated by specimens collected at Kure (about 34°N). Therefore, the differential selection of ovipositor is not affected by climatic conditions alone.

The lawn ground crickets (*Pteronemobius taprobanensis* and *P. mikado*) and the band-legged ground crickets (*P. fascipes* and *P. nigrofasciatus*) give further support to the above inference (Figure 2.5B and C). Each of these pairs comprises a temperate and a subtropical species (Masaki 1983). The temperate forms hibernate exclusively as eggs, whereas the subtropical forms more frequently hibernate as nymphs. The subtropical band-legged ground cricket *P. fascipes* is able to enter an egg diapause, but this does not seem to be the primary means of hibernation (Masaki 1978a). At the end of February (late winter), 1984, I collected 20 adults and 42 nymphs of this species on the subtropical island of Ishigaki. Both adults and nymphs of the subtropical lawn ground cricket *P. taprobanensis* were also common. There is therefore no clear period of developmental arrest at this subtropical latitude, although the nymphs of both species regulate their development by photoperiodic responses of long-day type

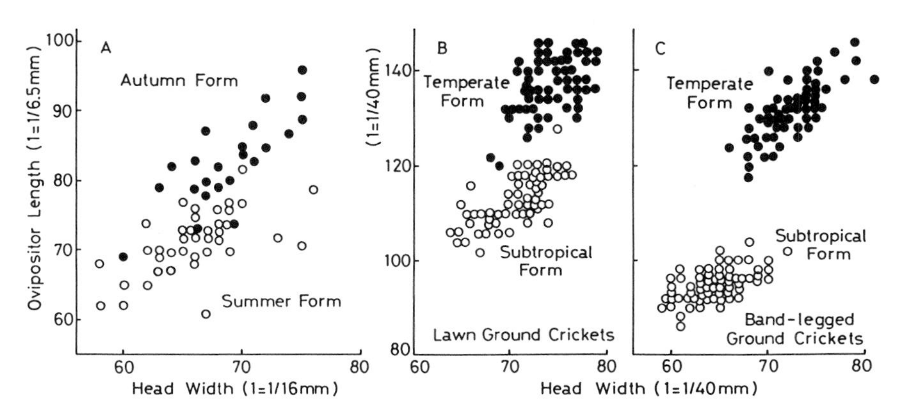

Figure 2.5. Comparisons of the relative ovipositor length between two closely related species of crickets mainly hibernating as eggs (●) and nymphs (○) respectively. (A) The two "seasonal forms" of *Velarifictorus micado*. (B) *Pteronemobius taprobanensis* (○) and *P. mikado* (●). (C) *Pteronemobius fascipes* (○) and *P. nigrofasciatus* (●).

(Masaki 1978a). On the other hand, neither adults nor nymphs of the temperate species *P. nigrofasciatus* and *P. mikado* survive winter. In both species egg diapause is the only possible means of hibernation, being obligatory in the univoltine populations and programmed by short, parental photoperiods in the bivoltine populations (Masaki 1973, 1979a, Kidokoro and Masaki 1978). The shift of life cycle from the subtropical to the temperate type is accompanied by a clear increase in ovipositor length (Masaki 1978a, 1979b, 1983), and the two climatic species in each group can be distinguished by ovipositor length (Figure 2.5B and C).

Similar associations between ovipositor length and egg diapause are also found in other parts of the world. In New Zealand, *Pteronemobius nigrovus* enters an intense diapause at the egg stage, and its ovipositor is longer than that of *P. bigelowi* whose egg diapause is short and unstable (McIntyre 1978). In Australia, *Teleogryllus commodus* with an egg diapause has a longer ovipositor than *T. oceanicus* without egg diapause (Otte and Alexander 1983). In North America, *Gryllus pennsylvanicus* and *G. firmus*, both overwintering as eggs, have longer ovipositors than the other members of the same genus overwintering as nymphs (Alexander 1957). This difference persists between *G. pennsylvanicus* and *G. veletis* throughout the wide range of their overlapping distributions (Alexander and Bigelow 1960).

Thus, the available comparisons of closely related species occurring in similar habitats with different seasonal cycles show a consistent tendency. The ovipositor is longer in the species with a stable egg diapause than in those without egg diapause or with brief diapause. An exception to this rule is a pair of marsh-inhabiting *Pteronemobius* (Figure 2.3). One of them, *P. ohmachii*, enters diapause as an egg (Masaki 1960) and the other, *P. nitidus*, as a nymph (Tanaka 1978), yet they do not show any significant difference in ovipositor length. It seems possible that the benefit in adaptation gained by the ovipositor becomes saturated within a very short range in the moist and stable soil of their habitats.

Latitudinal Adaptations

It is now clear that at least two different categories of factors are involved in the natural selection of ovipositor length: habitat factors and life cycle factors. The absence of egg diapause as well as moist and stable soil conditions select for a shorter ovipositor. The presence of egg diapause as well as dry and unstable soil conditions select for a longer ovipositor. If the association of egg diapause with a longer ovipositor is due to the greater cumulative risks in the longer egg stage, we can predict further adaptive variations to occur in ovipositor length.

Cricket life cycles vary along a latitudinal gradient, even when the same hibernating stage is maintained. If a latitudinal shift from a univoltine to a bivoltine cycle occurs in an egg-overwintering species, a substantial change in the selection pressure on ovipositor length should be expected. In the univoltine population the egg stage is consistently long, lasting several months. In the bivoltine population, the nondiapause generation with an egg stage of only a few weeks alternates with the diapause generation with an egg stage of several months. Therefore, the direction of selection would be reversed from one generation to the next. Furthermore, within the range of constant voltinism, the duration and severity of winter vary from place to place. The selection balance predicts that the ovipositor will be longer in the northern than in the southern population of the same voltinism.

To test these predictions, data are available for three species of crickets, *Pteronemobius nigrofasciatus, P. mikado* and *Teleogryllus emma*. All these species show geographical variation in adult size (Masaki 1967, 1973, 1978a), and the allometric regressions of ovipositor length on head width are all significant (Figure 2.6A, C, and E), accounting for about 60–80% of the interpopulation variation. The ovipositor length is therefore primarily determined by the species-specific allometric factor.

If the selection pressure on ovipositor length varies from place to place, this would be reflected in different deviations of the sample means from the regression equation. By plotting the deviations against the latitude of collecting sites, we can perceive geographical trends, if any, in the selection pressure. The data for *P. mikado* were previously analyzed in detail by a multiple regression technique (Masaki 1979b), but they are reexamined here by this simpler procedure to compare with the other two species.

The two ground cricket species show similar trends. Broadly speaking, the deviations from the allometric regressions show clines ascending northwards in parallel with the increasing duration of winter. From the photoperiodic responses controlling the egg diapause and nymphal development and also from the adult size cline, it has been inferred that the shift from the southern bivoltine to the northern univoltine cycle occurs at about 37°N in *P. nigrofasciatus* and at about 35°N in *P. mikado* (Masaki 1973, 1978a,b). To detect the effect of this

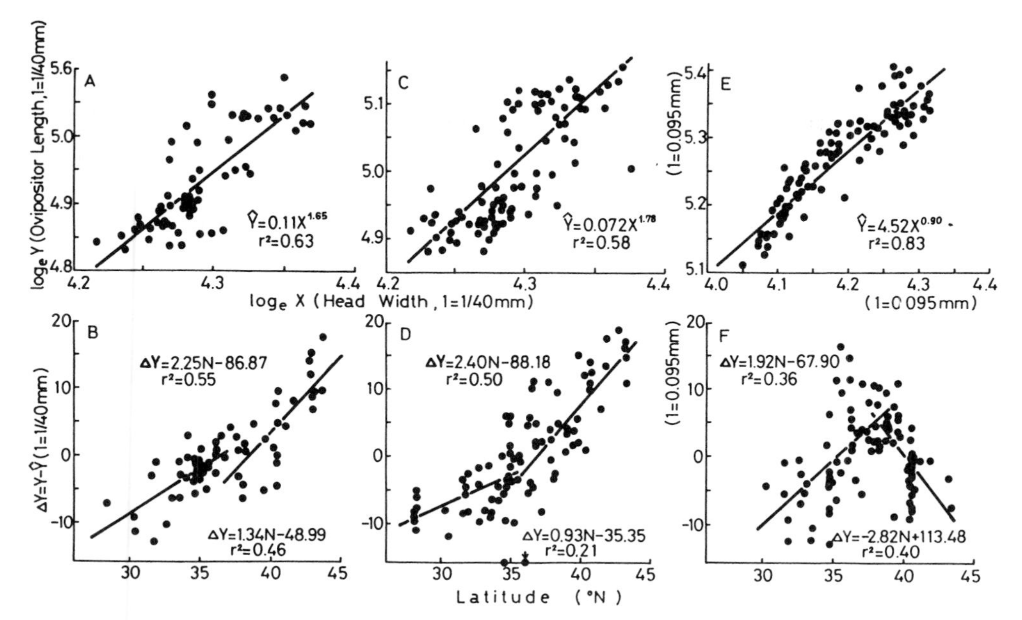

Figure 2.6. Allometric regressions of ovipositor length on head width (top) and the latitudinal trends of the deviation of ovipositor length from the allometric regression line (bottom) in three species of crickets. (A,B) *Pteronemobius nigrofasciatus* (62 local samples, 1067 specimens). (C,D) *Pteronemobius mikado* (84 local samples, 3835 specimens). (E,F) *Teleogryllus emma* (94 local samples, 7532 specimens).

life cycle shift, the regressions of the deviations (ΔY) on latitude were separately computed for the univoltine and bivoltine areas (Figure 2.6B and D). In both species the regression coefficients take a larger value in the univoltine area than in the bivoltine area (one-way analysis of variance, $F = 3.91$; df = 1, 59; $p < 0.05$ for *P. nigrofasciatus*; $F = 9.23$; df = 1, 83, $p < 0.01$ for *P. mikado*).

These results generally agree with the predictions. In the predominantly univoltine area, the selection pressure for a longer ovipositor is directly proportional to the duration of the hibernating period or the egg stage. In the bivoltine area, selection favors a shorter ovipositor for the parents of the nondiapausing summer generation but a longer ovipositor for the parents of the diapausing winter generation. The winter selection would be intensified northwards, but the summer selection would be almost constant everywhere because of the small variation in the duration of the egg stage in summer. The additive effects of selection would be the reduction of the regression coefficient for ΔY on latitude to about one-half of that in the univoltine area (Figure 2.6B and D).

A third example of latitudinal variation is afforded by the univoltine field cricket, *Teleogryllus emma*. This species forms a linear size cline that conforms to the converse of Bergmann's rule (Masaki 1967). Despite the simple life cycle, the deviations (ΔY) of mean ovipositor lengths from the allometric regression reveal an unexpected feature (Figure 2.6F). From the southern limit of distribution, ΔY increases northwards as predicted from the increasing duration of winter or the hibernating egg stage. However, a maximum is reached at about 38°N, and then a sharp drop occurs farther north.

To establish the reality of this reversion in a more objective way, a series of correlation coefficients between ΔY and latitude were computed by successively extending the sample area with 1°N increments southwards from the northern limit (44°N). The highest negative correlation (r- − 0.63) was obtained in the area between 44 and 38°N. As more southern samples were added, the correlation coefficient decreased gradually and reached 0.15 (not significant) for all local samples from 44 to 30°N. When computation was started from the southern limit in a similar way, a maximum positive correlation ($r = 0.60$) was obtained for the area between 30 to 39°N. More northern extensions of the sample area gradually decreased the correlation coefficient, again reflecting the reversion of the geographical trend of ΔY at around 38–39°N.

There may be two mutually exclusive interpretations of this unexpected tendency. It might be an artifact. The exponential function might not precisely describe the allometric relationship between the ovipositor length and head width, so that systematic deviations from the regression line occur. Since body size decreases northwards in a simple linear fashion, such size-related deviations would show a definite latitudinal trend. The other possibility is that the tendency reflects the real status of selection pressure. There are two factors that may cause this reversion. The northeastern part of Honshu and Hokkaido are normally covered with snow during winter. The deep snow cover may reduce the environmental stresses on the hibernating eggs and hence the selection pressure on ovipositor length. The selection balance would shorten the ovipositor when the stress decreases. Another possible factor is the heat available for the eggs in postdiapause development. After the snow cover has melted away, the ground surface absorbs radiant heat during the daytime, and a steep vertical gradient of temperature may be formed in the soil. A deeper placement of eggs might delay hatching, and the sum of heat may become an important limiting factor in the north.

A question naturally arises: Why doesn't a similar reversion of geographical trend occur in the other two species? The big difference in absolute length of ovipositor between *T. emma* and the ground crickets may be a possible answer. The short ovipositor of the ground crickets may not cause any substantial difference in heat availability within its range of variation. Moreover, the small ground crickets mature faster than the large field cricket so that they may be relatively free from this sort of selection. At present I have no evidence to conclude which one of these possibilities is closer to the truth.

Optimal Ovipositor Length: Concluding Remarks

The cost of the ovipositor increases in proportion to its length. The substance invested in the ovipositor is a function of its length. Its surface area also varies with length and may be related to evaporation and metabolic rate (Wigglesworth 1972). It is not easy to evaluate the relative importance of various costs, but

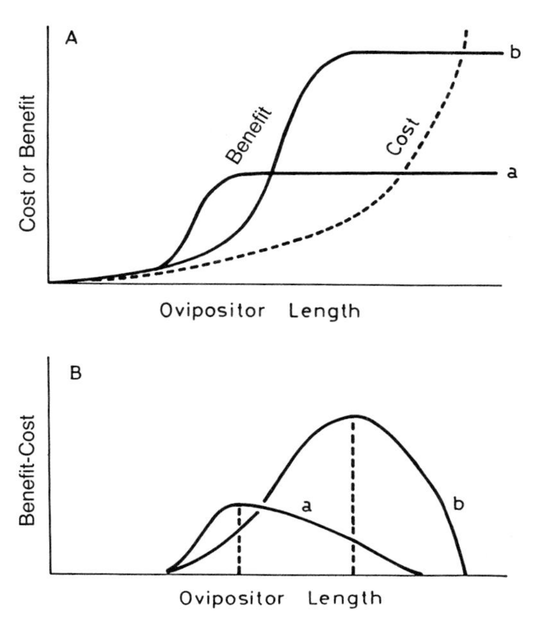

Figure 2.7. Graphic model for the natural selection of the cricket ovipositor. (A) Both the cost and the benefit that can be ascribed to the ovipositor vary as functions of its length. Cost items are mainly endogenous so that the cost curve is not highly variable. Benefit items are related to environmental stresses that vary in both space and time (see Figure 2.1). The protecting effect of laying eggs in the soil will become saturated with a short ovipositor in a mild environment (a), but only with a much longer ovipositor in a harsh environment (b). (B) Therefore, the optimal length (shown by vertical broken line) giving a maximum net benefit (benefit − cost) is smaller in the mild than in the harsh environment.

at least in terms of the invested substance the cost of ovipositor would be an exponential function of its length so far as the width and thickness of ovipositor are positively correlated with the length (Figure 2.7A).

On the other hand, the benefit may be a sigmoid function of ovipositor length. Eggs cannot enjoy any protecting effect unless they are entirely buried in the soil. If they are buried very close to the surface, they may easily be exposed to rain and wind, attacked by predators, or desiccated. The protecting effect therefore increases only slightly until a certain depth is reached, so that the benefit will increase with ovipositor length only at a low rate in the shorter range. However, the benefit increase would become larger as the effective depth for egg protection approaches. As their ovipositor lengthens further, it would reach a depth at which the protecting effect and therefore the benefit become saturated.

The saturation level of the benefit curve would vary with the life cycle and habitat, since the amount of benefit that can be derived is proportional to the size of risk. The larger the risk the larger is the benefit gained by avoiding it.

A benefit is of course an increase in survival rate of eggs by any means of protection. If eggs scattered around on the soil surface suffered no mortality, there would be no benefit gained by an ovipositor of any length. The same would be the case if eggs hatch as soon as they are laid. The saturation level of the benefit curve is therefore higher in a harsh environment than in a mild one, and also for a long egg stage than for a short. On the other hand, a lower degree of protection would be more significant under mild than under harsh conditions. Therefore, a change in saturation level is associated with a horizontal position shift of the benefit curve (Figure 2.7A).

Subtraction of the cost from the benefit gives a net-benefit curve (Figure 2.7B). The outcome of natural selection should be to produce the ovipositor length that maximizes the net benefit. It now seems evident that this optimal length of ovipositor varies with the way of life defined in both time and space, i.e., the life cycle pattern and habitat preference (Figure 2.8). If closely related species prefer habitats of clearly different types, the habitat factors should differentiate their ovipositor lengths. If they share the same habitat, the life cycle difference would be a key factor in differential selection of their ovipositor lengths. Geographical populations of the same species usually adhere to species-specific habitats. Therefore, the variation in length of their ovipositors would reflect their life cycle variations along the latitudinal gradient. As the key limiting factor shifts from one to another at a certain point in the latitudinal gradient, a reversion of the selection balance that determines the ovipositor length may occur. The cricket's ovipositor is thus a highly sensitive indicator of various aspects of its life cycle adaptation.

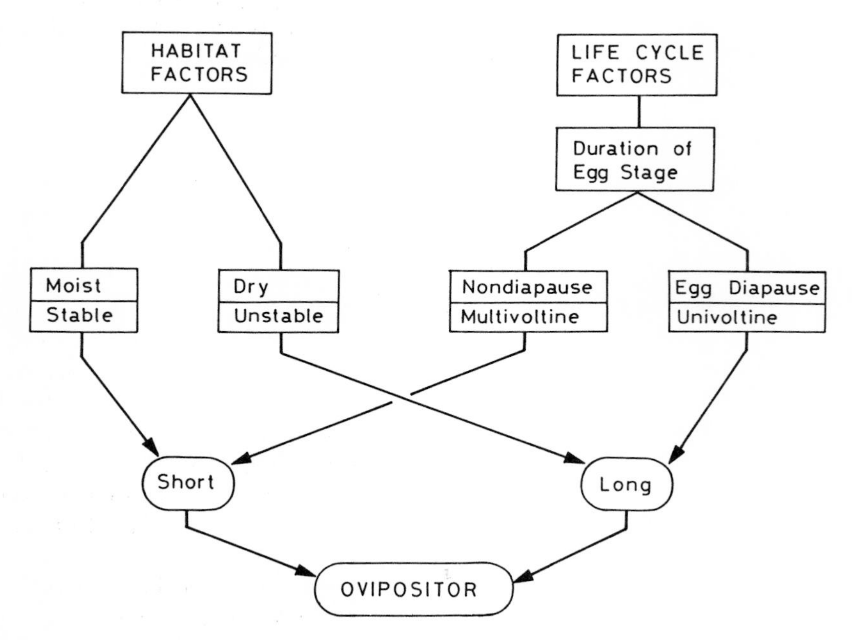

Figure 2.8. Summary of natural selection on ovipositor length in crickets.

The ovipositor, however, does not always bear all the responsibility for the protection of eggs. The female can also protect her eggs by giving them a harder chorion. The eggs themselves may be able to form a tougher serosal cuticle and a thicker wax layer. Different cricket species in fact show different degrees of resistance to desiccation in the egg stage (Hogan 1967; McIntyre 1978). The adaptive burden on the ovipositor would vary accordingly. As each organism is a highly integrated system, it is not surprising that various aspects of its adaptation are closely interlinked.

Acknowledgments I thank E. Funamizu and M. Sasaki for typing the manuscript, N. Sasaki for drawing the figures, and anonymous reviewers for suggesting improvements. This study is contribution No. 99 from the Laboratory of Entomology, Hirosaki University, and was partially supported by a grant (58480044) from the Ministry of Education, Science and Culture, Japan.

References

Alexander, R.D.: The taxonomy of the field crickets of the eastern United States (Orthoptera: Gryllidae). Ann. Entomol. Soc. Am. 50, 584–602 (1957).

Alexander, R.D.: Life cycle origins, speciation, and related phenomena in crickets. Quart. Rev. Biol. 43, 1–41 (1968).

Alexander, R.D., Bigelow, R.S.: Allochronic speciation in field crickets, and a new species, *Acheta veletis*. Evolution 14, 334–346 (1960).

Alexander, R.D., Thomas, E.S.: Systematic and behavioral studies on the crickets of the *Nemobius fasciatus* group (Orthoptera: Gryllidae: Nemobiinae). Ann. Entomol. Soc. Am. 52, 591–605 (1959).

Browning, T.O.: The influence of temperature and moisture on the uptake and loss of water in the eggs of *Gryllulus commodus* Walker (Gryllidae—Orthoptera). Aust. J. Sci. Res. B 5, 112–127 (1953).

Hogan, T.W.: The influence of diapause on the resistance to desiccation of eggs of *Teleogryllus commodus* (Walk.) (Orthoptera: Gryllidae). Proc. R. Soc. Victoria 80, 37–42 (1967).

Hubbell, T.H., Norton, R.H.: The systematics and biology of the cave crickets of the North American tribe Hadenoecini (Orthoptera: Saltatoria: Ensifera: Rhaphidophoridae: Dolichopodinae). Misc. Publ. Mus. Zool. Michigan 155, 1–124 (1978).

Kidokoro, T., Masaki, S.: Photoperiodic response in relation to variable voltinism in the ground cricket, *Pteronemobius fascipes* Walker . (Orthoptera: Gryllidae). Jpn. J. Ecol. 28, 291–298 (1978).

Maskai, S.: Thermal relations of diapause in the eggs of certain crickets (Orthoptera: Gryllidae). Bull. Fac. Agric. Hirosaki Univ. 6, 5–12 (1960).

Masaki, S.: Geographic variation of diapause in insects. Bull. Fac. Agric. Hirosaki Univ. 7, 66–98 (1961).

Masaki, S.: Geographic variation and climatic adaptation in a field cricket (Orthoptera: Gryllidae). Evolution 21, 725–741 (1967).

Masaki, S.: Climatic adaptation and photoperiodic response in the band-legged ground cricket. Evolution 26, 587–600 (1973).

Masaki, S.: Seasonal and latitudinal adaptations in the life cycles of crickets. In: Evolution of Insect Migration and Diapause. Dingle, H .(ed.). New York: Springer-Verlag, 1978a, pp. 72–100.

Masaki, S.: Climatic adaptation and species status in the lawn ground cricket. II. Body size. Oecologia 35, 343–356 (1978b).
Masaki, S.: Climatic adaptation and species status in the lawn ground cricket. I. Photoperiodic response. Kontyu 47, 48–65 (1979a).
Masaki, S.: Climatic adaptation and species status in the lawn ground cricket. III. Ovipositor length. Oecologia 43, 207–219 (1979b).
Masaki, S.: Climatic speciation in Japanese ground crickets. GeoJournal 7, 483–490 (1983).
Masaki, S., Ohmachi, F.: Divergence of photoperiodic response and hybrid development in *Teleogryllus* (Orthoptera: Gryllidae). Kontyu 35, 83–105 (1967).
McFarlane, J.E., Ghouri, A.S.K., Kennard, C.P.: Water absorption by the eggs of crickets. Can. J. Zool. 37, 391–399 (1959).
McIntyre, M.E.: Some aspects of diapause in the field cricket *Pteronemobius nigrovus* and *P. bigelowi* (Orthoptera: Nemobiinae), with notes on their ecology. Mauri Ora 6, 3–10 (1978).
Otte, D., Alexander, R.D.: The Australian crickets (Orthoptera: Gryllidae).Acad. Natl. Sci. Phil. Monogr 22, 1–477 (1983).
Peters, R.H.: The Ecological Implications of Body Size. Cambridge: Cambridge University Press, 1983.
Pianka, E.R.: Evolutionary Ecology. New York: Harper & Row, 1974.
Roff, A.D.: Size and survival in a stochastic environment. Oecologia 36, 163–172 (1978).
Roff, A.D.: On being the right size. Am. Natur. 118, 405–422 (1981).
Schmidt-Nielsen, K.: Scaling: Why is Animal Size So Important? Cambridge: Cambridge Unviersity Press, 1984.
Schoener, T.W., Janzen, D.H.: Notes on environmental determinants of tropical versus temperate insect size patterns. Am. Natur. 102, 207–224 (1968).
Sonan, J.: Studies on the giant Formosan cricket and its control. (In Japanese). Bull. Formosan Cent. Res. Sta. Dept. Agric. 86, 1–22 (1931).
Tanaka, S.: Effects of changing photoperiod on nymphal development in *Pteronemobius nitidus* Bolivar (Orthoptera, Gryllidae). Kontyu 46, 135–151 (1978).
Walker, T.J.: Systematics and acoustic behavior of United States and Caribbean short-tailed crickets (Orthoptera: Gryllidae: Anurogryllus). Ann .Entomol. Soc. Am. 66, 1269–1277 (1973).
West, M.J., Alexander, R.D.: Sub-social behavior in a burrowing cricket *Anurogryllus muticus* (de Geer). Ohio J. Sci. 63, 19–24 (1963).
Wigglesworth, V.B.: The Principles of Insect Physiology. [7th ed.] London: Chapman and Hall, 1972.

Chapter 3

Geographical Variation and Ecological Control of Diapause in Flies

E.B. VINOGRADOVA

Diapause is a characteristic feature of the life cycles in insects of temperate regions synchronizing development with cyclic, seasonal conditions and is important for the flies of interest to me, which belong to the families Calliphoridae, Muscidae, and Sarcophagidae. These flies are numerous in nature and include many species of importance to man. Although the spring-summer part of the annual cycle of development in these flies has been studied fairly well, our knowledge about their winter adaptations remains inadequate. The flies overwinter as larvae, adults, or pupae. Usually the hibernating individuals enter diapause but a few species (e.g., *Musca domestica*) hibernate without diapause. In Sarcophagids a pupal diapause is most common, in Calliphorids imaginal and larval, and in Muscids imaginal and pupal. As a rule, each species has only one diapause stage but some flies (e.g., *Calliphora vicina* and *C. vomitoria*) can diapause as a larva or as an adult or both.

In flies, as in other insects, diapause induction, maintenance, and termination are controlled by temperature and photoperiod (Danilevsky 1961; Tauber and Tauber 1976, 1978, 1981). There are some interesting peculiarities, however, concerning the stages of development that take part in the determination of larval diapause in Calliphorids and pupal diapause in Sarcophagids. In this chapter, I shall review experimental findings, largely from my lab, concerning geographical patterns of diapause stage, incidence, and duration among various species of flies.

Larval Diapause

Larval diapause is determined by maternal conditions. This phenomenon has been studied in detail in our experiments with *C. vicina,* which is a popular model organism for investigations of insect physiology and ecology (Vinogradova 1984). The reaction of the larvae to environmental factors is modified by means of parental conditioning and the primary diapause-inducing factor is maternal photoperiod. The progeny of short-day-treated females are likely to enter larval diapause (Zinovjeva and Vinogradova 1972, Vinogradova and Zinovjeva 1972b). Maternal photoperiod determined not only the proportion of diapause larvae in the progeny, but also the characteristics of the diapause; a more intensive diapause is typical for the progeny of short-day-treated flies (Vinogradova 1974). As to the factors acting directly on the larvae, the temperature fulfills the main role and the photoperiod a secondary one in diapause induction. For instance, in *C. vicina* from Aktuz, the incidence of diapause increases from 9 to 100% as temperature decreases from 17 to 12°C. Changes in photoperiod during the female's life induce some physiological changes, which determine corresponding changes in the state of the progeny. There is considerable intra- and interpopulation variation in the rate of female response to the new photoperiodic regimen (Vinogradova 1976a). For instance, it took 9–24 short-day cycles for females of the Gorky strain to switch from the production of nondiapausing to diapausing progeny when a long-day regimen was replaced by a short-day regimen (Vinogradova and Zinovjeva 1972b). This blowfly reacts not only to the single (one-step) change but also to gradually changing day length, as observed in nature. Progressive increasing of the day length (7 min per day) lowers the incidence of diapause in the progeny whereas shortening (4 min per day) increases diapause incidence (Vinogradova 1976a). These experimental results allow us to understand the reservoir of hibernating larvae in nature, which includes progeny of females belonging to several different generations.

Maternal participation in the determination of the physiological state of the progeny is probably typical for Calliphorids. This has been demonstrated for *Lucilia caesar* (Ring 1967a,b), *L. sericata* (Cragg and Cole 1952), *L.illustris, L. hirsutula,* and *C. uralensis* (Zinovjeva 1978). The degree of maternal effect can vary in different species, populations, and laboratory generations of flies. A review of the literature shows that the maternal influence on diapause of progeny is widely distributed among the insects though it has been studied insufficiently. In contrast to the complicated manner of larval diapause induction, completion of diapause in *C. vicina* and *L. caesar* (Ring 1968) is under direct control of photoperiod and temperature. Long days and an increase in temperature to 20°C or 25°C accelerate the breaking of diapause (Vinogradova 1974, 1984). A slow, spontaneous termination of diapause culminating in the puparia occurs at 6°C and 12°C. Temperatures of 0°C and −5°C are less favorable for reactivation. In the latter case, sizeable mortality of the larvae was observed (Vinogradova and Bogdanova 1984). Photoperiodic reactivation was effective in the laboratory but its occurrence in nature is doubtful (Vinogradova 1974).

In *C. vicina,* inheritance of larval diapause has been studied experimentally by hybridization of forms developing with or without larval diapause (Vinogradova and Tstutskova 1978). If the females of nondiapausing forms are crossed with the males of the diapausing forms, most of the F_1 hybrid larvae develop without diapause. In the reciprocal crossing, the results are not clear-cut but the tendency of the hybrid larvae to develop with maternal diapause traits is suggested. As a whole, the results agree with the widely held idea about polygenic inheritance of larval diapause.

Imaginal Diapause

The main peculiarities of imaginal (reproductive) diapause and its ecological control have been studied in *Phormia regina* (Stoffolano 1974, Stoffolano et al. 1974), *Musca autumnalis* (Stoffolano and Matthysse 1967, Valder et al. 1969), *Aldrichina grahami* (Vinogradova 1978a), *Protophormia terraenovae* (Vinogradova 1986, in press) and *C. vicina* (Vinogradova 1984). In flies, both the females and males overwinter. Imaginal diapause of the females is characterized by the delay of ovarian development phases I or II (Vinogradova 1978a). In some species, diapause is accompanied by great hypertrophy of the fat body (*P. regina, P. terraenovae*) but other species never exceed the second degree of fat body development (*C. vicina, C. vomitoria*). In diapausing flies the fat body composition is altered; in diapausing *M. autumnalis* synthesis of triglycerides and free fatty acids prevails (Valder et al. 1969) and the accumulation of the glycerol and lower respiration rates has been noted for *P. terraenovae* diapausing at low temperature (Wood and Nordin 1976). This is thought to play a defensive role. In diapausing *C. vicina* and *P. terraenovae*, behavior is also affected and the flies exhibit reduced activity levels and form compact clusters.

In most of the species that have been studied, temperature is the main diapause-inducing factor while photoperiodism has secondary significance; the physiological state of the flies depends presumably on the imaginal conditions. The range of temperatures that is optimal for diapause induction is species specific. In *P. terraenovae* at 25°C, there are some nondiapausing individuals, but at 20°C all the flies enter diapause. In this species, diapause is very intense and prolonged: at 20°C only about 20% of the females had terminated their diapause by the end of the fourth month. In *C. vicina* at 20°C few females enter diapause, at 12°C the incidence of diapause increases considerably, and at 6°C (darkness) all of the flies diapause. The spontaneous termination of diapause takes place in 75–90 days at 12°C (LD 12:12) and after 4–6 months at 6°C. In the latter case, the complete maturation of the females takes place at this low temperature. In *C. vicina* at 12°C the tendency to diapause is more frequently expressed under short-day than under long-day conditions (Vinogradova 1984).

In all the species, an increase in temperature usually terminates diapause. In some species, photoperiodic reactivation occurs. In *C. vicina*, most diapausing females terminated diapause 10–17 days after being transferred to 20°C. If the diapausing *P. terraenovae* are transferred from 20°C to 27°C or 30°C, the first oviposition takes place in 5–7 days but some females die without terminating diapause at this temperature. If the diapausing *M. autumnalis* are transferred into LD 16:8 and 27°C, the first mature flies appear after 15 days but about 33% remain in diapause after 3 months (Valder et al. 1969).

Pupal Diapause

In Sarcophagids, pupal diapause is expressed in the arrest of development at the stage of the young phanerocephalic pupa (Fraenkel and Hsiao 1968). In diapausing pupae of *Sarcophaga argyrostoma* and *S. crassipalpis* (Denlinger et al. 1972) and of *Haematobia irritans* (Hoelscher and Combs 1970), the rate of oxygen consumption is considerably lower compared to that of nondiapausing individuals. Cold-hardiness is a component of the diapause syndrome for pupae of *S. crassipalpis* (Adedokin and Denlinger 1984).

Ecological regulation of pupal diapause has been investigated in some Sarcophagids from the temperate regions of North America, Europe, and Asia (*S.argyrostoma, S. bullata, S. crassipalpis, S. peregrina, Ravinia striata, Parasarcophaga similis, P. semenovi, Bercaea haemorrhoidalis, Boettcherisca septentrionalis*) Fraenkel and Hsiao 1968; Denlinger 1971, 1972; Vinogradova and Zinovjeva 1972a, Vinogradova 1976b, 1978b; Kurahashi and Ohtaki 1979; Saunders 1984). from Australia (*Tricolioproctia impatiens*) (Roberts and Warren 1975) and in *Haematobia irritans* (Muscidae) (Depner 1961, 1962). Furthermore, the capacity to diapause as pupae is displayed by some tropical flesh flies from equatorial Africa and South America (*S. par, S. inzi, S. exuberans, S.monospila, S. ruficornis, Poecilometopa spilogaster,* and *P. punctipennis*)(Denlinger 1974, 1978, 1979). In these species, diapause can be induced by rearing in the laboratory under appropriate conditions. It should be noted, however, that whereas in Kenya in some years only few natural pupae of *P. spilogaster* enter a short diapause during the coldest season (June–August) (Denlinger 1978). The similarity of diapause patterns between the tropical and temperate region flesh flies suggests a common origin; on these grounds Denlinger (1979) developed the hypothesis about the tropical origin of pupal diapause in Sarcophagids. This question is discussed in detail by Tauber and Tauber (1981).

Diapause in flesh flies from the temperate region is generally induced by short-day photoperiods with the usual dependence on temperature. On the basis of numerous experiments with *S. argyrostoma,* Saunders (1984) worked out a model to explain the photoperiodic measurement of time in insects. In this flesh fly, diapause induction depends on the interaction of two varying values, namely the length of the period of sensitivity (larval development) and the number of short-day inductive cycles needed to raise the incidence of pupal diapause to 50%. The former value depends on the temperature ($Q = 2.7$), whereas the latter one is 13–14 short days and it is almost thermostable ($Q = 1.4$). Therefore, at high temperature, continuous development occurs because pupariation takes place before the accumulation of the necessary number of short-day cycles. On the contrary, at low temperatures the necessary number of inductive cycles is accumulated before the sensitive period finishes. This tendency to increase the incidence of diapause at lower temperatures is observed in many species of flesh flies (Denlinger 1972, Vinogradova and Zinovjeva 1972a, Vinogradova 1978b). For Sarcophagids in the tropical regions, diapause is determined by temperature exclusively and photoperiodism has no effect (Denlinger 1974). An interesting feature of pupal diapause induction in Sarcophagids is the early appearance of photoperiodic sensitivity in late embryos, which are still in the female's body. Denlinger (1971) collected direct evidence of independent embryonic perception of photoperiodic signals in *S. crassipalpis* by rearing embryos in artificial medium at different day lengths. He found that only 1–2 short-day cycles in late embryonic life induced pupal diapause.

Temperature is the main factor that terminates pupal diapause in Sarcophagids from the temperate region, whereas photoperiod is ineffective in this respect. Spontaneous reactivation occurs at all temperatures that induce diapause and the duration of diapause depends on temperature. Diapause duration decreases as temperature increases. For instance, in *S. argyrostoma* diapause continues for 227 ± 1.3 days at 17°C and 91.7 ± 1.1 days at 25°C, in *S. crassipalpis* for 117.9 ± 0.4 days at 17°C and 69.8 ± 0.7 days at 25°C (Denlinger 1972). Diapause in *P. similis* and *R. striata* terminates after 3–4 months of cooling at6°C (Vinogradova and Zinovjeva 1972a).

Intraspecific Variability of Diapause

Interpopulation Variability of Hibernation Adaptations

Intraspecific variability of insect diapause indicates ecological plasticity and reflects adaptation to a broad diversity of climatic conditions. These patterns are generalized in the summaries of Danilevsky (1961), Masaki (1978), and Tauber and Tauber (1976, 1978, 1981). Intraspecific variability of diapause in

Table 3.1. The incidence of female diapause (%) for various geographical strains of *P. terraenovae* in various photoperiodic conditions at 25°C

Strain	Geographical coordinates, °N, °E	Female age (days)	Day length, LD			
			12:12	16:8	18:6	20:4
1. Kirovsk	67, 36	20	100	100	—	—
		62	100	100		
2. Leningrad	60, 30	20	100	100	100	89
		62	100	100	95	96
3. Tumen	57, 65	30	100	100	100	100
		60	100	100	94	—
4. Moscow	56, 38	20	100	100	78	50
		62	100	96	100	95
5. B. Tserkov	50, 30	20	100	100	100	100
		62	100	100	100	98
6. Dnepropetrovsk	48, 36	20	100	100	91	68
		62	100	100	89	88
7. Stavropol	45, 42	30	100	100	100	100
		60	100	95	88	40
8. Ordgonikidze	43, 45	20	62	96	—	95
		40	60	85	—	—
9. Kasbegi	43, 45	20	12	33	18	—

The altitude of locality 9 is about 1700 m; that of locality 8, 700 m; and that of the other localities, below 500 m. In each case, 20–45 females were dissected.

flies remains poorly studied. We have examined interpopulation variability of diapause in two Calliphorids, *P. terraenovae* and *C. vicina*.

The blowfly, P. terraenovae, is distributed widely in Holarctic regions and is the typical representative of the dipterous fauna in the Arctic and Subarctic, where it is dominant. This fly is adapted very well to the severe conditions of this region. We have compared experimentally the imaginal diapause in nine geographical strains from 43° to 68°N and from 30° to 65°E (Table 3.1). At 20°C both long and short days force all individuals originating from all localities to enter prolonged diapause. At 25°C, on the 20–30th day of imaginal life some females of the strains 2, 4, and 6 reproduce in long-day regimens and in two southern strains, 8 and 9, there are nondiapause females in all regimens. Only strain 9 differs significantly from the other strains in its tendency to enter diapause ($p < 0.01$). Thus, in *P. terraenovae,* geographical variability of imaginal diapause is expressed weakly. Most strains have a tendency towards univoltinism throughout a large north-south distribution.

The blowfly, *C. vicina,* is a typical species widely distributed with man in temperate and subtropical regions of the Holarctic. We studied imaginal and larval diapause in 16 strains from different portions of the area of distribution from 38° to 69°N and from 20° to 77°E in USSR (Table 3.2). All of the populations can enter reproductive diapause, and the ranges of temperature that are optimal for diapause induction are similar, 12°C and below.

Table 3.2. Interpopulation variability in mean length of larval development (± SEM) for *C. vicina* at different temperatures

Strain	Geographic coordinates °N, °E	Mean length (days) 12°C LD 12:12	Mean length (days) 6°C LD 0:24
1. Leningrad	60, 30	79.4 ± 1.3	178.4 ± 1.5
2. Ribatchy	56, 20	58.0 ± 1.1	83.7 ± 1.9
3. Moscow	56, 38	69.8 ± 1.5	76.0 ± 3.3
4. Belaj Tserkov	50, 30	109.0 + 1.2	149.7 ± 1.4
5. Dnepropetrovsk	48, 36	50.4 ± 1.5	105.9 ± 1.6
6. Jalta	44, 34	18.4 ± 0.1	82.2 ± 0.6
7. Sukchumi	43, 41	93.8 ± 0.6	160.0 ± 1.2
8. Ordzonikidze	43, 45	60.7 ± 1.0	—
9. Aktuz	43, 77	93.6 ± 1.7	171.4 ± 1.9
10. Yerevan	40, 45	18.9 ± 0.4	121.6 ± 0.8
11. Baku	40, 50	17.5 ± 0	84.9 ± 1.7
12. Shafirkan	40, 64	17.5 ± 0	—
13. Dushanbe	39, 69	22.1 ± 0.1	29.5 ± 1.3
14. Kondara	39, 69	53.7 ± 1.0	105.9 ± 2.6
15. Ashhabad	38, 58	34.0 ± 1.2	60.8 ± 0.7
16. Kuljab	38, 69	19.7 ± 0.3	56.0 ± 0.8

The altitude of locality 9 is about 2000 m; that of locality 14, 1100 m; that of localities 8 and 10, 700 m; that of other localities, below 500 m. In each sample there were 200–800 larvae.

Interpopulation variability of larval hibernation adaptations has been investigated among the strains originating from the temperate region with continental (3–5, 12–16) and maritime (1, 11) climates and from the Black Seacoast subtropics (7). To investigate the capacity to diapause as larvae, all strains were reared under diapause-inducing conditions at two constant temperature and photoperiod regimens, 6°C (darkness) and 12°C (LD 12:12), and pupariation dates were recorded. On the basis of the response at 12°C, the strains were divided into two groups (Table 3.2). Larvae belonging to the first group develop without diapause and their mean length of development to pupariation varies from 17.5 to 34 days (at 6°C it varies from 29.5 to 121.6 days). This group includes strains from the southern flat country of the Crimea (6), Transcaucasia (10, 11), and Middle Asia (12, 13, 15, 16). In all these localities, except Yerevan (10), mean temperatures of the coldest month, January, are above 0°C (0.3'dg–5°C). The second group, showing a larval diapause, includes the more northern strains from European USSR and from Sukchumi (7) and mountainous localities of Middle Asia (9, 14). In all these localities except Sukchumi, mean temperatures of January are below 0°C (−3°–11°C). Larvae of this group develop with diapause at both temperatures. Mean length of larval development changes from 50 to 109 days at 12°C and from 76 to 178 days at 6°C.

The genetic basis and stability of these responses to changes in temperature between 12°C and 6°C are confirmed by laboratory experiments lasting 3–4 generations in strains 1, 2, 7, 13, 15, and 16. The various strains differ dramatically in the median and variability of the larval developmental period (Figure 3.1). The responses of larvae from different strains maintained first at 6°C and

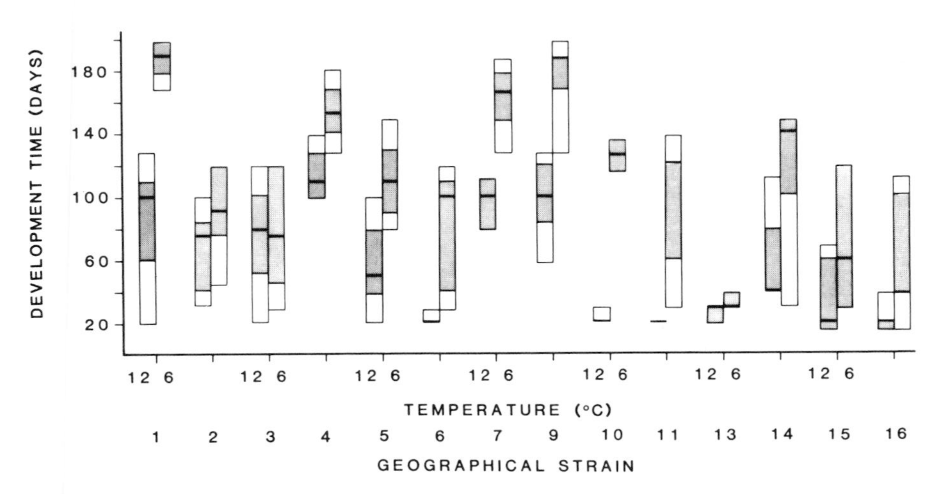

Figure 3.1. Geographical variability of larval diapause in the blowfly, *Calliphora vicina:* Pupariation of larvae at constant temperatures and photoperiods of 6°C (darkness) and 12°C (LD 12:12). Open columns represent q_{10}–q_{90}, the time required for the frequency of pupariated larvae to increase from 10% to 90%, and shaded columns represent q_{25}–q_{75}, the time required for the frequency to increase from 25% to 75%. The thick line shows the median time for each strain.

transferred to 12°C at 20-day intervals (transfers or treatments 1–5) served as a second index of diapause intensity (Figure 3.2). The higher the incidence of the puparia and the shorter the period of the pupariation, the less is the intensity of diapause. The results of transfer 1 (at a larval age of 26 days) is the most instructive because diapause is induced by this time, but reactivation has not occurred. From the first group, strain 13 from Tadzhik is the most extreme case since larvae develop without arrest at both 12°C and 6°C. Pupariation occurs early during a short period and both medians coincide (Figure 3.1), and in transfer 2 about 85% of the larvae pupariate during the first 6 days (Figure 3.2). These results indicate that larval diapause is absent in this strain. The other strains of this group pupariate very rapidly at 12°C except for strain 15. In transfer 1, about 35–45% of larvae of strains 6 and 10 and 75–100% in the other strains pupariated during the first 15 days. Therefore, the delay is unstable and development resumes easily in response to an increase in temperature. These data allow us to hypothesize that among the strains from the first group, only some individuals are able to enter larval diapause at 6°C and diapause is weak. The tendency to diapause is most pronounced in strain 10, its q_{10}–q_{90} is 120–140 days.

In strains of the second group, diapause appears at both 12°C and 6°C. But in some strains some larvae develop without diapause, especially at 12°C (Figure 3.1, strains 1, 2, 3, 5). At first, nondiapause larvae pupariate and then the diapause individuals complete diapause. From the temperate region, strains 1 and 4 have the most intense diapause. At 6°C their medians and q_{10}–q_{90} are large (130–200 days) and their diapause does not break easily. Within 15 days only 4–18% of the larvae pupariate in transfer 1 and 26–37% in transfer 2. The strains from Belomorsk, North Karelia (65°N, 34°E) (Vinogradova 1975) and the northern most one from the Barentsovo sea coast (69°N, 34°E) have diapause of similar intensity. Unexpectedly, strain 7 (Sukchumi), originating from the subtropics, also has intensive larval diapause, which probably prevents the premature development of larvae during the relatively warm winter. An analogous function of intense diapause in southern populations has been emphasized in other insects (Masaki 1978).

Altitudinal variability of larval diapause has been found in *C. vicina*. The Tadzhik strain from Dushanbe (13) has no diapause whereas a neighboring strain (14) inhabiting higher elevations (1100 m) does diapause. Strain 9 from Kirgizia (2000 m) has an especially intense diapause.

Mean length of larval development at 12°C correlates with the latitude where the strain was collected ($r = 0.74$; $p < 0.01$). It also correlates with mean temperature in January ($r = 0.73$; $p < 0.05$) and with mean frost period ($r = 0.86$; $p < 0.01$). This latter factor is probably most important in determining the pattern of hibernation, namely the capacity to diapause as a larva ($r = 0.68$; $p < 0.05$). The mean length of development at 6°C is correlated both with mean January temperature ($r = 0.93$; $p < 0.01$) and with mean frost period ($r = 0.83$;$p < 0.01$). The degree of the larval developmental arrest (the incidence of larvae pupariating within 6 days in transfer 1) is correlated with mean frost period, also ($r = 0.76$; $p < 0.05$).

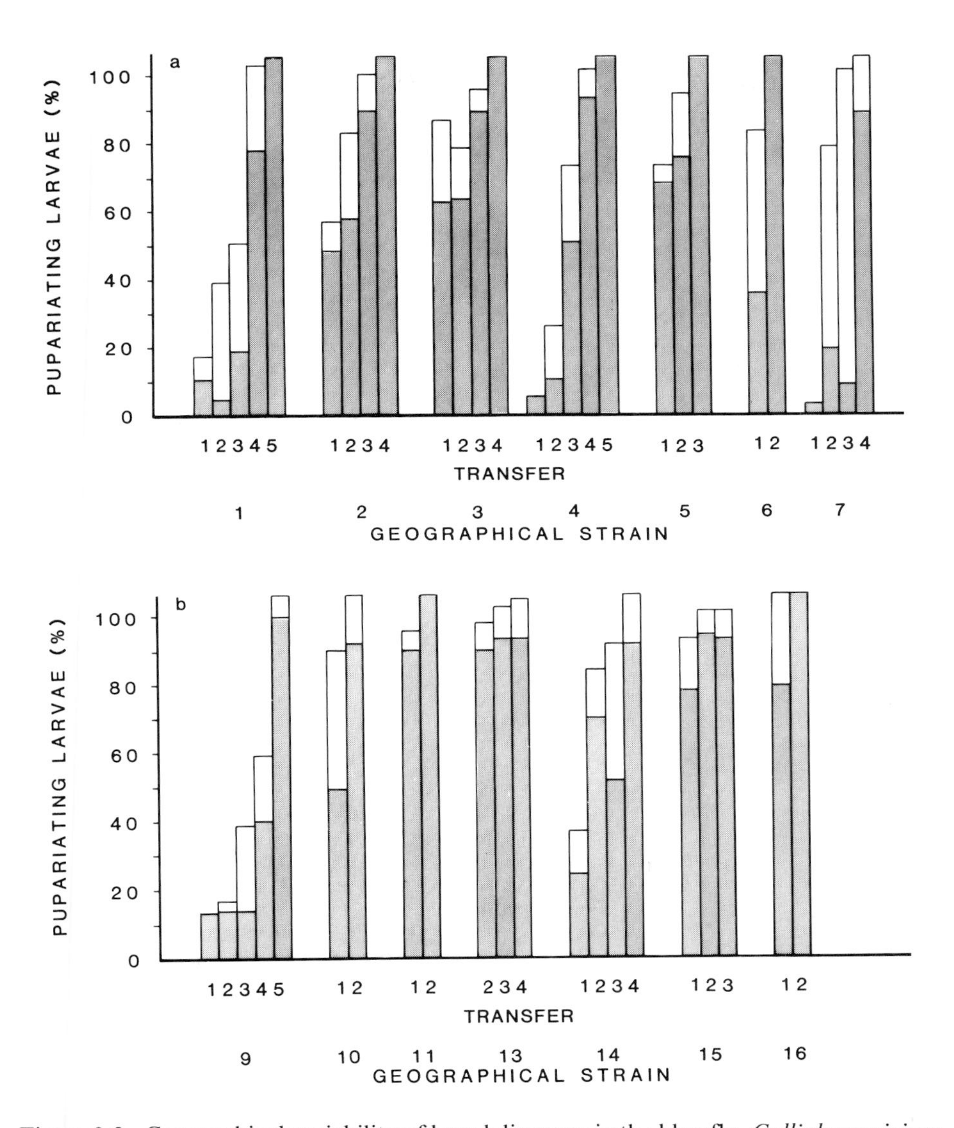

Figure 3.2. Geographical variability of larval diapause in the blowfly, *Calliphora vicina:* Pupariation of larvae that have been transferred from 6°C to 12°C. The following variants of the experiments were conducted: larvae were maintained at 6°C for 26 (1), 46 (2), 66 (3), 86 (4), and 106 (5) days and then were transferred to 12°C. Larvae that pupariated from day 1 to 6 after being transferred to 12°C are represented by shaded columns, those that pupariated from day 7 to day 15 after being transferred are represented by open columns. (n = 200).

The experiments show that the blowfly *C. vicina* is a good model organism for the investigation of winter adaptations, which include imaginal and larval diapause. The ontogeny of this fly is characterized by great ecophysiological plasticity. It has been shown experimentally that larval diapause does not exclude imaginal diapause, i.e., each individual has the potential capacity to diapause twice. Imaginal diapause is a characteristic of all populations that have been studied from 38° to 69°N while the capacity to enter larval diapause varies with distribution. Larval diapause is expressed very distinctly in populations from the north and weakens southwards to complete disappearance in the southern strain from Dushanbe.

Populations differ in the upper range of temperatures favorable for induction of larval diapause, the incidence of diapause, and the intensity of diapause. Geographical differentiation of larval diapause behavior is related to the duration of the frost period, which correlated positively with winter temperatures. Besides the tendency for diapause to weaken to the south, there is a tendency for larval diapause to strengthen in mild southern winters. But it is still unknown how widespread this phenomenon is for *C. vicina;* it is revealed now only in one strain (Sukchumi). Similar trends in geographical variation of diapause intensity have been noted in other species (Tauber and Tauber 1976; Masaki 1978). In the southern portion of the distribution, where the winter temperatures are presumably above 0°C, *C. vicina* can overwinter safely as active, reproductive females as well. It has been shown experimentally that a temperature of 6°C cannot induce diapause in reproducing females, which live successfully under these conditions for up to 3–4 months and then begin oviposition if the temperature rises.

Intrapopulation Variability

As a final note, I shall discuss some interesting features of intrapopulation variability of larval diapause in *C. vicina*. In the northern strain from Belomorsk about 98–100% of the larvae produced by most females enter diapause but in the progeny of a few females only 30–36% of the larvae diapause (Vinogradova 1975). In another strain from Gorky under long-day conditions, the majority of the females (21) produced nondiapause progeny but three females produced entirely diapause progeny. By collecting these larvae for four generations a new race was selected. This new race differed from the initial strain by having a higher incidence of larval diapause that was also more intense. A "univoltine" race was selected from the progeny of one female (from Leningrad) and its obligate diapause was very intense (Vinogradova 1975).

The incidence of pupal diapause in *Parasarcophaga similis* was increased from 53% in F_1 to 97% in F_4 by selecting under diapause-inducing conditions (Vinogradova and Zinovjeva 1972a). In another flesh fly, *Pseudosarcophaga affinis,* a nondiapausing laboratory stock has been obtained as a result of 200 generations of unplanned selection (House 1967). Selection of nondiapause pupae of *P. spilogaster* at 15°C resulted in the dropping of the level of diapause from 80 to 10% in 10 generations (Denlinger 1979).

Acknowledgments I am very grateful to Drs. W. E. Bradshaw and H. Dingle for the invitation to take part in this interesting symposium and Drs. R. Karban and F. Taylor for their help in the preparation and review of the manuscript. I express my sincere thanks to T. P. Bogdanova for assistance in the experimental investigations and Dr. S. J. Reznik for valuable advice.

References

Adedokin, T.A., Denlinger, D.L.: Cold-hardiness: a component of the diapause syndrome in pupae of the flesh flies, *Sarcophaga crassipalpis* and *S. bullata*. Physiol. Entomol. 9, 361–364 (1984).

Cragg, J.B., Cole, P.: Diapause in *Lucilia sericata* Mg., Diptera. J. Exp.Biol. 29, 600–604 (1952).

Danilevsky, A.S.: Photoperiodism and Seasonal Development of Insects. Leningrad: Leningrad State University Publishing House, 1961 (in Russian).

Denlinger, D.L.: Embryonic determination of pupal diapause in the flesh fly, *Sarcophaga crassipalpis*. J. Insect Physiol. 17, 1815–1822 (1971).

Denlinger, D.L.: Induction and termination of pupal diapause in *Sarcophaga* (Diptera: Sarcophagidae). Biol. Bull. 142, 11–24 (1972).

Denlinger, D.L.: Diapause potential in tropical flesh flies. Nature 252, 5480, 223–224 (1974).

Denlinger, D.L.: The developmental response of flesh flies (Diptera, Sarcophagidae) to tropical seasons: variation in generation time and diapause in East Africa. Oecologia 35, 105–107 (1978).

Denlinger, D.L.: Pupal diapause in tropical flesh flies: environmental and endocrine regulation, metabolic rate and genetic selection. Biol. Bull. 156, 31–46 (1979).

Denlinger, D.L., Willis, J.H., Fraenkel, G.: Rates and cycles of oxygen consumption during pupal diapause in Sarcophaga flesh flies. J. Insect Physiol. 18, 871–882 (1972).

Depner, K.R.: The effect of temperature on development and diapause of the horn fly *Siphons irritans* (Diptera, Muscidae). Can. Entomol. 93, 855–859 (1961).

Depner, K.R.: Effects of photoperiod and of ultraviolet radiation on the incidence of diapause in the horn fly *Haematobia irritans* L. (Diptera, Muscidae). Int. Biometeorol. 5, 68–71 (1962).

Fraenkel, G., Hsiao, C.: Manifestation of a pupal diapause in two species of flies *Sarcophaga argyrostoma* and *S. bullata*. J. Insect Physiol. 14, 689–705 (1968).

Hoelscher, E.E., Combs, R.L.: The horn fly. II. Comparative physiological studies of reproductive and diapaused pupae. Physiol. Zool. 43, 241–248 (1970).

House, H.I.: The decreasing occurrence of diapause in the fly *Pseudosarcophaga affinis* through laboratory-reared generations. Can. J. Zool. 45, 149–153 (1967).

Kurahashi, H., Ohtaki, T.: Induction of pupal diapause and photoperiodic sensitivity during early development of *Sarcophaga peregrina* larvae. Jpn. J. Med. Sci. Bull. 32, 77–82 (1979).

Masaki, S.: Seasonal and latitudinal adaptations in life cycles of crickets.
In: Evolution of Insect Migration and Diapause. Dingle, H. (ed.). NewYork: Springer-Verlag, 1978, pp. 72–100.

Ring, R.A.: Photoperiodic control of diapause induction in the larva of *Lucilia caesar* L. (Diptera, Calliphoridae). J. Exp. Biol. 46, 117–122 (1967a).

Ring, R.A.: Maternal induction of diapause in the larva of *Lucilia caesar* L. (Diptera, Calliphoridae). J. Exp. Biol. 46, 123–136 (1967b).

Ring, R.A.: Termination of diapause in the larva of *Lucilia caesar* L. (Diptera, Calliphoridae). Can. J. Zool. 46, 335–344 (1968).

Roberts, B., Warren, M.A.: Diapause in the flesh fly *Tricholioproctia impatiens* (Diptera: Calliphoridae). Aust. J. Zool. 23, 563–567 (1975).

Saunders, D.S.: Photoperiodism in insects (Russian translation). In: Biological Rhythms. Aschoff, J. (ed.). Moscow: "Mir," 1984, v. 2, pp. 81–129.

Stoffolano, J.G.: Influence of diapause and diet on the development of the gonads and accessory reproductive glands of the black blowfly, *Phormia regina* (Meigen). Can. J. Zool. 52, 981–988 (1974).

Stoffolano, J.G., Matthysse, J.G.: Influence of photoperiod and temperature on diapause in the face fly, *Musca autumnalis* (Diptera, Muscidae). Ann. Entomol. Soc. Am. 60, 1242–1246 (1967).

Stoffolano, J.G., Greenberg, S., Calabrese, E.L.: A facultative imaginal diapause in the black blowfly, *Phormia regina*. Ann. Entomol. Soc. Am. 67, 518–519 (1974).

Tauber, M.J., Tauber, C.A.: Insect seasonality: diapause maintenance, termination, and postdiapause development. Annu. Rev. Entomol. 21, 81–107 (1976).

Tauber, M.J., Tauber, C.A.: Evolution of phenological strategies in insects: a comparative approach with eco-physiological strategies and genetic considerations. In: Evolution of Insect Migration and Diapause. Dingle, H. (ed.). New York: Springer-Verlag, 1978, pp. 53–71.

Tauber, C.A., Tauber, M.J.: Insect seasonal cycles: genetics and evolution. Annu. Rev. Ecol. Syst. 12, 281–308 (1981).

Valder, S.M., Hopkins, T.L., Valder, S.A.: Diapause induction and changes in lipid composition in diapausing and reproducing face flies, *Musca autumnalis*. J. Insect Physiol. 15, 1199–1214 (1969).

Vinogradova, E.B.: Maternal influence on the progeny diapause in insects. In: Annual Reports on Memory of N.A. Cholodkovsky in 1972. Leningrad: "Nauka,"1973, pp. 39–66 (in Russian).

Vinogradova, E.B.: The pattern of reactivation of diapausing larvae in the blowfly, *Calliphora vicina*. J. Insect Physiol. 20, 2487–2496 (1974).

Vinogradova, E.B.: Intraspecific variability of reactions controlling the larval diapause in *Calliphora vicina* R.-D. (Diptera, Calliphoridae). Entomol. Rev. USSR 54, 720–735 (1975) (in Russian, English summary).

Vinogradova, E.B.: Influence of the change of photoperiodic regime during the female life on the progeny state in the blowfly, *Calliphora vicina* (Diptera, Calliphoridae). Entomol. Rev. USSR 55, 790–799 (1976a) (in Russian, English summary).

Vinogradova, E.B.: Embryonic photoperiodic sensitivity in two species of flesh flies *Parasarcophaga similis* and *Boettcherisca septentrionalis*. J. Insect Physiol. 22, 819–822 (1976b).

Vinogradova, E.B.: Influence of the photothermic conditions on the reproduction and imaginal diapause in *Aldrichina grahami* Aldrich (Diptera:Calliphoridae). In: Photoperiodic Reactions of the Insects. Zaslavsky,V.A. (ed.). Leningrad: "Nauka," 1978a, pp. 187–194 (in Russian).

Vinogradova, E.B.: Feature of the photothermic induction of the pupal diapause in flesh fly, *Boettcherisca septentrionalis* (Diptera, Sarcophagidea) .Zaslavsky, V.A. (ed.). Leningrad: "Nauka," 1978b, pp. 147–166 (in Russian).

Vinogradova, E.B.: Blowfly *Calliphora vicina* as a model insect for the physiological and ecological investigations (in Russian). Leningrad: "Nauka," 1984.

Vinogradova, E.B.: Peculiarity and ecological control of the imaginal diapause in blowfly, *Protophormia terraenovae* (Diptera, Calliphoridae). (in press).

Vinogradova, E.B., Bogdanova, T.P.: Correlation between the larval and imaginal diapause during the ontogenesis of the blowfly, *Calliphora vicina*. Dokladi Acad. Nauk SSSR 278, 505–507 (1984) (in Russian).

Vinogradova, E.B., Tsutskova, I.P.: Inheritance of larval diapause in the crossing of intraspecific forms of *Calliphora vicina* (Diptera, Calliphoridae). Entomol. Rev. USSR 57, 242–255 (1978) (in Russian).

Vinogradova, E.B., Zinovjeva, K.B.: The control of seasonal development in parasites of flesh and blowflies. I. Ecological control of pupal diapause in Sarcophagidae (Diptera). In: Host-Parasite Relationships in Insects. Zaslavsky, V.A. (ed.). Leningrad: "Nauka," 1972a, pp. 77–89 (in Russian, English summary).

Vinogradova, E.B., Zinovjeva, K.B.: Maternal induction of larval diapause in the blowfly, *Calliphora vicina*. J. Insect Physiol. 18, 2401–2409 (1972b).

Wood, F.E., Nordin, J.H.: Studies on the low temperature induced biogenesis of glycerol by adult *Protophormia terraenovae*. J. Insect Physiol. 22, 1665–1673(1976).

Zinovjeva, K.B.: Maternal induction of the larval diapause in *Lucilia hirsutula* Gr., *L. illustris* Meig., *Calliphora uralensis* Vill. (Diptera, Calliphoridae). In: Photoperiodic Reactions of the Insects. Zaslavsky,V.A. (ed.). Leningrad: "Nauka," 1978, pp. 80–94 (in Russian).

Zinovjeva, K.B., Vinogradova, E.B.: The control of seasonal development in parasites of flesh and blowflies. II. Ecological control of winter adaptations in *Calliphora vicina* R.-D. (Diptera, Calliphoridae). In:Host-parasite Relationships in Insects. Zaslavsky, V.A. (ed.). Leningrad: "Nauka," 1972, pp. 90–99 (in Russian, English summary).

Chapter 4
Geography of Density-Dependent Selection in Pitcher-Plant Mosquitoes

WILLIAM E. BRADSHAW and CHRISTINA M. HOLZAPFEL

Two questions appear to us to be fundamental to the understanding of life history evolution. First, what sorts of selective forces are impinging on a species over its range or a given environmental gradient? Second, what is the response by this species to this gradient? As has often been pointed out, there is abundant theory relating to general (many assumptions) or special (very specific) circumstances but fewer direct answers to the two questions posed above. Before one can predict the outcome of selection, one must first know how selection is operating; only then can one test the predictions about what organisms ought to do by comparing these predictions with what one actually finds them doing.

A few investigators have now sought to test theory in the laboratory by imposing artificial selection designed to distinguish between alternative predictions in *Drosophila* (Giesel and Zettler 1980; Taylor and Condra 1980; Mueller and Ayala 1981), bacteria (Luckinbill 1978), protozoa (Luckinbill 1979), and copepods (Bergmans 1984). Although we believe these experiments to be revealing and important tests of theoretical predictions, one must examine the experimental designs carefully to determine what is being selected. For instance, Taylor and Condra (1980) reared *Drosophila* on a small, fixed amount of medium allowing adults continuous opportunity to oviposit (supposed K selection) or on a generous supply of medium, but allowing only the first 100 flies that emerged to oviposit and allowing only 2 days in which to oviposit (supposed r selection). Because of the design of their experiment, they were primarily selecting for rapid development instead of r selection directly; instead of a K-selected strain, they selected for high adult survivorship instead of a higher equilibrium density. It is not, therefore, surprising that their primary result was that "r-selected" strains showed rapid development and "K-selected" strains showed high adult survivorship. Their strains did not differ in either capacity for increase or carrying capacity. Did these latter traits not respond to selection because the selection was not what it superficially appeared to be or because these traits were manifestations of a relatively inflexible genotype?

Instead of setting up a selection experiment in the laboratory, one can also examine life history changes along a gradient of selection. One is equally responsible for identifying the factors actually impinging on a population in the field as well in the laboratory. In this chapter, we define a geographical gradient in density-dependent, juvenile mortality in the pitcher-plant mosquito, *Wyeomyia smithii* (Coq.), and describe what we believe to be some of the potentials and pitfalls of such a basis from which to test theory.

Background on *W. smithii* and Its Host Plant

The mosquito *W. smithii,* is a member of the tribe Sabethini, most of which are tropical mosquitoes that oviposit into and develop in small, circumscribed (container) habitats such as bamboo internodes, bromeliad axils, and pitcher plants. *W. smithii* is no exception as it develops exclusively in the leaves of insectivorous plants belonging to the genus *Sarracenia*. Within *Sarracenia* there are only two species that retain water throughout the year; one, *S. psittacina* Michx., has a cryptic opening, has large internal hairs, and, in occupying the wettest subhabitat of the genus, is submerged much of the time. The other *Sarracenia* with persistent pitchers, and the only one to contain standing water regularly, is *S. purpurea* L., the principal habitat of *W. smithii*. *S. purpurea* is found from the Gulf of Mexico to Canada, where it ranges from Newfoundland (the provincial flower) to the McKenzie Territory and British Columbia. The range of *W. smithii* follows that of its host from about 30–55°N latitude.

Prey captured by the host leaf constitute the resource base of the community of living insect inhabitants (inquilines). Although prey decompose and disarticulate, arthropod head capsules or cephalothoraces tend to remain intact and identifiable. The number of prey captured is then quantifiable by examining the "gut contents" of a leaf and counting the number of head capsules of prey contained therein. We have shown previously that both mosquito biomass sustained by and pupae produced in a leaf are directly proportional to the number of prey it has captured (Bradshaw 1983a; Bradshaw and Holzapfel 1983). Although a leaf may live for over a year, it is most active in capturing prey only for a short period 2–4 weeks after it has opened. Thereafter, prey capture declines exponentially and becomes highly sporadic. Thus, resource availability is predictable but transient. Ovipositing females of *W. smithii* select only the very youngest leaves and, even though prey capture is proportional to leaf size, females oviposit preferentially into a smaller, younger leaf than into an older, larger one (Bradshaw 1983a). Attractability of a leaf to *W. smithii,* unlike to prey, is immediate on opening and declines exponentially from day zero. *W. smithii* will oviposit into a leaf that has opened the same day, has no standing water, and no prey, and is not yet fully open or hardened. The adaptive significance of this behavior seems straightforward: since mosquitoes hatch into a leaf prior to maximal prey capture, resources will peak as they develop and resource demand increases. Further, where mosquito resources are limiting, early cohorts may have a competitive advantage over later ones (Livdahl 1982). Plant and mosquito behavior combine to create an ideal field experimental system: both mosquito recruitment and resource input are transient and occur early in the life of a leaf. Densities or food levels may be manipulated soon after a leaf opens, populations censused with replacement for the duration of the experiment, and resource level (head capsules) determined by destructive sampling at a later date. Finally, *W. smithii* enter dormancy (diapause) as larvae under the influence of short-day photoperiod (Bradshaw 1976; Bradshaw and Lounibos 1977) and overwinter inside the pitcher plant leaves. Thus, during the winter and early spring, 100% of the population can be censused and resource availability during the previous year may be quantified at the same time.

On a geographical scale, *S. purpurea* along the Gulf Coast and northwards into the Carolina coastal plane live in wet, pine savannahs and the wet areas surrounding cypress domes and titi pocosins. In the mid-Atlantic states, *S. purpurea* may be found in sphagnous marshes, cedar swamps, or bogs. Further north and at higher elevations in the south, they occur in bogs, in sphagnous cattail (*Typha*) marshes, in marl pits, and in moist areas around kettle holes, ponds, and lakes, usually associated with ericaceous shrubbery and tamarack (*Larix larcina* DuRoi). At 54°N, a single meristem of *S. purpurea* may produce two leaves per year; at 30°N, it may produce a dozen leaves or more but seven or eight leaves are about average. It is important to emphasize that, although the macrohabitatof the host plant varies over this range, *W. smithii* is a single species of mosquito whose principal habitat is the water-filled leaves of a single species of plant. Along the Gulf Coast and at low elevations in North Carolina, *W. smithii* diapause as fourth instars although overwintering larvae may be found regularly as second, third, or fourth instars; further north and at high southern elevations, *W. smithii* diapause as third instars and 98–100% of the overwintering larvae are found in that stage.

In the present chapter, we start with a close examination of density-dependent development in a single wet pine savannah at the southern extent of *W. smithii*'s range in north Florida (30°N). We then consider density-dependent development at other localities along the Gulf Coast, at low elevations in North Carolina, and at higher latitudes and altitudes. Finally, we shall argue that the degree of density-dependent selection is measured by the concentration of *W. smithii* per unit resource and that this concentration declines continually with latitude throughout the range of this species.

Phenology of *W. smithii* and Its Host Leaves in North Florida

At Wilma in north Florida, pupation among the 1977–1978 overwintering generation of *W. smithii* (Figure 4.1A) commenced in mid-March, peaked the first week in April, and declined but continued thereafter. Starting in mid-April, small numbers of first instars appeared in the overwintered leaves. The first summer generation remained distinguishable until mid-May when the overlap in their developmental stages made it difficult to tell the two generations apart. We ceased censusing of the last year's leaves and the overwintering generation at this time. New leaves of the host plant first appeared in early to mid-April

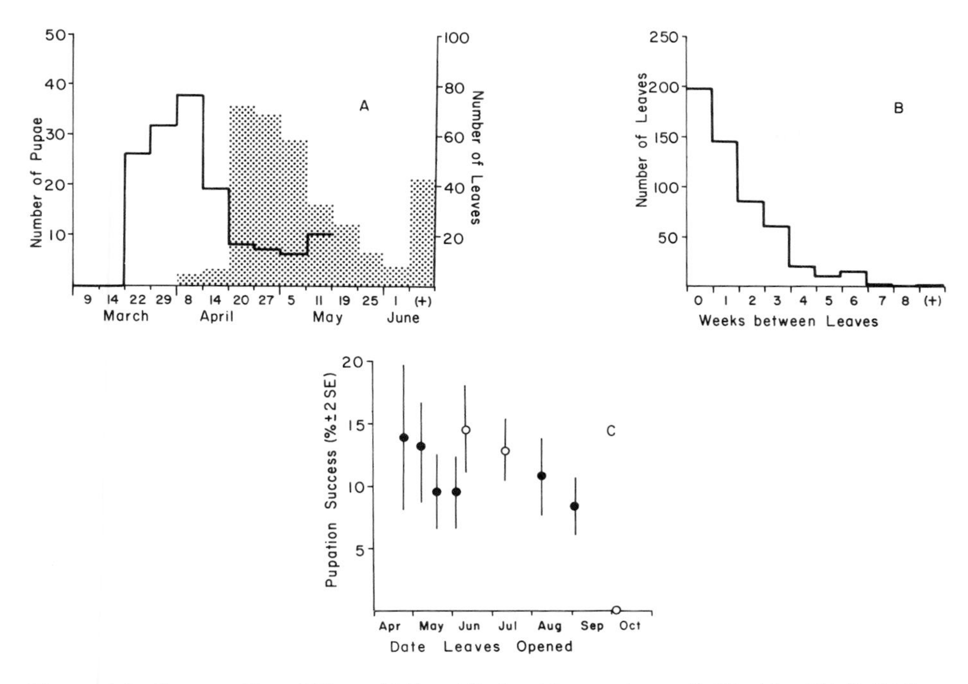

Figure 4.1. Seasonality of *W. smithii* and its host leaves in north Florida. (A) Solid line: number of *W. smithii* pupating in 55 leaves censused weekly from 14 March to 11 May, 1978. Shaded columns: number of plants putting on the first new leaf of the year among 158 plants observed weekly from 29 March until 1 June; (+) represents plants initiating their first leaf after 1 June. (B) Interval (weeks) between opening of 532 leaves on 119 plants after the opening of the first leaf from weekly observations 29 March to 1 August. If two leaves opened on the same plant during a given week, the older leaf was scored as the number of weeks since the previous leaf had opened and the younger leaf as zero weeks. (C) Pupation success among leaves opening from March to October. Pupation success is measured as (the cumulative number of pupae found in weekly censuses from the third to the eighth week after a leaf opened) ÷ (number of *W. smithii* found in the leaf four weeks after it had opened). Heterogeneity among ratios was observed (χ^2 = 84.47; $p < 0.001$; Snedecor 1956, Chap. 9) and significant deviations ($p < 0.01$) from mean expected pupation are indicated by open circles.

(Figure 4.1A) and appearance of the first leaf on plants peaked in late April. The influx of eggs and resultant first instars into the new leaves was immediate and, unlike the previous year's leaves, abundant. After the initial leaves opened, further leaves opened at a median interval of 0.5 weeks (Figure 4.1B) at a rate of about 8 new leaves per 100 plant days (Bradshaw and Holzapfel 1983). *W. smithii* continued to oviposit into and to complete development in these leaves through October. Pupation success averaged 8–15% in leaves opening between April and September but no larvae pupated in leaves that opened after September (Figure 4.1C). Since *W. smithii* diapause as larvae, the decline in pupation late in the season probably resulted from larvae entering diapause as a consequence of declining photoperiod (Bradshaw and Lounibos 1972, 1977; Evans and Brust 1972), food (Istock et al. 1975), or both. These observations suggest (1) that the growing season for *W. smithii* in north Florida is about 7.5 months long, (2) that *W. smithii* emerge about 1–3 weeks before new habitats become available, (3) that new leaves open rapidly and continually once a plant starts producing leaves, and (4) that *W. smithii* readily occupy and develop in these leaves as they appear.

Development During the Summer

Biomass Accumulation

The above observations indicate that in the spring when suitable habitats appear, they are rapidly exploited by an abundance of previously emerged adults. Do these adults saturate the resources and, if so, do their offspring produce sufficient numbers of fertile adults to maintain this saturation throughout the year? To answer this question, we set up an experiment to determine whether leaves could sustain either greater mosquito biomass or pupal production than was already being utilized by the resident mosquito population. On nine separate dates (5 May, 11 May, 19 May, 2 June, 6 June, 10 July, 8 August, 2 September, and 3 October) we selected recently opened leaves on a number of plants and added either (1) 100 first instar larvae "recruited" the same day from local pitcher plant leaves or (2) no mosquitoes but only the pitcher plant water equivalent to that required for transferring 100 first instar larvae. We censused each leaf one at a time in the field, commencing the third week after the experiment was started and continuing weekly through the eighth week. At each census, we counted and removed pupae, counted the number of larvae in each instar, and returned the larvae immediately to the leaf. We determined biomass by ascribing to each individual the mean dry weight for its respective stage of development (Bradshaw 1983b). At the conclusion of each experiment, we harvested the leaves destructively, measured the volume by filling each empty leaf to capacity with water and emptying it into a graduated cylinder, and quantified prey capture by the leaf as the number of head capsules of prey present in the base of the leaf. Among experiments started 5 and 11 May, the number of leaves was not as great as in the other experiments and there was considerable overlap in opening dates of the constituent leaves; consequently, we combined these two treatments to give eight starting dates and two experimental densities totalling, at each density, 76 leaves, which remained intact throughout the duration of their respective treatments.

As a measure of biomass sustained by each leaf, we calculated the biomass of larvae remaining in a leaf on the 8th week and added to it the mass of the cumulative sum of pupae produced up through the 8th week. We found that biomass was positively correlated with prey capture by the host leaf (Figure 4.2A). ANCOVA revealed heterogeneous regression coefficients ($F_{1,148} = 5.50$; $P < 0.05$) between leaves that received 0 or 100 larvae so that it was not realistic

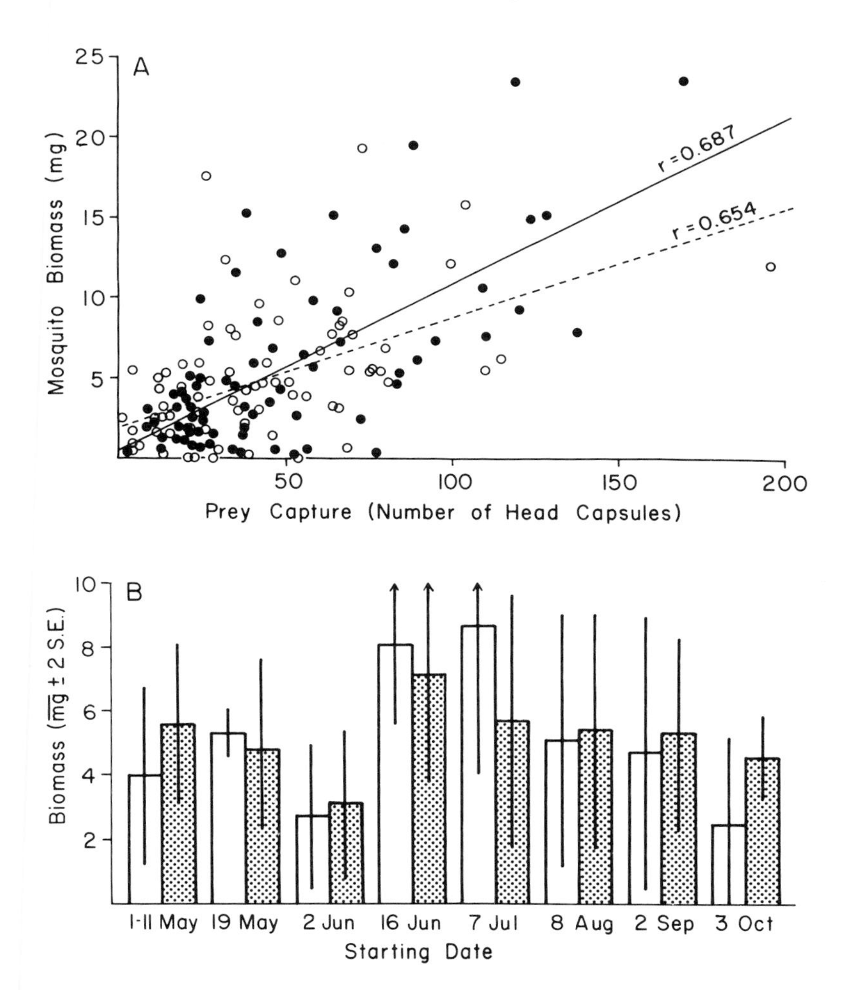

Figure 4.2. Biomass of *W. smithii* supported by leaves of *S. purpurea* in north Florida. (A) Comparison of mean aestival biomass accumulated in leaves to which either zero (——, (●) or 100 (—–, (○) supplemental larvae had been added. Both correlations are very highly significant ($p < 0.001$). (B) Comparison of mean aestival biomass accumulated in leaves to which either zero (open columns) or 100 (shaded columns) larvae had been added in experiments starting at eight different times. All leaves had opened during the previous week except 1–11 May when leaves may have been open for up to 2 weeks at the start of the experiment.

to test for differences between adjusted means. If we ignored the correlation between biomass and prey capture, two-way ANOVA (Figure 4.2B) of biomass with number added and starting date as treatments revealed no heterogeneity among cells ($F_{15,136} = 1.28$; $P > 0.05$). When, despite apparent heterogeneity among cells, we examined the least-squares ANOVA, we observed a barely "significant" effect of time of year ($F_{7,136} = 2.17$; $p < 0.05$) but no significant effect of *W. smithii* added ($F_{1,136} = 0.05$; $p >> 0.05$) or interaction between these treatments ($F_{7,136} = 0.53$; $p > 0.05$). Both an examination of the pattern of points about the regression lines in Figure 4.2A and the results of the ANOVA in Figure 4.2B provide no evidence that more biomass is potentially sustainable by leaves than is already being produced. Thus, leaves are, on the average, saturated with mosquito biomass throughout the growing season.

Pupation Success

The accumulation of biomass has little significance if this accumulation does not result in adult emergence. As an index of adult emergence, we counted the number of pupae in our weekly censuses in the convergence experiment above and expressed pupation success as the cumulative sum of pupae through week

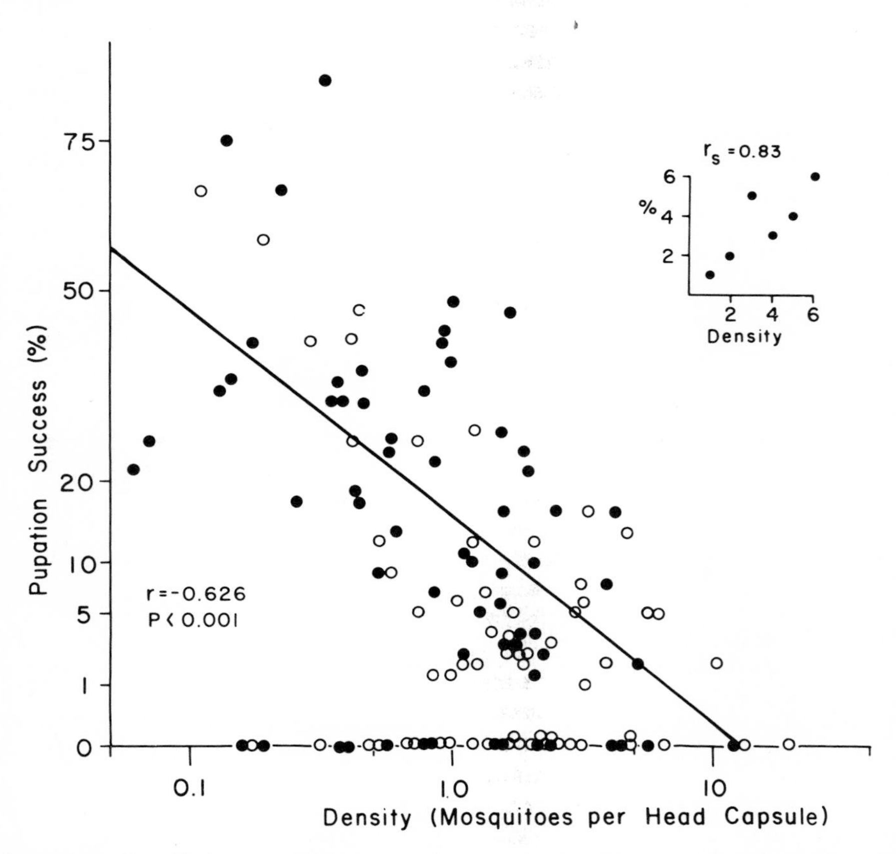

Figure 4.3. Cumulative pupation success for experiments illustrated in Figure 4.2 (omitting those started 3 October). Density is measured as *W. smithii* per head capsule of prey captured by the host leaf. (●), Leaves to which no supplemental larvae were added; (○), leaves to which 100 first instars were added. The solid line plots the regression when zero percent pupation is omitted. Inset: rank correlation between percentage of leaves sustaining no pupation (%) and density.

8 (after pupation had declined) divided by the number of mosquitoes (larvae + pupae) present on week 4 (by which time few new first instars were appearing in leaves and few larvae had started to die). In our analyses, we omitted consideration of the experiment started in October since we ascribed the absence of pupae to dormancy, not limitations of available resources. Since pupation success is essentially a percentage datum, we used an angular transformation prior to any analysis. Figure 4.3 shows that pupation success was inversely correlated with density. The addition of 100 larvae to leaves as described above resulted in a decline in pupation success (ANCOVA: regression coefficients, $F_{1,127} = 0.04$; $p > 0.05$; adjusted means, $F_{1,128} = 5.55$; $p < 0.05$). If the leaves in which there were no pupae are deleted from the ANCOVA, however, then the treatment effects do not differ (ANCOVA: slopes, $F_{1,86} = 0.16$, $p > 0.05$; adjusted means, $F_{1,87} = 1.05$, $p > 0.05$). Of the 64 leaves receiving 100 larvae, 15 failed to produce pupae and, of the 68 leaves receiving no supplemental larvae, 25 failed to produce pupae but these numbers are not significantly different ($\chi^2 = 3.14$; $p > 0.05$). To see if there was a density-dependent component to the lack of pupation in these leaves, we divided densities into six intervals (larvae per unit prey = 0.25 or less, 0.25–0.50, 0.51–1.00, 1.1–2.0, 2.1–4.0, and 4.1 or greater). Percentage of leaves not producing any pupae within each interval was then positively correlated with density interval ($r_s = 0.829$; $p = 0.029$). Thus, both the probability of producing any pupae as well as the incidence of pupation in productive leaves show a close correlation with density. These results support the previous general conclusion; namely, that there appear to be little or no resources available to convert into additional mosquito biomass or pupae during the aestival growing season.

Overwintering Ecology

The overwintering population of *W. smithii* in north Florida is comprised entirely of larvae in the diapausing (fourth) or in earlier instars (Bradshaw and Lounibos 1977; Bradshaw and Holzapfel 1983; Bradshaw 1983a). It is this population of larvae from which adults are recruited in the spring. In this section, we describe the establishment of the overwintering population and then factors that affect its survivorship and eventual pupation success.

Among the leaves in the convergence experiment to which no supplemental larvae were added, those that opened during the week prior to 3 October produced no pupae and those that opened during the week prior to 2 September produced pupae from 27 September to 24 October (Figure 4.4A) but none thereafter. We attribute the complete lack of pupation during November and December to the combined effects of density and the shortening day length that induce larval diapause. These larvae thus represent members of the population destined to overwinter in the leaves and not pupate until the following spring.

To see to what extent density affects the age structure of larvae at this time of year, we regressed mean age (average instar number) on density (*W. smithii* per prey captured by the host leaf). In this chapter, we shall be concerned with

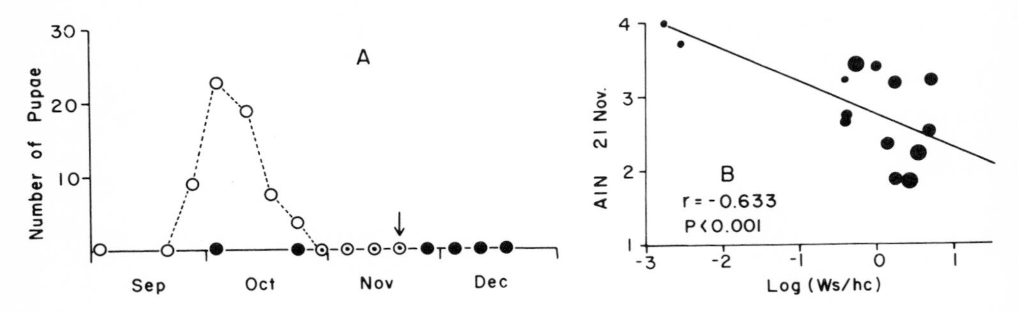

Figure 4.4. Autumnal development of *W. smithii* in north Florida in leaves to which no supplemental larvae had been added. (A) Pupation in leaves that opened on or during the week prior to 2 September (○) or 3 October (●). The latter produced no pupae. (B) Dependence of age structure (AIN, average instar number) on density measured as *W. smithii* per head capsule (*Ws/hc*) on 21 November (arrow in A) in the combined census of leaves shown in A. The number of larvae in each leaf is proportional to dot size: 1–10, 11–25, 26–50; >50 larvae.

what is happening to populations of mosquitoes, not populations of habitats. It is density that is encountered by the average individual that is important. This density is not necessarily that which prevails in the average habitat. Consequently, in calculating regressions and correlations, we summed the sums of squares and cross products over individuals. Because individuals in the same leaf may not be considered independent observations, we made error estimates and tested for significance using the number of leaves to generate degrees of freedom. After summing over individuals, there was a negative correlation between average instar number and density in leaves (Figure 4.4B). Thus, density at the end of the growing season has a depressing effect on the age structure of the overwintering population. This result is especially significant since pupation success in the spring is positively correlated with the average instar number of the overwintering population (Bradshaw andHolzapfel 1983).

To examine the effects of autumnal resource levels and mosquito densities further, we marked leaves that had opened from late October to mid-November and on 21 November added 25, 50, and 100 *W. smithii* recruited on the same day from nearby leaves, 50 or 100 ants (which constitute the main prey captured by *S. purpurea* at Wilma) from a nearby anthill, or nothing (control). We censused these leaves from 30 January through 6 February to determine the overwintering population size and then weekly from 9 March through 27 April. Figure 4.5 shows that pupation success in the spring was adversely affected by increased larval density and greatly enhanced by increased resources the previous fall.

Once established, at least two further factors affected survivorship and pupation success of overwintering larvae: (1) flooding and flushing of leaves by winter and spring rains, and (2) predation of host leaves by the noctuid moth, *Exyra* sp. These effects are illustrated by comparing pupation success in 97

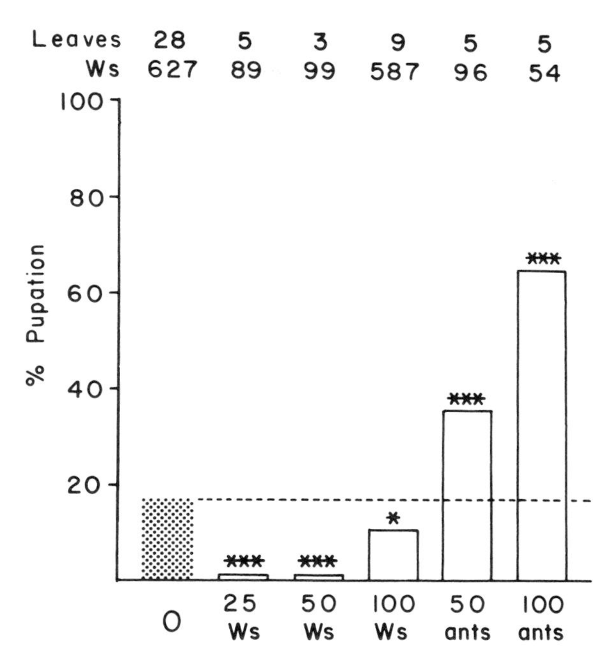

Figure 4.5. Percentage pupation during April and May in leaves that received nothing (0), 25, 50, or 100 first instar *W. smithii* (*Ws*) or 50 or 100 ants (predominant prey of *S. purpurea*) the previous November. χ^2 tests the null hypothesis that percentage pupation is the same in all treatments as in the control (0, screen and dashed line); significance is denoted by $^*p < 0.05$; $^{***}p < 0.001$.

leaves marked in November, 1977, and followed through the ensuing winter and spring (Table 4.1). Hereafter, we define the overwintering population as those individuals we found by censusing leaves on 30 January and 6 February, 1978. In the winter, these 97 leaves contained 1974 larvae of which only 8.3% pupated by 27 April. Moths ate three of these leaves; all filled with frass and never produced any pupae. On 6 and 9 March, heavy rains flooded 39 of the leaves while 56 remained above water on these dates. Flooding resulted in a precipitous decline in *W. smithii* populations. Survivorship through the spring was 30.7% in unflooded leaves and only 8.8% in flooded leaves. Among survivors of the flooding, however, pupation success was the same as among individuals in leaves that had not flooded. Thus, flooding and moth predation have a substantial effect on survivorship of the overwintering population but mosquitoes that survive flooding do not appear either to suffer adverse consequences as a result of the flooding or to obtain relief from any density-dependent factors that affect them.

Taken together, these results and observations show that among larvae that survive the hazards of flooding and habitat destruction, overwintering density is a substantial determinant of eventual pupation success. This density de-

Table 4.1 Survivorship and pupation of overwintering *W. smithii* in north Florida

	Number of leaves	Larval population 30 Jan–6 Feb	Percent surviving or pupating by 27 April	Percent total population that pupated	Percent post-flood survivors that pupated
All *W. smithii*	97[a]	1974	18.9	8.3	22.9
W. smithii in nonflooded leaves	56	922	30.7	14.5	23.0
W. smithii in flooded leaves	39	1024	8.8	2.9	22.5
W. smithii in moth-eaten leaves	3	28	0.0	0.0	—

[a]Sum of individual leaves = 98 since 1 leaf was both moth-eaten and flooded.

pendence is influenced by the age structure of the previous fall; these effects may then be reinforced or relaxed by an increase in density or resources, respectively. Density-dependent age structure in the autumnal population is in turn the final manifestation of density-dependent development, which has continued from the first summer generation. Thus, at Wilma in north Florida, *W. smithii* encounter density-dependent constraints year round; not even diapause provides an "escape in time" from the exigencies of resource depletion and density-dependent limitations to development.

Geographical Variation of Density-Dependent Development

The results and observations of the previous section indicate that resources and consequential density dependence at Wilma in north Florida may be measured by some function of prey captured by host leaves. In this section, we address two extensions of this conclusion. First, we ask to what extent prey capture is a reasonable measure of resource availability at other localities and then examine variation in resource-dependent densities encountered by *W. smithii* throughout its range from the Gulf of Mexico to northern Manitoba (30–54°N latitude). To answer these questions, we analyze the interaction between density and age structure of the overwintering population. We chose this population as our basis for comparison because *W. smithii* diapause exclusively as larvae; 100% of the population is thereby confined to leaves during the winter and spring when they may be readily sampled.

From the Gulf of Mexico north through low elevations in the Carolinas, *W. smithii* diapause primarily in the fourth instar. As at Wilma, the average instar number of the overwintering population is somewhere around three, not four (Bradshaw and Lounibos 1977), and is inversely correlated with mosquitoes per head capsule of prey captured by the host leaf (Bradshaw and Holzapfel, 1983).

At northern latitudes or higher altitudes, *W. smithii* diapause as third instars and 98–100% of the population overwinter in this stage (Bradshaw and Lounibos 1977). Istock et al. (1976), running convergence experiments similar to those we ran at Wilma, concluded that *W. smithii* in New York State experience density-dependent development in the spring and fall but escape these constraints during the summer. We have already reported (Bradshaw and Holzapfel 1983) that at the same latitude, average instar number of *W. smithii* during the spring is inversely correlated with mosquitoes per leaf. We now provide data from vernal censuses at six northern and one high elevation, southern locality where *W. smithii* diapause as third instars. As above, since we are interested in the conditions affecting the average individual, not conditions prevailing in the average habitat, we summed regressions over individuals but tested for significance using degrees of freedom based on the number of leaves sampled. We found a significant inverse correlation between average instar number and density (Figure 4.6) at four of the six northern localities but not the high-elevation, southern locality.

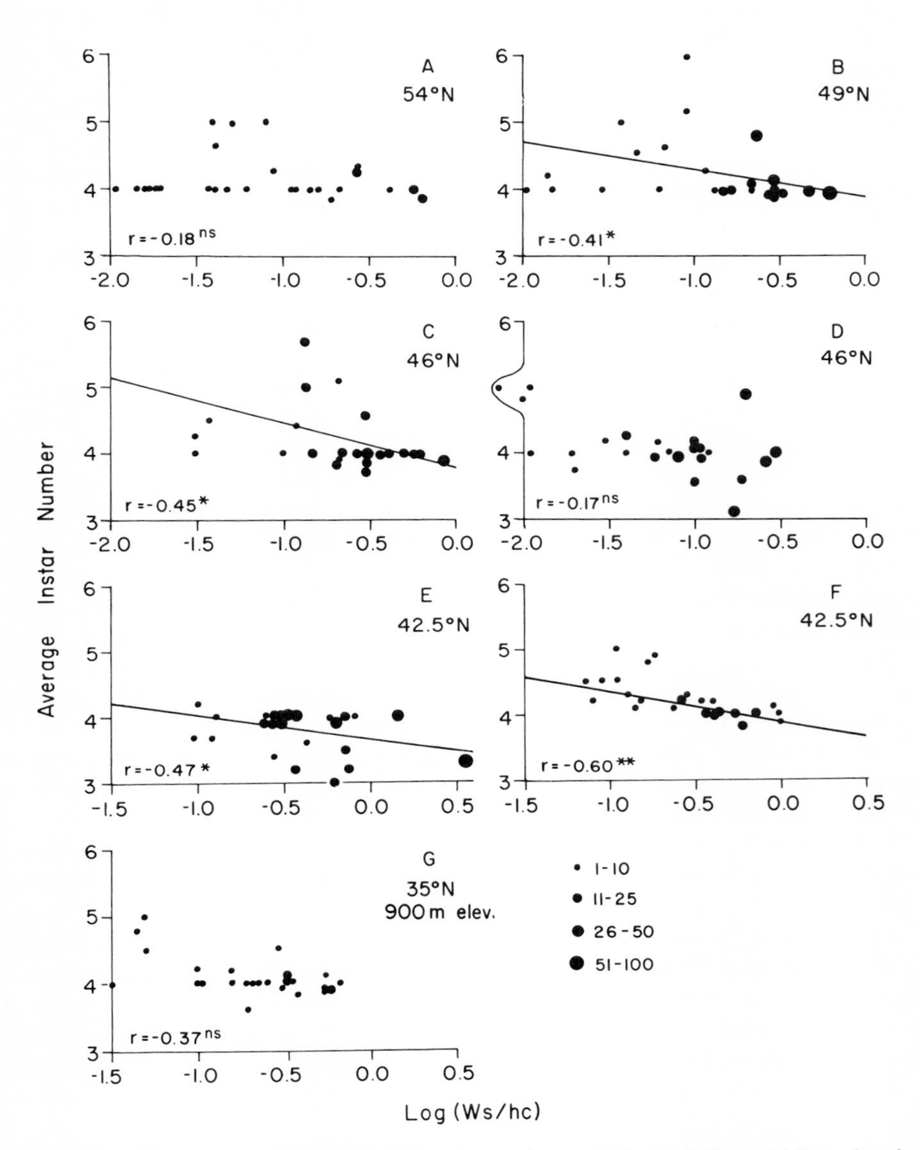

Figure 4.6. Dependence of age structure of vernal populations of *W. smithii* on density at northern (A–F) or high-elevation, southern (G) localities. Pupae are scored as 5 and pupal exuviae (adults) as 6. Density is measured as *W. smithii* per head capsule of prey in the leaf (*Ws/hc*). (A) Northern Manitoba, 2 July, 1979; (B) northern Ontario, 29 June, 1979; (C) northern Wisconsin, 28 June, 1979; (D) northern Maine, 22 June, 1979; (E) southern Michigan, 17 May, 1976; (F) eastern Massachusetts, 12 June, 1976; (G) western North Carolina, 31 May, 1976. Significance of regressions is designated by NS, $p > 0.05$; $^{*}p < 0.05$; $^{**}p < 0.01$.

A clue to the reason for the lack of correlation between development and density in these three localities emerges from consideration of relative resource availability. We continue to be interested in conditions experienced by the average individual at a locality. Consequently, we calculated the mean intraspecific crowding of *W. smithii* per unit resource (Hurlbert 1978):

$$\frac{\overset{*}{Ws}}{hc} = \frac{\Sigma \dfrac{Ws_i\,(Ws_i - 1)}{hc_i}}{\Sigma Ws_i}$$

where Ws_i is the number of *W. smithii* in the *i*th leaf and hc_i is the number of head capsules of prey captured by that leaf. When we plotted census data from four localities along the Gulf Coast, two from a low elevation in North Carolina and nine from more northern localities, we found an inverse correlation between mean crowding per unit resource of the overwintering generation and latitude (Figure 4.7). The average of three censuses at two localities at high elevation in North Carolina produced a mean crowding per unit resource equivalent to about 43°N latitude (Figure 4.7, open circles). The three regressions in Figure 4.6 that did not yield significant correlations between average instar number and density represent the two lowest mean crowdings per unit resource along the latitudinal gradient and the lowest of the three censuses at high elevation in North Carolina. Thus, when resources become sufficiently abundant, there no longer is a significant correlation between development and density.

A more complete manifestation of this trend may be seen by examining the correlation coefficients relating average instar number to density (number of *W. smithii* per head capsule of prey) along this same gradient. We have such data for 13 of the 17 separate localities shown in Figure 4.7. Because the censuses at southern localities were taken in midwinter and at northern and high elevation localities in the spring at different times of different years, we used rank rather than least-squares correlation. As seen in the inset to Figure 4.7, there was a very close correlation ($r_s = 0.84$; $p < 0.01$) between the correlation coefficient relating age structure to density and the mean crowding per unit resource at that locality. This result means that as the number of individuals that must share a given resource increases, their development more clearly becomes dependent on the quantity of those resources.

Conclusions

Many of our analyses have considered what is happening to the average individual in a population. We believe that this type of analysis is, unfortunately, greatly underutilized. It is far easier to sum regressions, ANOVA, and other statistics over samples, be they pitcher-plant leaves or quadrats, particularly now that canned computer programs make for statistics with fewer tears.

There are two main considerations here. Summing over individuals avoids giving undue emphasis to aberrant subhabitats with few larvae in them but it

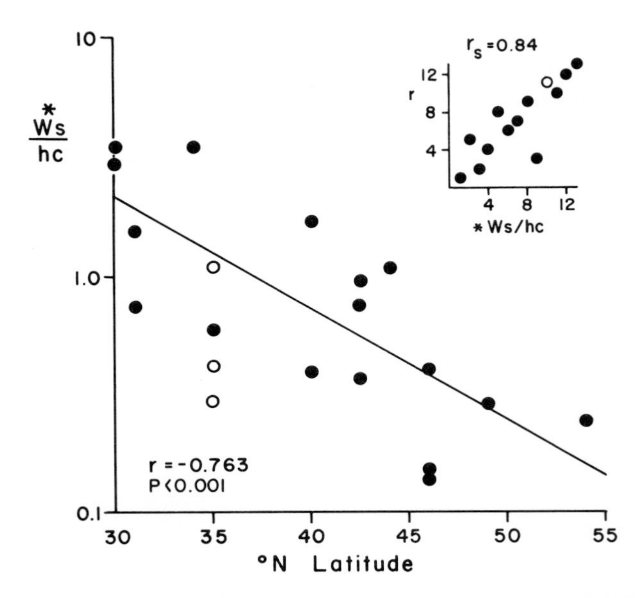

Figure 4.7. Resource availability to the average larva over the range of *W. smithii*. The regression was run on the latitudinal gradient (●); the mean of three censuses at 900 m in North Carolina (○) intercepts the regression line at 43°N. Resource availability is measured as mean crowding of *W. smithii* per head capsule of prey in the leaf during January and February at 30–35°N and during May to July at high elevation at 35°N and all points 40–54°N. Inset: rank correlation between the correlation coefficient (*r*) for the regression of average instar number on density as shown in Figure 4.6 and resource availability (**Ws/hc*), for the 13 localities for which we have such data.

does give much more weight to the subhabitats with many larvae. One's uneasiness here stems from the apprehension that somehow something important is being missed by deemphasizing what is going on in the peripheral or sparsely populated habitats. The underlying assumption in the former analysis is that all habitats and all individuals are equally important. This assumption and uneasiness loses track of two points. First, one is generally interested in what is happening to a population of organisms, not a population of habitats, and second, a leaf with more larvae produces absolutely more pupae than a leaf with fewer larvae. What happens within the more heavily populated leaf will have a greater impact on which mosquitoes will comprise the breeding population than what happens within the more sparsely populated leaf. In Figure 4.3 we presented analyses both including leaves that produced no pupae and excluding these leaves. We believe the latter to be the more important comparison as, in this case, we are dealing with the subset of mosquitoes that will produce the next generation. The proximal causes of developmental failure in leaves not producing pupae has little adaptive significance. The important selective forces are those impinging upon and overcome or exploited by the mosquitoes that eventually reproduce. In the best of all possible worlds, density-dependent effects in the parental generation might then be summed over offspring produced, not the parents themselves, and certainly not over all the subhabitats in which the parental generation might have lived.

Flooding and predation of leaves are severe causes of mortality (Table 4.1). Along with freezing further north, these aspects of seasonal harshness probably constitute the greatest sources of density-independent mortality among the overwintering generation. Yet, unlike the scenario envisioned by Roughgarden (1971), these aspects of seasonal harshness do not relieve the effects of density-dependent constraints in either the overwintering (Table 4.1) or subsequent (Figure 4.2) generation. In *W. smithii,* density dependence exerts its effects in the fall during the establishment of the overwintering population. Once resources have been depleted during the fall and have either been taken up by the plant or incorporated into mosquito bodies, there are no resources remaining for survivors of environmental harshness. Unless density-independent mortality occurs before the density-dependent bottleneck during the life cycle of an organism, it is not necessarily going to relieve that density-dependence. The very important point is that to assess the relative importance of density-dependent versus independent causes of mortality as selective factors molding life history traits of a species, the timing of events must be taken into account.

We are quick to point out that the correlation shown in Figure 4.7 is clearly not perfect, even though measurements were always made on the overwintering generation when 100% of the population was confined to the leaves. Local conditions vary in both space and time; these variations affect measurements of mean crowding as well as measurements of density-dependence using regressions of age, biomass, pupation success, or other life history traits on density. The perception of a gradient depends on the number of observations along the gradient. The perception of geographical variation in density-dependent development among populations of *W. smithii* might have appeared very different had we sampled only two or three localities drawn at random from those shown in Figures 4.6 and 4.7. The literature is full of examples of studies using two points, one insular, one continental; one southern, one northern; one low elevation, one high elevation; one wet, one dry, and so on; but, apart from the problem of controls, two points do not indicate whether the response to the gradient is linear or nonlinear, continuous or discontinuous, or provide an idea of within- versus between-region variation. If the gradient is carefully chosen, one usually wants to know the quantitative relationship between the environmental gradient and the organismic or populational response. Three or even five points are precious few upon which to base a correlation, a regression, or an error estimate; for a two-dimensional gradient (e.g., latitude and altitude) they are simply inadequate.

In the case of *W. smithii*, we concluded that there is a continual, linear decline in density-dependent juvenile mortality from the Gulf of Mexico to northern Canada. To the best of our knowledge, *W. smithii* is the first case for which this statement can be made for the entire latitudinal range of a species. The average mean crowding per unit resource (Figure 4.7) from three censuses of two localities at 900 m in North Carolina intercepts the regression line at 43°N, providing an altitude/latitude conversion factor of 113 m/°N. This value falls between the altitude/latitude conversion factors of 99 m/°N for length of growing season (mean number of frost-free days) over this same range and 122m/°N for Hopkins "bioclimatic law" relating phenology to geography (Bradshaw 1976). More studies of the present kind will tell us whether these comparative values are coincidence or represent a general ecogeographical pattern.

Only when one understands the selective forces actually impinging upon populations of a species will one be able to make predictions about the consequesnces of this selection on life history. The quantitative perception of the selection gradient will then provide the basis for quantitative testing of our models. Initially, it makes no difference whether these populations exist in the laboratory or are spread over several continents. Ultimately, however, scholars of life history evolution will have to formulate hypotheses and to identify species and selection gradients that permit us to test theory in the crucible of natural, not artificial, selection.

Acknowledgments Much of the field data in this paper were collected while the authors were research fellows at the Tall Timbers Research Station. We thank Ms. Lee Szyska, Dr. David Schleimer, Dr. Robert Godfrey, Dr. James Farr, Dr. Thomas Gibson, Mr. John Leydon, Jr., Dr. Kent Fiala, and Ms. Barbara and Laura Kittredge for help in the field. Mrs. Ada Simons, Mrs. Bridget Gullikson, and Mr. Clyde Belcher gave permission to do field work on their land; the staff of the Appalachicola National Forest, Dr. F. C. Evans, Dr. Richard Bruce, and Dr. William Peters, gave permission to use land under their supervision. Much welcome logistical support and encouragement were provided by Dr. and Mrs. Donald Strong, Dr. and Mrs. Irving Cantrall, Mr. and Mrs. Carl Holzapfel, Mr. and Mrs. John Bradshaw, Jr., Dr. and Mrs. Carroll Williams, Dr. Byron Ingram, Dr. and Mrs. John Lundberg, Dr. D. Bruce Means, Dr. and Mrs. E. V. Komarek, and Mr. and Mrs. Lyell Porter. Dr. Fritz Taylor read and provided many helpful suggestions for this manuscript. This research was supported by the National Science Foundation Grants GB-41753 and DEB-00918-A01 and by the Tall Timbers Research Station.

References

Barclay, H.J., Gregory, P.T.: An experimental test of models predicting life-history characteristics. Am. Natur. 117, 944–961 (1981).

Bergmans, M.: Life history adaptation to demographic regime in laboratory-cultured *Tisbe furcata* (Copepoda, Harpacticoida). Evolution 38, 292–299(1984).

Bradshaw, W.E.: Geography of photoperiodic response in a diapausing mosquito. Nature 262, 384–386 (1976).

Bradshaw, W.E.: Interaction between the mosquito, *Wyeomyia smithii,* the midge, *Metriocnemus knabi,* and their carnivorous host, *Sarracenia purpurea*. In: Phytotelmata: Terrestrial Plants as Hosts of Aquatic Insect Communities. Frank, J.H., Lounibos, L.P. (eds.). Medford: Plexus, 1983a, pp. 161–189.

Bradshaw, W.E.: Estimating biomass of mosquito populations. Environ. Entomol. 12, 779–781 (1983b).
Bradshaw, W.E., Holzapfel, C.M.: Life cycle strategies in *Wyeomyia smithii:* Seasonal and geographic adaptations. In: Diapause and Life Cycle Strategies in Insects. Brown, V.K., Hodek, I. (eds.). The Hague: Junk, 1983, pp. 169–187.
Bradshaw, W.E., Lounibos, L.P.: Photoperiodic control of development in the pitcher-plant mosquito, *Wyeomyia smithii.* Can. J. Zool. 50, 713–719 (1972).
Bradshaw, W.E., Lounibos, L.P.: Evolution of dormancy and its photoperiodic control in pitcher-plant mosquitoes. Evolution 31, 546–567 (1977).
Evans, K.W., Brust, R.A.: Induction and termination of diapause in *Wyeomyia smithii* (Diptera: Culicidae), and larval survival studies at low and subzero temperatures. Can. Entomol. 104, 1937–1950 (1972).
Giesel, J.T., Zettler, E.E.: Genetic correlations of life historical parameters and certain fitness indices in *Drosophila melanogaster*: r_m, v_s, diet breadth. Oecologia 47, 299–302 (1980).
Hurlbert, S.H.: The measurement of niche overlap and some relatives. Ecology 59, 67–77 (1978).
Istock, C.A., Vavra, K.J., Zimmer, H.: Ecology and evolution of the pitcher-plant mosquito: 3. Resources tracking by a natural population. Evolution 30, 548–557 (1976).
Istock, C.A., Wasserman, S.S., Zimmer, H.: Ecology and evolution of the pitcher-plant mosquito: 1. Population dynamics and laboratory response to food and population density. Evolution 29, 296–312 (1975).
Livdahl, T.P.: Competition within and between hatching cohorts of a treehole mosquito. Ecology 63, 1751–1760 (1982).
Luckinbill, L.S.: r- and K-selection in experimental populations of *Escherichia coli.* Science 202, 1201–1203 (1978).
Luckinbill, L.S.: Selection and the r/K continuum in experimental populations of protozoa. Am. Natur. 113, 427–437 (1979).
Mueller, L.D., Ayala, F.J.: Trade-off between r-selection and K-selection in *Drosophila* populations. Proc. Natl. Acad. Sci. USA 78, 1303–1305 (1981).
Roughgarden, J.: Density-dependent natural selection. Ecology 52, 453–468(1971).
Snedecor, G.W.: Statistical Methods. Ames: The Iowa State University Press, 1956.
Taylor, C.E., Condra, C.: r- and K-selection in *Drosophila pseudoobscura.* Evolution 34, 1183–1193 (1980).

Chapter 5

Geographical Patterns in the Photoperiodic Induction of Hibernal Diapause

FRITZ TAYLOR and JOHN B. SPALDING

By studying geographical variation in the characteristics of plant and animal species, biologists have gained valuable insights as to the potential for phenotypic and genotypic variability within a species and variability in the interactions among traits in different populations, as well as to the relationships between phenotypes and important variables in the environment (Clausen et al. 1940; Heisey 1965; Teeri and Stowe 1976; Berven and Gill 1983). Comparisons among populations occurring at different latitudes and elevations have proven especially useful because one can assume that there are underlying gradients in the physical environment, particularly in day length, temperature, and duration of the growing season. An extensive literature deals with geographical, or clinal (Endler 1977), patterns in the use of day length by arthropods to time their preparations for the onset of winter conditions (Masaki 1961, Danilevskii et al. 1970; Beck 1980; Saunders 1982) and we feel that this isan appropriate time to assess what has been learned from these studies.

In the most often repeated experiment to measure the photoperiodic response of a population, samples of an arthropod population are reared under a series of constant photoperiods at a constant temperature and the percentage diapausing is recorded at each photoperiod. Arthropods typically exhibit a long-day response for the induction of hibernal diapause, meaning that short-day conditions, signalling the onset of winter, induce diapause. In this chapter, we summarize the responses of 27 species of insects, 2 species of mites, and 2 species of *Daphnia*. These 31 species represent a distillation using stringent criteria for inclusion of data sets from well over 100 papers. In addition to analyzing latitudinal patterns in the median (or critical photoperiod [CPP]) and variability of the population response to photoperiod, we consider the importance of changes in day length within the sensitive period during which the-diapause decision is made and we suggest directions for further research on the timing of diapause induction.

Analyses to Be Performed

The data sets included in these analyses are summarized in Table 5.1. Two different sets of criteria were used for inclusion of photoperiodic response data depending on whether the data were to be used to estimate the median or the spread of the response. A photoperiodic response curve consists of a series of points, each representing the percentage of diapausing individuals in a sample reared at a particular day length. To examine environmentally relevant photoperiodic responses, we applied the following criteria in determining which points contributed useful information about the median photoperiodic response: (1) A day length giving 0 or 100% response was kept if it was within 1 hr of the adjacent point showing a less extreme response. More distant points provide misleading information as to the day lengths first producing 0 or 100% diapause. (2) Additional points showing 0 or 100% responses at longer or shorter day lengths, respectively, were not used. (3) Points in the decreasing section of the photoperiodic response curve, in which longer day lengths produce lower percentage diapause, were kept; points showing the opposite relationship may occur at unnaturally long or short day lengths and these were omitted. After the above criteria were applied, a photoperiodic response curve was used to estimate the median of the response if there were three or more points whose range included 50% diapause. Photoperiodic response curves based on two points, which were not common, were also used if the day lengths were separated by no more than 0.5 hr or if neither point showed 0 or 100% diapause.

Photoperiodic response curves used to estimate the spread of the response were subjected to further requirements because of the need for more information to describe a shape rather than a central tendency (Copenhaver and Mielke 1977). One or more points had to occur within each of the ranges 0–15% and 85–100% diapause and at least one point within 15–85%. A photoperiodic response curve of only three points, but not meeting these criteria, was used if the points were nomore than 0.5 hr apart and only one showed a response of 0 or 100% diapause.

Those data sets that met the above criteria were subjected to a probit analysis, which fits a cumulative normal distribution to the photoperiodic response curve, or to a logit analysis, which fits a cumulative logistic distribution (Finney 1978). In most cases the fit of the probit curve was superior. For 10 photoperiodic response curves (representing seven species) a logistic curve more closely matched the points; in these cases diapause responses of 0 and 100% were changed to 0.01 and 99.99%, respectively, because the transformation used to linearize the logistic function is undefined for these extreme values.

To compute the day length for a particular latitude and date, we used a formula provided by the Nautical Almanac Office (1980).

Table 5.1. Data sets included in the analysis

Order	Family	Species	Symbol[a]	Range[b] of Latitudes	Number of locations	Source
Psocoptera	Psocidae	*Amphigerontia bifasciata*[c]	A	60	1	Glinyanaya 1975
Orthoptera	Gryllidae	*Pteronemobius taprobenensis*	B	28–34	4	Masaki 1978, 1979
Homoptera	Psyllidae	*Psylla pyricola*[c]	C	50	1	McMullen and Jong 1976
		Psylla pyricola	C	33–38	2	Oldfield 1970
Coleoptera	Coccinellidae	*Epilachna varivestris*	D	34	1	Taylor and Schrader 1984
Neuroptera	Chysopidae	*Chrysopa carnea*	E	33–42	2	Tauber and Tauber 1972a
		Chrysopa nigricornis[c]	F	42	1	Tauber and Tauber 1972b
		Chrysopa oculata	G	42	1	Propp et al. 1969
Lepidoptera	Arctiidae	*Hyphantria cunea*	H	36	1	Masaki et al. 1968
	Gelechiidae	*Pectinophora gossypiella*	I	32	1	Ankersmit and Adkisson 1967
	Lymantriidae	*Orgyia thyellina*[c]	J	40	1	Kimura and Masaki 1977
	Noctuidae	*Acronycta rumicis*[c]	K	50	1	Danilevskii 1965
		Barathra brassicae[c]	L	42–50	2	Danilevskii 1965
	Nymphalidae	*Limenitis archippus*	M	39–44	2	Hong and Platt 1975
	Pieridae	*Pieris brassicae*	N	43–52	3	Danilevskii 1965
	Pyralidae	*Chilo suppressalis* (1967, 1968 data)	O	34–41, 32–43	7, 10	Kishino 1970
		Diatraea saccharalis[c]	P	32–43	1	Fuchs et al. 1979
		Homoeosoma electellum[c]	R	27	1	Teetes et al. 1969
		Ostrinia nubilalis	S	43	1	Danilevskii 1965 (data of Du Chzhen-ven)

	Tortricidae	*Grapholitha molesta*[c]	T	34	1	Dickson 1949
		Laspeyresia pomonella	U	34	1	Dickson 1949
		Laspeyresia pomonella	U	39	1	Shel'deshova 1967
Diptera	Culicidae	*Aedes atropalpus*	V	30–42	2	Anderson 1968
		Aedes atropalpus[c]	V	34–45	2	Beach 1978
		Aedes atropalpus	V	44	1	Kalpage and Brust 1974
		Aedes sierrensis	W	42		Jordan and Bradshaw 1978
		Culex pipiens	2	42	1	Spielman and Wong 1973
		Toxorhynchites rutilis[c]	3	40	1	Bradshaw and Holzapfel 1977
		Toxorhynchites rutilis	3	30–40	2	Trimble and Smith 1979
		Wyeomyia smithii	4	34–49	7	Bradshaw and Lounibos 1977
		Wyeomyia smithii	4	42	1	Bradshaw and Phillips 1980
		Wyeomyia smithii	4	50	1	Evans and Brust 1972
	Drosophilidae	*Drosophila littoralis*[c]	6	64	1	Lumme and Oikarinen 1977
		Drosophila phalerata	7	61–65	2	Muona and Lumme 1981
Acari	Tetranychidae	*Metatetranychus ulmi*	9	52	1	Lees 1953
		Tetranychus urticae	*	41–45	2	Bondarenko and Kuan Khai-Yuan[1] 1958
Cladocera	Daphnidae	*Daphnia middendorfiana*		71	1	Stross 1969
		Daphnia pulex		46	1	Stross and Hill 1965

[a]Used in figures.
[b]To nearest degree, analyses used nearest 1/10 degree.
[c]Data adequate only to estimate critical photoperiod.

Geographical Patterns

Median Response versus Latitude

The signal that arthropods generally use to induce diapause is day length, which varies predictably with latitude and over time (Figure 5.1). Because of the high sensitivity of photoreceptors involved in diapause induction, it is necessary to include civil twilight in the determination of day length (Beck1980; Saunders 1982). When civil twilight is included, day lengths of 24 hrs occur around the summer solstice (June 21 or day 172) at latitudes above 62°. Days are longer at higher latitudes until around October 5 (day 278), which is actually after the autumnal equinox on September 23 (or day 266) because of the effects of civil twilight; thereafter, days are shorter at higher latitudes.

Generally, it is expected, and has been found, that median, or critical, photoperiods (CPP) occurring before October 5 will increase with latitude for local populations from a particular species. This relationship is expected because, with an earlier end of season at higher latitudes, diapause should be initiated at an earlier date corresponding to a longer day length. After the equinox, the CPP should generally decrease with latitude (but see below). Ten out of the 12 latitudinal gradients, for which the CPPs occur before the autumnal equinox, show positive slopes (Figure 5.2a). The exceptions are *Pteronemobius tapro-*

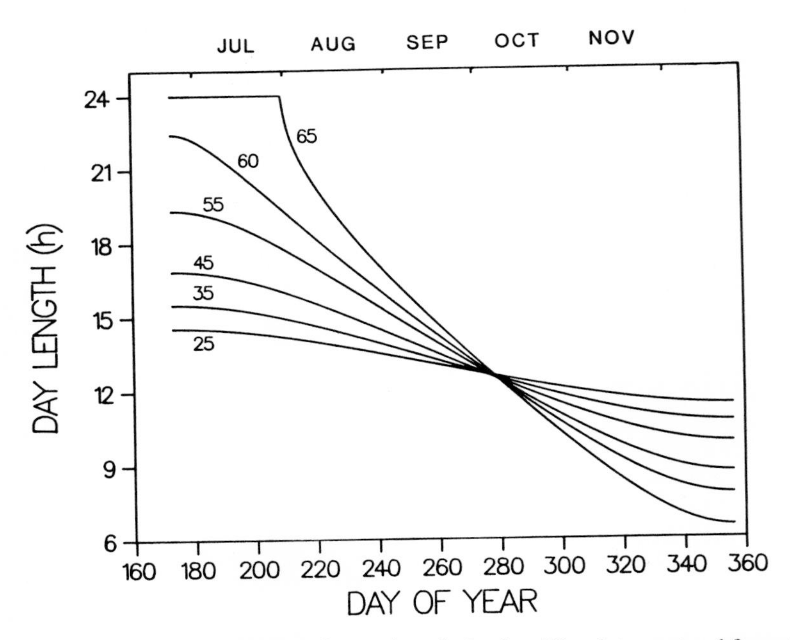

Figure 5.1. Photoperiod regimens for various latitudes. The dates extend from the summer solstice on June 21 (day 172) to the winter solstice on December 23 (day 357). Day lengths include civil twilight. The numbers indicate the latitudes of the curves.

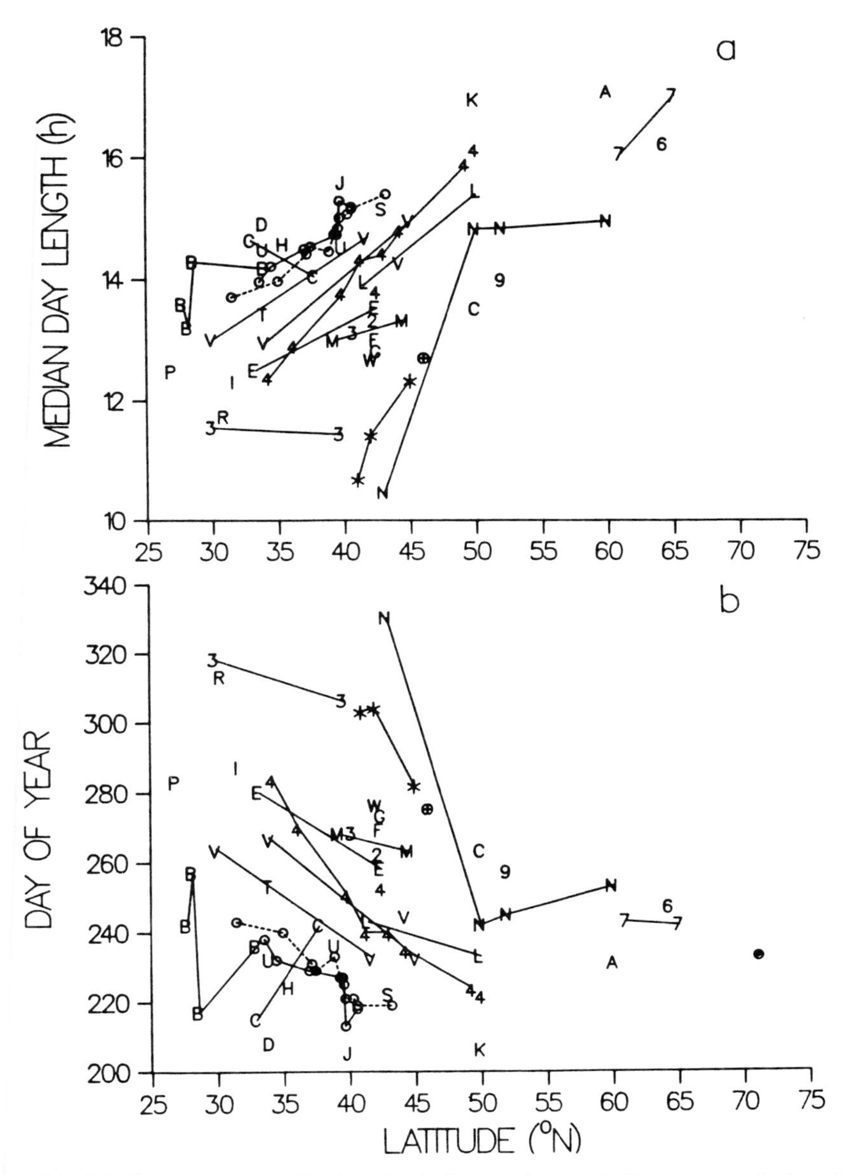

Figure 5.2. Median responses. Each point indicates the result for one population. Populations considered in the same study at the same time are connected with straight lines. The symbols refer to species listed in Table 5.1. (The dashed line for *Chilo suppressalis* [o] represents data from 1968). (a) Median photoperiodic response. (b) Dates corresponding to median photoperiods.

banensis (B) and *Psylla pyricola* (C). (Letters, numbers, and symbols are those used for a particular species in the figures and are listed in Table 5.1. We shall refer to species without citing papers when these are provided in Table 5.1). As discussed below, the CPPs for *Toxorhynchites rutilis* (3, Trimble and Smith 1979) occur after the autumnal equinox. The slopes for data sets containing more than two points are similar, ranging from 0.16 to 0.25 (Table 5.2). *Pteronemobius taprobanensis* (B) was not included in the regression analysis because a linear hypothesis was clearly inappropriate (Figures 5.2 and 5.3). The two species for which two gradients have been sampled show consistent results: two slopes of 0.16 for two different years in *Chilo suppressalis* (o) and slopes of 0.14 and 0.18 for separate studies on *Aedes atropalpus* (V). The single points lie close to the scatter of points produced by the gradients, but the CPP for the extreme high latitude population of *Daphnia middendorfiana* (22.9 hr, but not shown in Figure 5.2a because it required that the ordinate be extended too far) was higher than a linear extrapolation from the points at lower latitudes, as might be expected since day length for a given date changes faster with increasing latitude at higher latitudes (Figure 5.1).

It is generally assumed that changes in CPP with latitude will translate into earlier median times of diapause induction at higher latitudes. We tested this assumption by computing the dates corresponding to the points in Figure 5.2a and replotting the results in Figure 5.2b. This expectation was confirmed for 11 out of 14 gradients, but the results have several interesting features. Most noticeable is that the fairly tight relationship with latitude seen in Figure 5.2a

Table 5.2. Geographical patterns for species occurring at two or more locations

Species	Symbol	Critical day length		Critical date		n	Range of latitude
		Slope	r^2	Slope	r^2		
Psylla pyricola	C	−0.125		5.2		2	33–38
Chrysopa carnea	E	0.11		-2.4		2	33–43
Barathra brassicae	L	0.13		−0.94		2	42–50
Limenitis archippus	M	0.06		−0.95		2	39–45
Chilo suppressalis (1967 data)	o	0.16	0.88*[b]	−2.6	0.72*	7	34–41
Chilo suppressalis (1968 data)	o	0.16	0.93***	−2.4	0.85***	10	32–43
Aedes atropalpus (Anderson 1968)	V	0.14		−2.7		2	30–42
Aedes atropalpus (Beach 1978)	V	0.18		−3.1		2	34–45
Toxorhynchites rutilis	3	−0.011		−1.2		2	30–40
Wyeomyia smithii	4	0.21	0.74*	−3.5	0.54 NS	7	34–47
Wyeomyia smithii[a]	4	0.23	0.99***	−4.0	0.93**	7	34–49
Drosophila phalerata	7	0.24		−0.25		2	61–65
Tetranychus urticae	*	0.23	0.90 NS	−5.7	0.92 NS	3	41–45

[a]Elevation-corrected latitude.

[b]Levels of statistical significance: *$p<0.05$; **$p<0.001$; ***$p<0.0002$.

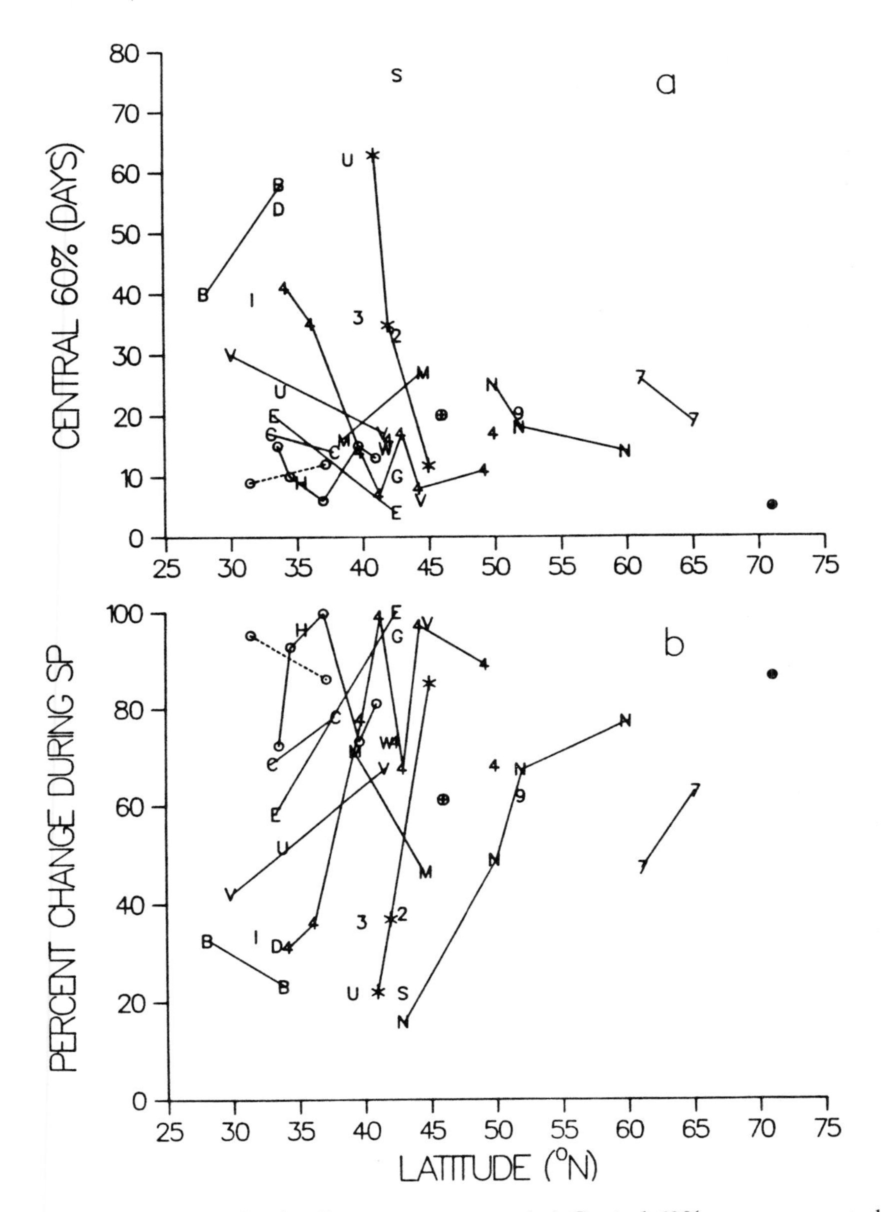

Figure 5.3. Variability in the diapause response. (a.) Central 60% response on a date scale. For each population, this is the range of dates between the dates corresponding to the 20% and 80% responses on the photoperiodic scale. (b.) Percentage difference in the photoperiodic response over a 20-day interval centered at the critical photoperiod. The dates were determined 10 days before and after the CPP and the corresponding day lengths computed for the appropriate latitude. When these photoperiods are projected onto the photoperiodic response curve, they provide a measure of the sensitivity of the population to changes in day length during a SP of 20 days centered at the CPP. Populations considered in the same study at the same time are connected with straight lines. The symbols refer to species listed in Table 5.1.

is lost. The slopes for the various gradients differ considerably (Table 5.2) and there is no tendency for diapause to be induced earlier at high than at low latitudes when all data sets are considered at once. The slope is highly significant for *Wyeomyia smithii* (4), which is the only case where elevational data were available, when elevation is included, but it is not significant when latitude is considered alone. (The latitudes for this species in Figures 5.2 and 5.3 are corrected for elevation based on the results from a multiple regression using the median photoperiods [see Bradshaw and Lounibos 1977].)

For increase in CPP at a higher latitude before October 5 (day 278) to correspond to an earlier date, it must be greater than the increase of day length that occurs with latitude on the same date. Thus, in one instance, *Pieris brassicae* (N), longer CPPs at higher latitudes actually translated into later median dates. On the other hand, one species that showed a shorter critical day length at higher latitudes, *Toxorhynchites rutilus* (3), now is seen to diapause at an earlier date at the higher latitude. In this case, because both dates follow October 5, the opposite relationship between the CPP and latitude is expected. *Tetranychus urticae* (*), on the other hand, which showed a positive slope for the two CPPs well after October 5, still showed a negative slope on the calendar date scale. Thus, the population at the higher latitude shows a longer CPP, but this corresponds to an earlier date. This result can occur after October 5 because, although the day length on the same date is shorter at higher latitudes, the day length still decreases with time at each latitude until the winter solstice (Figure 5.1). So the day length at an early date at a high latitude may be longer than that at a lower latitude on a later date. Finally, the longest CPP, which belonged to *Daphnia middendorfiana* (⊙) sampled from a very high latitude, is transformed into a centrally located date, day 232 or August 20, rather than a particularly early date.

Variability of Response versus Latitude

We chose the central 60% of the population response (30% on either side of the median) as a measure of variability. It is easy to interpret and is preferable to a broader region because the estimates of the response curve around the median are better than those approaching the tails (Copenhaver and Mielke 1977). Because the distributions (normal and logistic) that we used to fit the responses are symmetrical, the central 60% will give the same qualitative results as would a measure based on the standard deviation. We also used a range of days over which the central 60% population response would occur instead of day length to make interpretation easier.

Seven out of 11 curves for the range of dates in the central 60% response versus latitude show a negative slope, indicating that variability in the population response tends to decrease with latitude (Figure 5.3a). The four exceptions are *Pteronemobius taprobanensis* (B), *Limenitis archippus* (M), and*Chilo suppressalis* (o, two curves). *Wyeomyia smithii* (4) is an exception over part of its range and this will be discussed later. In 31 out of 48 populations the central

60% response occurs in 20 or fewer days and in 10 populations it occurs in 10 or fewer days. The two populations having the most precise responses are *Chrysopa carnea* (E) at 42.5°N with a central 60% response of 4 days and *Daphnia middendorfiana* (⊙) at 71°N with a central 60% response of 5 days. Five populations show responses spread over more than 50 days. The most extreme population in this respect is *Ostrinia nubilalis* (S) with a central 60% response over 76 days.

The central 60% of the photoperiodic response, in contrast to the range of days, shows no consistent pattern as a function of latitude. Four of the seven data sets with two locations (Table 5.2) show a negative slope and the other three show a positive slope. Of the four data sets with three or more points, only that for *Tetranychus urticae* (*) exhibits any consistent pattern, having a much less variable response at higher latitudes, as would be expected from the extremely negative slope in Figure 5.3a.

Changing Photoperiod During the Sensitive Period

Most treatments of the photoperiodic induction of diapause do not include naturally changing day lengths during the sensitive period (SP), when the decision whether to diapause is made. A notable exception is the study of Butler et al. (1978). The following analyses are designed to indicate when changing day lengths may be important. One way to use the photoperiodic response curve to approach this problem is to compute the difference in percent diapause that would occur if the day length at the start of a SP were determining the percent diapause as compared to the day length at the end of the SP. To do this, for each population we supposed there was a 20-day SP centered at the date corresponding to the CPP and we computed the day length at the start of the SP (date of the CPP - 10 days) and at the end of the SP. Each of these day lengths was then translated into a percent diapause by referring to the photoperiodic response curve for that population. The difference between these two percentages measures the sensitivity of the population to naturally changing day lengths near the CPP. The 20-day SP was chosen as reasonable for many insects at ambient temperatures (Saunders 1982), although the duration of the SP has not been accurately estimated for many species (Taylor 1985).

For most species, the difference in percent diapause over the duration of the 20-day SP is > 60% (Figure 5.3b). Some photoperiodic response curves are so steep that the difference approaches 100% whereas in others the difference is only about 20–30%. Since this analysis depends on the steepness of the photoperiodic response curves, this figure is close to a mirror image about the abscissa of Figure 5.3a, which gave the dates spanning the central 60% response. A steep photoperiodic response curve gives a low central 60% response (Figure 5.3a) and a high level of sensitivity to changing day length during a 20-day SP (Figure 5.3b). The locations of the points along the ordinate in Figure 5.3b depend on the duration of the SP; a shorter SP will make the differences smaller and a longer SP will make them greater.

How much the day length changes during a SP depends on the time of year and the latitude. We addressed this problem by moving a 20-day window across dates starting at the summer solstice on day 172 (June 21) and ending at the winter solstice on day 356 (December 22). The 20-day interval is centered at each date. As would be expected, the day length changes most rapidly over a fixed interval of time at high latitudes (Figure 5.4). For example, for a SP centered at August 15 (day 227), the change in day length at 25°N is 0.45 hr whereas at 60°N it is 2 hrs. It might not have been anticipated, however, that because of the inclusion of civil twilight the most rapid change at a particular latitude does not occur at the autumnal equinox on day 266 (September 23). At high latitudes around 60° the slope of the day length curve (Figure 5.1) increases toward the summer solstice because of the increase in civil twilight. Especially abrupt changes occur in the difference in day length between the beginning and end of the SP for latitudes approaching and above 62°, where 24-hr day lengths are first achieved. During the time of continuous daylight at these latitudes, there may be no change in day length in the SP, but immediately thereafter the day length decreases precipitously (Figure 5.1). Because continuous daylight extends later in the year with increasing latitude above 62°, the date of maximum change during the SP again moves to later dates (maximum curve in Figure 5.4).

It is of interest to know the maximum potential change in day length that may occur for particular lengths of the SP at various latitudes. To find this value, we computed the day length change during the SP of each length at each latitude over the sequence of dates when the maximum change is likely to occur

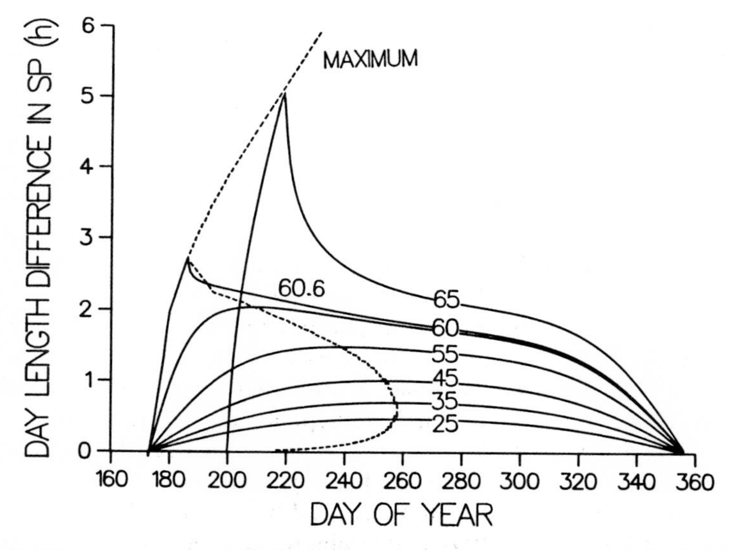

Figure 5.4. Change in photoperiod during a 20-day interval. The range of dates at the center of the interval is as in Figure 5.1. The S-shaped curve connects the dates of maximum change at successively higher latitudes. The numbers next to the curves indicate latitude. Increments in computing the maximum curve were 0.1° latitude.

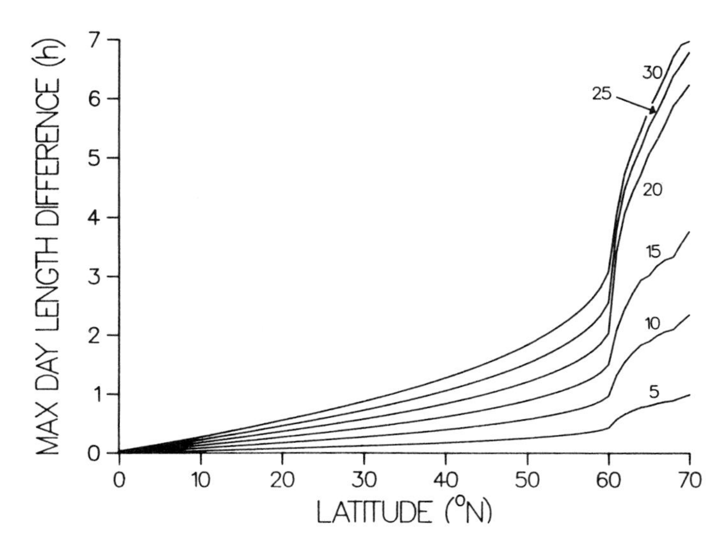

Figure 5.5 The maximum differences in photoperiod occurring over intervals of various lengths as a function of latitude. For example, the curve for a 20-day interval consists of the day length differences joined by the S-shaped curve in Figure 5.4 and plotted as a function of latitude. The numbers next to the curves indicate the lengths of the intervals. Values were computed for each 1° latitude.

and recorded the maximum value. As shown in Figure 5.5, the maximum change in day length during the SP ascends dramatically after 60°N. This rise is most significant for SPs of longer duration. The maximum day length change for a SP of 20 days, for example, is just over 2 hr at 60°N latitude and increases to about 5¼ hr at 65°N.

Discussion

Median Diapause Response

A striking feature of the median responses on both the photoperiod and calendar date scales is the similarity of the slope among latitudinal clines. On the photoperiod scale, nine gradients show similar slopes ranging from 0.11 to 0.24 (Figure 5.2, Table 5.2) as follows: *Chrysopa carnea* (E), *Barathra brassicae*(L), *Chilo suppressalis* (o, two gradients), *Aedes atropalpus* (V, two gradients), *Wyeomyia smithii* (4), *Drosophila phalerata* (7), and *Tetranychus urticae* (*). The consistency of results within a species is supported by *Chilo suppressalis* (o) in 2 years and for *Aedes atropalpus* (V) in two separate studies. On the calendar date scale, six gradients show similar slopes ranging from −4 to −2.4 (Figure 5.2b, Table 5.2): *Chrysopa carnea* (E), *Chilo suppressalis* (o, two gradients), *Aedes atropalpis* (V, two gradients), and *Wyeomyia smithii* (4). Three other gradients have shallower slopes and their slopes are very close to one another, lying between −1.20 and −0.94: *Barathra brassicae* (L), *Limenitis archippus* (M), and *Toxorhynchites rutilis* (3). The latter species is interesting because on the photoperiod scale it shows a slightly negative slope, but, because the day lengths correspond to dates after October 5, the slope remains negative on the calendar date scale and close to that of the other two species. Overall, some patterns emerge, but since the preponderance of studies provide only two points (after our criteria for acceptance of data sets were applied), not much can be concluded from most of the individual gradients. Furthermore, the necessity of including elevation in the analysis when populations have been sampled at markedly different elevations is emphasized in the results from the gradient for *Wyeomyia smithii*. When elevation was removed as a variable, highly significant regressions were reduced to low significance for the CPPs and became nonsignificant for the critical dates (Table 5.2). We cannot assess the influence of elevation in the other studies because this information was not provided.

A plausible explanation for the similarity of slopes across latitude is that the optimal time to induce diapause varies in a similar way with latitude. A possible candidate for the event selecting for diapause induction is the occurrence of the first hard frost, which may be lethal to individuals not in diapause and cold-hardened or, in the case of phytophagous arthropods, may kill the leaves of the host plant. We are testing for geographical patterns in the occurrence of the first hard frost using long-term temperature records (Spalding and Taylor in preparation).

Pteronemobius taprobanensis (B) and *Psylla pyricola* (C) are exceptional on both the photoperiod and calendar date scales. Short of accepting the possibility that the median responses of local populations in these instances are simply random processes not representing adaptive responses to their environments, three plausible explanations can be proposed for the deviant behavior. First, elevational differences between populations, which were not available to be included in our analyses, may have overridden the latitudinal effects. Second, the selection regimens for these species may vary with latitude in ways quite different from the others. Third, the median responses of these two species may show patterns similar to those of the others in nature, but their responses to laboratory conditions were aberrant. The second explanation applies to *Pteronemobius taprobanensis*, which is changing from a univoltine to a bivoltine life history over the range of latitudes being considered, and associated with this change are radical alterations in its photoperiodic response (Masaki 1978, 1979). Testing these hypotheses for *Psylla pyricola* requires detailed information not available to us.

Variability in the Diapause Response

As a rule, the median photoperiod increases with latitude before October 5 (10 out of 12 gradients in Figure 5.2a) and the median date decreases with latitude (11 out of 14 gradients in Figure 5.2b). It is also true that the variability of the photoperiodic response translated to a calendar date scale usually decreases with latitude (7 out of 11 gradients in Figure 5.3a). We shall argue that two of the four exceptions may behave consistently with an explanation for this trend. In addition, the codling moth, *Cydia pomonella,* which was not included in our analyses because photoperiodic response curves were not available, shows a decrease with latitude in the variance of the time of diapause on a calandar date scale in populations in nature (Riedl and Croft 1978). Three explanations can be forwarded for this phenomenon. (1) The catastrophic event selecting for the induction of diapause may be more variable in the south. Analysis of the occurrence of first hard frosts over many years along a latitudinal gradient in the Great Plains from Texas to North Dakota and another gradient on the East Coast from South Carolina to Maine demonstrates no consistent latitudinal trend (Spalding and Taylor, in preparation). It is conceivable that variability in some other event related to diapause induction does decrease from south to north. (2) Variability in survivorship and fecundity affecting the position of the optimal time to induce diapause (Taylor 1980) may decrease from south to north. This possibility seems unlikely because the latitudinal change in variability would have to be very dramatic to produce a measurable effect in the median time of diapause induction (Taylor and Spalding, in preparation), but there are no data from nature relating to this point.

(3) Arthropods can tell time (date) better at high latitudes and, therefore, can display a more precise response. This should be true because day length changes more rapidly over time at higher latitudes (Figures 5.1 and 5.4). At low latitudes, where day lengths change little from day to day, e.g., a maximum of about ½ hr over a 20-day interval at 30°N (Figure 5.4), arthropods may have difficulty telling time (or date) and this imposes some relatively high minimal variance on the population response. At higher latitudes, where day lengths change more rapidly, e.g., 1¼ hr over a 20-day interval at 50°N, arthropods can tell time better and the variability in the timing of diapause induction declines. Thus, if two samples of individuals having broadly differing ages (so individuals would pass through the sensitive period at different times) from a population with a particular photoperiodic response curve were reared under two different day length regimens, one corresponding to low and the second to high latitude, the latter should show diapause induction over a narrower range of days. Thus, even if a population maintained the same level of variability on the photoperiod scale with increasing latitude, it would show a decreasing variability on the calendar date scale. Eventually, at high latitudes, it could have an extremely precise response. But yearly variation in the position of the optimum (Slatkin and Lande 1976; Roughgarden 1979), as well as dispersal, may maintain variability in the response on the calendar date scale. According to this hypothesis, at low latitudes, where photoperiod changes slowly from day to day, the lower limit for the variability of the response will be set by the limitation of the diapause induction mechanism to determine the date, whereas at high latitudes the lower limit will be set by variation in the selection regimen and by dispersal. It should be recalled in evaluating this hypothesis that variability in the photoperiodic response shows no consistent patterns as a function

of latitude. Therefore, any pattern of variability observed on the date scale is not simply a by-product of an underlying pattern on the photoperiod scale.

To consider the agreement between this hypothesis and the available data, we shall focus on the four sets of data providing information on the central 60% response at three or more latitudes (Figure 5.3A). The results for *Wyeomyia smithii* (4) are particularly instructive in this context in that variability decreases steadily with latitude and then remains low, varying around 12 days, above 39.5°N. This pattern could be explained by the above hypothesis: the central 60% decreases as the ability to tell time increases until a lower limit is reached that is maintained by variability in the selection regimen and by dispersal. Also in this context, *Chilo suppressalis* (o) may not be an exception to this pattern. This species may simply be able to tell time well at low latitudes and, over the range of latitudes observed, may show levels of variability that are maintained by variation in the selection regimen and dispersal. These quantities are not expected to display a trend with latitude. The gradients for *Pieris brassicae* (N) and *Tetranychus urticae* (*), both of which have three points, also show negative slopes. Unfortunately, data are not available for either species at higher latitudes where the variances may be approaching a lower limit. An aspect of the results for *T. urticae* that appears to be at odds with the above hypothesis is the dramatic decrease in the central 60% for populations separated by only 1° (41 and 42°) with very similar critical dates. It is hard to believe that this decrease in the variability of the population response could be caused by improved ability to distinguish dates using day length, but this problem requires further analysis. In summary, 10 out of 12 gradients (including that for the codling moth) are potentially consistent with the third hypothesis. Latitudinal comparisons of local populations from species occurring over broad ranges of latitude directed at this question would be informative.

Effect of Changing Photoperiod During the Sensitive Period

The foregoing results argue forcefully for a need to understand better how arthropods decide to diapause under natural conditions of changing photoperiod and temperature. For as long as we do not understand this, we shall never be certain how the findings from the laboratory relate to nature. Here we focus on the following problem: If individuals are passing through the sensitive period over a range of times around the CPP, which of them will enter diapause? Or, stated another way, how does one go about predicting the time course of diapause induction for a population in nature?

For the sake of discussion, we assume that the physiological mechanism for diapause induction involves the required day number (RDN)–sensitive period (SP) interaction model proposed by Saunders (Saunders 1981, 1982; Veerman and Vaz Nunes 1984). The RDN is the number of light-dark cycles (days) that must be experienced within the SP for diapause to be induced. It is mainly a function of day length and has been shown to be temperature-compensated (Saunders 1971, 1982). The duration of the SP is the time elapsed between two developmental events and, thus, for well-fed arthropods is mainly a function of temperature (Taylor 1981). The interaction between the RDN and the SP gives a population of long-day arthropods its characteristic response to different combinations of day length and temperature as follows. The RDN increases with increasing day length, so at a constant temperature the proportion of a sample of arthropods entering diapause is less at longer day lengths. The duration of the SP decreases with increasing temperature more rapidly than does the RDN; so at a given day length the proportion entering diapause decreases as temperature increases (Saunders 1982).

According to this model, variability in the photoperiodic response at a constant temperature will be caused by four parameters: the variance of the RDN and the variance of the duration of the SP, the rate of response of the median RDN to changes in photoperiod, and the length of the SP. While this is not the place to develop this problem in detail, it is clear that the interactions among these variables are important for our ability to predict and understand the evolution of the timing of diapause induction. Since populations at all latitudes can show a high degree of sensitivity to changing day length within the SP (Figure 5.3b), this is a problem of general significance. More specifically, it would be interesting to know, for example, why populations of *Chilo suppressalis* (o at 40.6° in Figure 5.3a) and *Ostrinia nubilalis* (S at 43°) with similar CPP values (15.2 and 15.6 hrs, respectively) at similar latitudes, display such markedly dissimilar variabilities in their diapause responses (central 60% of 13 and 76 days, respectively). One also wonders how, as well as why, a high variability is maintained at very high latitudes. *Drosophila phalerata* (7 in Figure 5.3a), for example, maintains a central 60% response falling in the middle of the range for populations at far lower latitudes (Figure 5.2a) when the change in photoperiod during a SP of 20 days (around its CPP on day 342 at 65°N latitude) is $>$ 2.5 hrs compared to $<$ 1 hr at 45°N. A particularly informative approach to such problems would be to compare the parameters of the RDN-SP model, or whatever alternative model is more appropriate, in populations within a species or between closely related species over a broad range of latitudes giving greatly differing rates of change in day length during the SP (Figure 5.4 and 5.5). Some work dealing with this problem has already been presented by Beach (1978) for *Aedes atropalpus* and by Reed et al. (1978) for *Ostrinia nubilalis*.

Summary and Concluding Remarks

Most studies of geographical patterns in the photoperiodic induction of hibernal diapause have emphasized the observation that the median response (or critical photoperiod) increases with latitude before the autumnal equinox (actually October 5 when civil twilight is included) and decreases with latitude thereafter. We confirmed this observation for 11 out of 14 cases in which the photoperiodic response was measured for two or more local populations along a latitudinal gradient. When these critical photoperiods are converted to the corresponding dates, the median date of response decreases with latitude, as expected, in 11 out of 14 cases. By considering the range of dates over which the central 60% of the population will be induced to enter diapause, we showed that in 7 out of 11 cases the variability of the response decreases with latitude. Nine out of the 11 cases were consistent with the following hypothesis. At low latitudes, where day lengths change little from day to day, arthropods may have difficulty telling time (or date) and this imposes some relatively high minimal variance on the population response. At higher latitudes, where day lengths change more rapidly, arthropods can tell time better and the variability in the timing of diapause induction declines. Above some latitude, variability is maintained by yearly variation in the optimal time to enter diapause and by dispersal, so it will not necessarily decline further with latitude. We found also remarkable differences in variability of the diapause response among different species occurring at similar latitudes and having similar critical photoperiods. Finally, populations occurring at very high latitudes, where day length changes extremely rapidly, e.g., more than 2½ hr over a 20-day interval at 65°N, can maintain a surprisingly high level of variability on the date scale.

These observations suggest the following conclusions. (1) Studies directed solely at examining latitudinal trends in the median day length or date of the diapause induction response are redundant, for the patterns are already clear. It remains of great interest, however, to understand the evolutionary mechanisms maintaining the median response at even a single location. (2) Geographical patterns in the variability of this response are much less clear and warrant a great deal of further consideration at both the physiological and evolutionary levels. A first step is to compare populations from the same species over a broad geographical range to see what the geographical patterns really are. In addition, at the proximate level, we need to elucidate the physiological mechanisms that produce variability in the diapause response and, at the ultimate level, we must learn what evolutionary forces influence this variation. (3) Considering the frequent occurrence and the potential importance of changing day lengths during the sensitive period, it is essential that we learn how natural populations respond to the conditions actually experienced during the sensitive period. This includes conducting laboratory experiments that simulate conditions found in nature as well as working out the details of the physiological response to such conditions. We have not considered here the influence of changing temperatures on diapause induction, but this is important and may be especially difficult to understand in nature because periodicity per se in the temperature signal alters the diapause response (Chippendale et al. 1976; Bradshaw 1980; Masaki and Kikukawa 1981; Beck 1982). Finally, comparisons among populations of diverse geographical origin should continue to provide a powerful tool in studying the ecological and evolutionary questions stimulated by the results of earlier geographical comparisons.

Acknowledgments We appreciate the comments of Bill Bradshaw, Ollar Fuller, Jaakko Lumme, and Ted Nusbaum on earlier drafts of this chapter. Ollar Fuller assisted in the graphics. This research was supported by the following two National Science Foundation grants to Fritz Taylor: DEB-8104698 and DEB-8208998.

References

Anderson, J.F.: Influence of photoperiod and temperature on the induction ofdiapause in *Aedes atropalpus* (Diptera, Culicidae). Entomol. Exp. Appl. 11, 321–330 (1968).

Ankersmit, G.W., Adkisson, P.L.: Photoperiodic responses of certain geographical strains of *Pectnophora gossypiella* (Lepidoptera). J. Insect Physiol. 13, 553–564 (1967).

Beach, R.: The required day number and timely induction of diapause in geographical strains of the mosquito, *Aedes atropalpus*. J. Insect Physiol. 24, 449–455 (1978).

Beck, S.D.: Insect Photoperiodism. Second Edition. New York: Academic Press, 1980.

Beck, S.D.: Thermoperiodic induction of larval diapause in the European corn borer, *Ostrinia nubilalis*. J. Insect Physiol. 28, 273–277 (1982).

Berven, K.A.: The genetic basis of altitudinal variation in the wood frog *Rana sylvatica*. I. An experimental analysis of life history traits. Evolution 36, 962–983 (1982).

Berven, K.A., Gill, D.E.: Interpreting geographic variation in life-history traits. Am. Zool. 23, 85–98 (1983).

Bondarenko, N.V, Kuan Khai-Yuan': The characteristics of the onset of diapause in different geographical populations of the spider mite. Rep. Acad. Sci. USSR 119, 1247–1250 (1958).

Bradshaw, W.E.: Thermoperiodism and the thermal environment of the pitcher-plant mosquito, *Wyeomyia smithii*. Oecologia 46, 13–17 (1980).

Bradshaw, W.E., Holzapfel, C.M.: Interaction between photoperiod, temperature, and chilling in dormant larvae of the tree-hole mosquito, *Toxorhynchites rutilus* Coq. Biol. Bull. 152, 147–158 (1977).

Bradshaw, W.E., Lounibos, L.P.: Evolution of dormancy and its photoperiodic control in pitcher-plant mosquitoes. Evolution 31, 546–567 (1977).

Bradshaw, W.E., Phillips, D.L.: Photoperiodism and the photic environment of the pitcher-plant mosquito, *Wyeomyia smithii*. Oecologia 44, 311–316 (1980).

Butler, G.D., Hamilton, A.G., Guttierez, A.P.: Pink bollworm: diapause induction in relation to temperature and photophase. Ann. Am. Entomol .Soc. 71, 202–204 (1978).

Chippendale, G.M., Reddy, A.S., Catt, C.L.: Photoperiodic and thermoperiodic interactions in the regulation of the larval diapause of *Diatraea grandiosella*. J. Insect Physiol. 22, 823–828 (1976).

Clausen, J., Keck, D.D., Heisey, W.M.: Experimental studies on the nature of species. I. Effect of varied environments on western North American plants. Carnegie Institution Wash. Publ. 520, 1940.

Copenhaver, T.W., Mielke, P.W.: Quantit analysis: a quantal assay refinement. Biometrics 33, 175–186 (1977).

Danilevskii, A.S.: Photoperiodism and Seasonal Development of Insects. Edinburgh: Oliver & Boyd, 283 pp., 1965.

Danilevskii, A.S., Goryshin, N.I., Tyshchenko, V.P.: Biological rhythms in terrestrial arthropods. Annu. Rev. Entomol. 15, 201–244 (1970).

Dickson, R.C.: Factors governing the induction of diapause in the oriental fruit moth. Ann. Entomol. Soc. Am. 42, 511–537 (1949).

Endler, J.A.: Geographic Variation, Speciation, and Clines. Princeton: Princeton University Press, 1977.

Evans, K.W., Brust, R.A.: Induction and termination of diapause in *Wyeomyia smithii* (Diptera, Culicidae), and larval survival studies at low and subzero temperatures. Can. Entomol. 104, 1937–1950 (1972).

Finney, J.D.: Statistical Method in Biological Assay. Third Edition. London: Charles Griffing and Company, 1978.

Fuchs, T.W., Harding, J.A., Smith, J.W.: Induction and termination of diapause in the sugarcane borer. Ann. Entomol. Soc. Am. 72, 271–274 (1979).

Glinyanaya, Ye. I.: The importance of day length in regulating the seasonal cycles, and diapause in some Psocoptera. Entomol. Rev. 54, 10–13 (1975).

Havelka, J.: Some aspects of photoperiodism of the aphidophagous gallmidge *Aphidoletes aphidimyza* Rond. Bull. S.R.O.P/W.P.R.S III/3, 75–82 (1980).

Heisey, W.M., Milner, H.W.: Physiology of ecological races and species. Annu Rev Plant Physiol. 16, 203–216 (1965).

Hong, J.W., Platt, A.P.: Critical photoperiod and day length threshold differences between northern and southern populations of the butterfly *Limenitus archippus*. J. Insect Physiol. 21, 1159–1165 (1975).

Jordan, R.G., Bradshaw, W.E.: Geographic variation in the photoperiodic response of the western tree-hole mosquito, *Aedes sierrensis*. Ann. Entomol .Soc. Am. 71, 487–490 (1978).

Kalpage, K.S.P., Brust, R.A.: Studies on diapause and female fecundity in *Aedes atropalpus*. Environ. Entomol. 3, 139–145 (1974).

Kimura, T., Masaki, S.: Brachypterism and seasonal adaptation in *Orgyia thyellina* Butler (Lepidoptera, Lymantriidae). Kontyu 45, 97–106 (1977).

Kishino, K.: Ecological studies on the local characteristics of seasonal development in the rice stem borer, *Chilo suppressalis* Walker. II. Local characteristics of diapause and development. Jpn. J. Appl. Entomol. Zool .14, 1–11 (1970).

Lees, A.D.: Environmental factors controlling the evocation and termination of diapause in the fruit tree red spider mite *Metatetranychus ulmi* Koch (Acarina, Tetranychidae). Ann. Appl. Biol. 40, 449–486 (1953).

Lumme, J., Oikarinen, A.: The genetic basis of the geographically variable photoperiodic diapause in *Drosophila littoralis*. Hereditas 86, 129–142 (1977).

Masaki, S.: Geographic variation of diapause in insects. Bull. Fac. Agric. Hirosaki Univ. 7, 66–98 (1961).

Masaki, S.: Seasonal and latitudinal adaptations in the life cycles of crickets. In: Evolution of Insect Migration and Diapause. Dingle, H .(ed.). New York: Springer-Verlag, 1978, pp. 72–100.

Masaki, S.: Climatic adaptation and species status in the lawn ground cricket. I. Photoperiodic response. Kontyu 47, 47–65 (1979).

Masaki, S., Kikukawa, S.: The diapause clock in a moth: response to temperature signals. In: Biological Clocks in Seasonal Reproductive Cycles. Follett, B.K., Follett, D.E. (eds.). New York: John Wiley and Sons, 1981, pp. 101–112.

Masaki, S., Umeya, K., Sekiguchi, Y., Kawasaki, R.: Biology of *Hyphantria cunea* Drury (Lepidoptera, Arctiidae) in Japan. III. Photoperiodic induction of diapause in relation to the seasonal life cycle. Appl. Entomol. Zool. 3, 55–66 (1968).

McMullen, R.D., Jong, C.: Factors affecting induction and termination of diapause in pear psylla (Homoptera, Psyllidae). Can. Entomol. 108, 1001–1005 (1976).

Muona, O., Lumme, J.: Geographical variation in the reproductive cycle and photoperiodic diapause of *Drosophila phalerata* and *D. transversa* (Drosophilidae, Diptera). Evolution 35, 158–167 (1981).

Nautical Almanac Office: Almanac for computers 1980. U.S. Naval Observatory, Washington, D.C., 1980.

Oldfield, G.N.: Diapause and polymorphism in California populations of *Psylla pyricola* (Homoptera, Psyllidae). Ann. Entomol. Soc. Am. 63, 180–184 (1970).

Propp, G.D., Tauber, M.J., Tauber, C.A.: Diapause in the neuropteran *Chrysopa oculata*. J. Insect Physiol. 15, 1749–1757 (1969).

Reed, G.L., Showers, W.B., Guthrie, W.D., Lynch, R.E.: Larval age and the diapause potential in northern and southern ecotypes of the European corn borer. Ann. Entomol. Soc. Am. 71, 928–930 (1978).

Riedl, H., Croft, B.A.: The effect of photoperiod and effective temperatures on the seasonal phenology of the codling moth (Lepidoptera: Tortricidae). Can. Entomol. 110, 455–470 (1978).

Roughgarden, J.: Theory of Population Genetics and Evolutionary Ecology: An Introduction. New York: Macmillan Publishing, 1979.

Saunders, D.S.: The temperature-compensated photoperiodic clock 'programming' development and pupal diapause in the flesh-fly, *Sarcophaga argyrostoma*. J .Insect Physiol. 17, 801–812 (1971).

Saunders, D.S.: Insect Clocks. Second Edition. New York: Pergamon Press, 1982.

Saunders, D.S.: Insect photoperiodism—the clock and the counter: a review. Physiol. Entomol. 6, 99–116 (1981).

Shel'deshova, G.G.: Ecological factors determining distribution of the codling moth *Laspeyresia pomonella* L. (Lepidoptera, Tortricidae) in the northern and southern hemispheres. Entomol. Rev. 46, 349–361 (1967).

Slatkin, M., Lande, R.: Niche width in a fluctuating environment–density independent model. Am. Natur. 119, 31–55 (1976).

Spielman, A., Wong, J.: Environmental control of ovarian diapause in *Culex pipiens*. Ann. Entomol. Soc. Am. 66, 905–907 (1973).

Stross, R.G.: Photoperiod control of diapause in *Daphnia*. II. Induction of winter diapause in the arctic. Biol. Bull. 136, 264–273 (1969).

Stross, R.G., Hill, J.C.: Diapause induction in *Daphnia* requires two stimuli. Science 150, 1462–1464 (1965).

Tauber, M.J., Tauber, C.A.: Geographic variation in critical photoperiod and in diapause intensity of *Chrysopa carnea* (Neuroptera). J. Insect Physiol. 18, 25–29 (1972a).

Tauber, M.J., Tauber, C.A.: Larval diapause in *Chrysopa nigricornis*, sensitive stages, critical photoperiod, and termination (Neuroptera, Chrysopidae). Entomol. Exp. and Appl. 15, 105–111 (1972b).

Taylor, F.: Optimal switching to diapause in relation to the onset of winter. Theor. Pop. Biol 18, 125–133 (1980).

Taylor, F.: Ecology and evolution of physiological time in insects. Am. Natur. 117, 1–23 (1981).

Taylor, F.: Estimating the ends of the sensitive period for diapause induction in arthropods. J. Theor. Biol. 117, 319–336 (1985).

Taylor, F., Schrader, R.: Transient effects of photoperiod on reproduction in the Mexican bean beetle. Physiol. Entomol. 9, 459–464 (1984).

Teeri, J.A., Stowe, L.G.: Climatic patterns and the distribution of C4 grasses in North America. Oecologia 23, 1–12 (1976).

Teetes, G.L., Adkisson, P.L., Randolph, N.M.: Photoperiod and temperature as factors controlling the diapause of the sunflower moth, *Homoeosoma electellum*. J. Insect Physiol. 15, 755–761 (1969).

Trimble, R.M., Smith, S.M.: Geographic variation in the effects of temperature and photoperiod on dormancy induction, development time, and predation in the tree-hole mosquito, *Toxorhynchites rutilus septentrionalis* (Diptera: Culicidae). Can. J. Zool. 57, 1612–1618 (1979).

Veerman, A., Vaz Nunes, M.: Photoperiodic reception in spider mites: photoreceptor, clock and counter. In: Photoperiodic Regulation of Insect and Molluscan Hormones. Porter, R., Collins, G.M. (eds.). London: Pitman Publishing, 1984, pp. 48–59.

Part II
Diversity of Life Cycle Patterns

Chapter 6

Diapause Strategies in the Australian Plague Locust (*Chortoicetes terminifera* Walker)

K.G. Wardhaugh

Insects that live in highly variable environments repeatedly encounter problems of survival in both time and space. The exploitation of these transient habitats is often achieved through elaborate behavioral or physiological responses, such as migration (e.g., Johnson 1969) or diapause (e.g., Andrewartha 1952). Migration results in the colonization of new habitats and hence allows escape in space, whereas diapause regulates seasonal phenology and thus confers escape in time. Numerous, diverse examples of the importance of migration and diapause in regulating the life histories of insects are given in Dingle (1978) and Brown and Hodek (1983).

The Australian plague locust (*Chortoicetes terminifera* Walker) is a highly migratory, multivoltine species with a facultative egg diapause. Its distribution covers most of the Australian continent (Key 1945), but areas known, or thought, to have supported frequent swarm formation (Key 1945; Casimir 1962; Magor 1970) are confined largely to the eastern states of Queensland, New South Wales, Victoria, and South Australia (Figure 6.1a). This extensive region varies from subtropical to temperate. Habitat conditions tend to be highly ephemeral, due to the extreme temporal and spatial variability of rainfall (Figure 6.1b) (Clark et al. 1969). Episodic migration (Clark 1969; Farrow 1975; Drake and Farrow 1983) and the ability of eggs to survive drought conditions for 2–3 months (Davidson 1936; Swan 1955) are the principal adaptive traits that enable the species to exploit these temporary habitats. During winter, however, most areas are either too dry and/or too cool to permit the normal development of the active stages. This chapter considers the role of diapause and a number of closely related survival strategies, which together govern the seasonality of *C. terminifera* and enable it to persist in winter environments that might otherwise be unsuitable for permanent habitation.

Diapause and Quiescence

The developmental period of eggs of *C. terminifera* was long thought to depend solely on temperature and moisture conditions experienced during development (Key 1945; Clark, L.R. 1947; Clark et al. 1969). Studies on the survival and

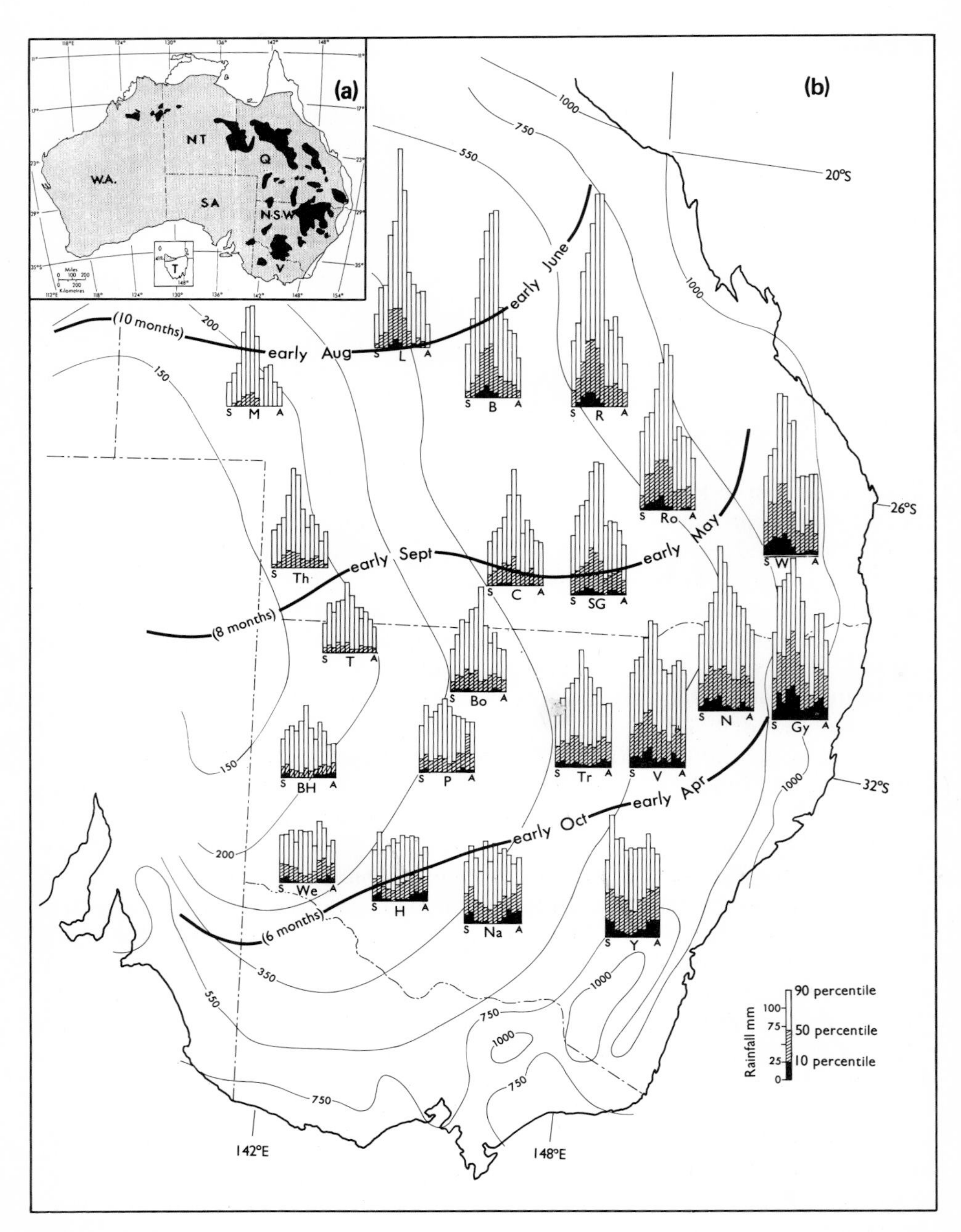

Figure 6.1. (a) Distribution area (stippled) and major breeding areas (solid) of *C. terminifera* in Australia. (b) Rainfall and temperature conditions in the major breeding areas of eastern Australia; bold curves indicate southern limit and period during which mean monthly maximum air temperature exceeds the 25°C level. Histograms show percentile rainfall variability by month from September through August for locations indicated by abbreviations. Annual isohyets (———) in mm.

growth rates of overwintering embryos, however, revealed the existence of an embryonic diapause (Wardhaugh 1972).

Diapause supervenes during late anatrepsis, at which stage the egg is almost fully hydrated and the embryo has completed about 45% of its development (Wardhaugh 1978). Hydration begins during early anatrepsis, but, in eggs subjected to moisture stress at this stage, embryogenesis ceases and resumes only when adequate moisture becomes available. In this prediapause, quiescent state (Quiescent Stage 1), the egg is highly resistant to desiccation and, because of low winter temperatures, can remain viable until spring (Wardhaugh1973).

Postdiapause development is also dependent on the moisture status of the egg. Diapausing eggs that lose as little as 9–14% of their moisture content are unable to resume development until this moisture is replenished (Wardhaugh 1973). This second quiescent stage (Q_2), which is morphologically indistinguishable from that at which diapause occurs, is particularly important in the warmer, more northerly parts of the species' range, which typically receive little or no winter-spring rainfall. When invoked, the arrest of development at Q_2 serves to delay hatching until the occurrence of adequate rainfall.

Seasonal and Regional Incidence of Diapause

The breeding season of *C. terminifera* normally extends from September-October (spring) to April-May (autumn) (Key 1945). Regular, biweekly sampling of adult populations in the central-west of New South Wales during three consecutive breeding seasons (Wardhaugh 1973) showed that diapause was confined largely to eggs laid during the autumn months of March-April (Figure 6.2). In 1969, the first diapausing eggs were recorded on 9 March as compared to 28 February in 1970. In 1971, diapausing eggs were produced by small numbers of females in both spring (2 from 38 females) and summer (5 from 354), with the main period of production of diapausing eggs beginning on 3 March.

Similar procedures were used to examine the spatial variation of diapause, females being collected along a north-south transect between Blackhall (24°S) and Canberra (35°S). The results showed that during the first 3 weeks of March most females in the cooler, southern areas were laying diapausing eggs, whereas those in the north were still laying nondiapausing eggs (Figure 6.3). By late March-early April, diapausing eggs were laid by females from all areas.

Photoperiodic Induction of Diapause

The fact that diapause began about the same time each year suggested that photoperiod may be the principal stimulus inducing diapause in *C. terminifera*. In March, the duration of day length is about 13.5 hr and decreases to about 12.5 hr by early April at which time the incidence of diapause is at a maximum. Since similar day lengths occur in spring, however, when few or no diapausing eggs are laid, it seemed likely that the directional change in day length might also be important. Furthermore, the regional study of diapause indicated that diapause began earlier in the south than in the north. In autumn, however, the

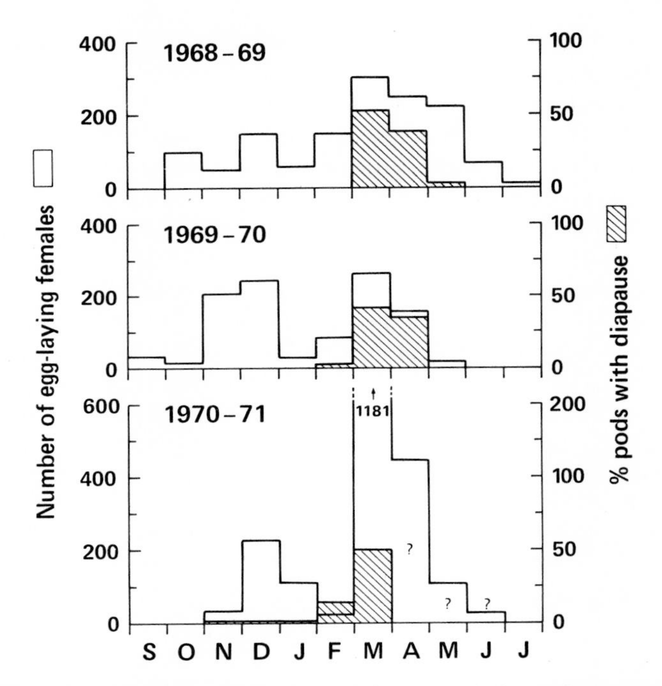

Figure 6.2. Seasonal occurrence of diapause in field populations of *C. terminifera*. Open columns indicate numbers of egg-laying females collected. Hatched columns show proportion laying diapause eggs; (?) indicates no data on diapause incidence. (Data from Wardhaugh 1973.)

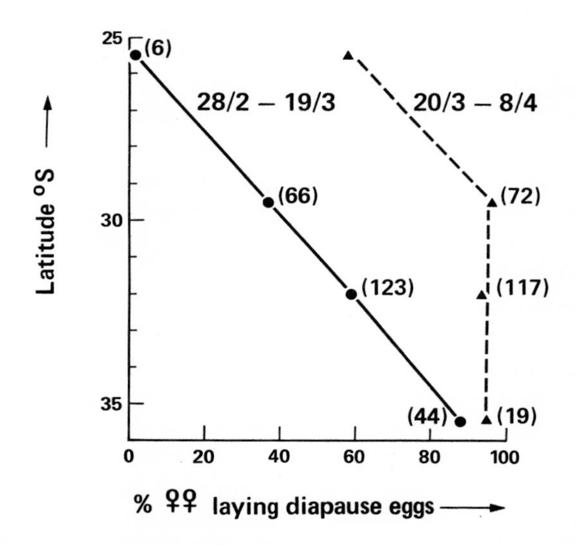

Figure 6.3. Regional and temporal variation in the incidence of diapause in field populations of *C. terminifera*. (Data from Wardhaugh 1973.)

processional change of day length with latitude is such that a photoperiod of, say 13.5 hr, occurs earlier in the north than in the south. Since it has been established for many insects [see, for example, Lees (1955), Beck (1968)] that the critical photoperiod for diapause induction is temperature-dependent, it seemed reasonable to suppose that the observed regional differences in the time of onset of diapause in *C. terminifera* may be due to differences in temperature, as well as day length. Accordingly, a series of experiments was undertaken to test these hypotheses (Wardhaugh 1980b).

Insects were held throughout their five nymphal stages at one of three light regimens (long, 15.5L:8.5D; medium, 12.5L:11.5D; and short, 11L:13D) and two temperature regimens (high temperature, 15.5 hr, 35°C: 8.5 hr, 29°C and low temperature, 15.5 hr, 29°C: 8.5 hr, 22°C), giving six combinations of temperature and photoperiod. At the end of the last nymphal stage, the insects in each temperature-photoperiod treatment were divided into six equal groups. Insects in one group were always left in situ in each treatment and thus spent their adult life at the same temperature and photoperiod as that experienced in the nymphal stage. The remaining five groups of insects in each treatment were transferred to one of the other five combinations of temperature-photoperiod and so experienced a sudden change in temperature and/or photoperiod. Such manipulation resulted in insects being reared at one or another of 36 combinations of temperature and photoperiod (Wardhaugh 1980b).

Insects held continuously at either long or short photophase laid few or nodiapausing eggs. At a photophase of 12.5 hr (medium), which is equivalent to the day length associated with maximal diapause production in the field, there was a significant increase in the incidence of diapause, especially from cultures held at low temperature, or those experiencing a decrease in temperature during development (Table 6.1). The overall response, however, was far below that expected on the basis of the field data.

In contrast, high percentages of diapausing eggs were laid by insects that experienced a decrease in photophase during development (Table 6.1). In the long/medium photophase treatments, insects that experienced either low temperature or a decrease in temperature during development laid more diapausing eggs than those subjected to high temperatures or an increase in temperature. In the medium/short photophase treatments, however, temperature effects were reversed, high temperature or an increase in temperature being more inductive than low temperature or a decrease in temperature. In the long/short photophase treatments, in which there were no apparent effects due to temperature, large numbers of diapausing eggs were produced in each treatment despite the fact that separately, long and short photophases were largely noninductive.

Among insects that experienced an increase in photophase during development, diapause was almost totally inhibited (Table 6.1). Even in cultures in which insects spent their nymphal or adult stages at the inductive photophase of 12.5 hr, few diapausing eggs were laid. It was concluded, therefore, that *C. terminifera* is capable of responding not only to absolute differences in photophase, but also to the direction in which day length changes.

Since the above experiments indicated that the inductive response of *C. ter-*

Table 6.1. The percentage of diapause in egg pods of *C. terminifera* reared at different combinations of temperature and photophase

Photophase and temperature during nymphal stages		Photophase and temperature during adult stage					
		Long		Medium		Short	
		HT	LT	HT	LT	HT	LT
LONG	HT	0.0	1.3	43.4	80.2	56.0	53.8
	LT	0.0	4.8	62.2	74.3	52.3	49.6
MEDIUM	HT	0.0	1.0	8.0	29.8	30.0	16.5
	LT	0.6	0.0	3.0	11.8	28.1	3.3
SHORT	HT	0.0	0.0	0.0	0.0	5.1	0.0
	LT	0.0	0.0	0.8	0.0	0.8	0.0

Long, 15½ L; medium, 12½ L; short, 11 L. HT, High temperature; LT, low temperature; see text.

minifera was confined to a relatively narrow range of photoperiod, similar to that experienced in the field, further experiments were carried out at several additional constant photoperiods to determine the optimum for diapause induction.

These experiments confirmed the inhibitory influence of long and short photophase and suggested that the maximum response occurred at a photophase of ca. 13.5 hr (Figure 6.4). In the field, however, day lengths of this duration occur during late February-early March at which time few or no diapausing eggs are laid.

In experiments to determine the stage(s) sensitive to photophase, in which insects were transferred from long to short photophase either at the end of each of the five nymphal instars or as mature adults, it was found that all except the mature adult stage were sensitive to a decrease in photophase (Wardhaugh 1980a). The maximum response occurred in insects transferred at the end of the final nymphal instar. Moreover, in insects reared initially at a photophase of 15.5 hr and subjected to successive hourly decreases in day length at the end of each nymphal instar, 93% of the 43 eggpods laid contained eggs that entered diapause (Wardhaugh, unpublished). The average photophase experienced by these insects was 13.5 hr.

Under field conditions, *C. terminifera* normally takes 5–6 weeks to complete its nymphal development and an additional 1–2 weeks for adult maturation (Key

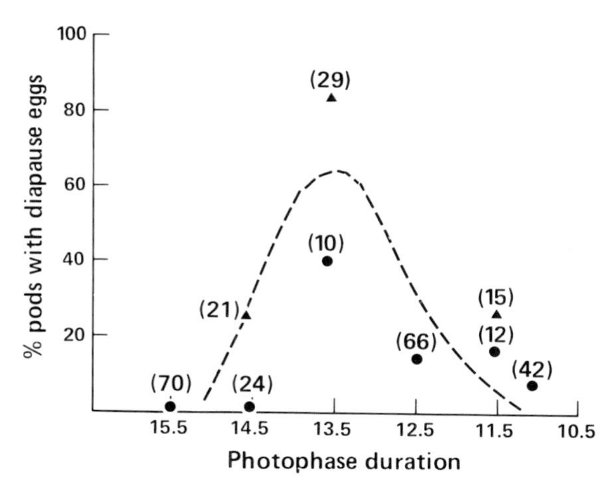

Figure 6.4. Incidence of diapause in egg pods laid by *C. terminifera* when reared at different constant photophases. Numbers in parentheses indicate sample sizes; (●), parent population from first-generation diapausing eggs; (▲), parent population from second-generation diapausing eggs. (Data from Wardhaugh 1980b.)

1938). Thus, insects laying eggs in late February (13.5 hr day length) can be expected to have hatched in early January (15 hr). Accordingly, the average photophase experienced during the sensitive stages would be about 14.3 hr. Such conditions would not be expected to elicit more than a weak inductive response (Figure 6.4). In contrast, insects laying in late March-early April (12.5 hr) would have hatched in early February (14.5 hr) and thus experienced an average photophase of 13.5 hr, i.e., the optimum photophase for diapause induction. It would appear, therefore, that the strictly seasonal nature of diapause in *C. terminifera* can be explained partly in terms of the average photophase experienced over the sensitive nymphal-early adult stage, and partly by their ability to discriminate between increasing and decreasing day lengths.

Inception of Diapause in Response to Temperature and Humidity

For experiments requiring large numbers of diapausing eggs, field populations of adult *C. terminifera* were sampled over a 6-week period during autumn. Newly laid eggpods were stored in air-dry soil at 15°C so that development would be arrested at the prediapause stage of Q_1. When subsequently moistened and incubated at 32.5°C, none of these eggs entered diapause (Wardhaugh 1973). Evidently, diapause had been averted in response to the low temperature and/or dry conditions experienced during storage.

To examine this phenomenon further, eggpods laid by field-collected females were transferred immediately after laying to either moist (12.9% moisture content) or dry (5.8%) soil and were then held for 40 days in either warm(32.5°C) or cool (15.5°C) temperatures. The proportion of eggs entering diapause in the cool, moist treatment, conditions that normally favor rapid embryogenesis and hatching, was used as a standard for assessing the efficacy of the other treatments in averting diapause.

The results showed that whereas low temperature resulted in a significant reduction of diapause, irrespective of the moisture content of the soil, dry conditions were effective in reducing the incidence of diapause only when associated with high temperature (Table 6.2, Wardhaugh 1980a). It was concluded, therefore, that the diapause is expressed most fully in eggs that undergo rapid development in the early stages of embryogenesis and that a combination of warm, moist conditions during this period is an essential element in the overall inductive process.

Recent work by Hunter and Gregg (1984), however, indicates that the process governing entry into diapause is even more complex. By collecting eggpods from the field and incubating them at various constant temperatures, they found that the incidence of diapause was consistently higher in eggs held at 20°C than in those kept at 26°C or 32°C. Moreover, they also showed that the propensity to enter diapause was often markedly different in newly laid eggs than in eggs left to develop in the field for a further 3–5 weeks. Hunter and Gregg (1984) concluded, therefore, that "diapause potential" was not fixed and may vary according to the environmental conditions experienced in the early stages of embryogenesis. More recent experiments (Gregg, personal communication), which support this conclusion, confirmed the inhibitory effects of low temperature and suggest that the optimum temperature for the inception of diapause is 20°C. Gregg has also shown, however, that newlylaid eggs that are incubated for various periods at 32°C before being transferred to 20°C always enter diapause if the transfer is effected at or before the stage at which diapause normally occurs, i.e., there appears to be no loss of diapause potential associated with high temperature exposure during the prediapause stages. In eggs incubated for longer periods at 32°C, however, there was an abrupt decrease in the numbers entering diapause.

Thus, it appears that there are at least two temperature-mediated processes controlling entry into diapause, namely, a high temperature effect, which is invoked at the time of onset of diapause and which affects only a portion of the population; and a low temperature effect, which operates during the prediapause stage and which can result in the circumvention of diapause. Thus, whereas it was formerly thought that in autumn *C. terminifera* laid a mixture of eggs, some with and some without diapause potential (Wardhaugh 1972), it now appears that all eggs laid at this season may have the potential to enter diapause and that the proportion that actually does so depends largely on the temperature experienced in the early stages of development. According to this view, the progressive seasonal changes in diapause incidence observed under field conditions (Wardhaugh 1973) may simply reflect systematic variation in the proportion of the population responding to the inhibitory influence of high temperature at the time when eggs would normally enter diapause. Further seasonal studies, involving the incubation of eggs at a range of temperatures that includes the optimum for diapause induction (20°C), are needed to examine this hypothesis.

Table 6.2. Comparison of numbers of diapause, nondiapause and nonviable eggs in egg pods of *C. terminifera* exposed to various temperature-moisture treatments at the time of laying (after Wardhaugh 1980a)

	Initial treatment after laying										
	Period	Temperature	Moisture	Total	Total	Diapause		Nondiapause		Nonviable	
Type	(days)	(°C)	(% dry wt.)	pods	eggs	Total	%	Total	%	Total	%
Cool-moist	40	15.5	12.9	22	734	13	1.8[b]	576	78.5	145	19.8
Cool-dry	40	15.5	5.8[a]	24	625	52	8.3[b]	454	72.6	119	19.0
Warm-dry	40	32.5	4.9[c]	22	608	210	34.5[b]	377	61.9	21	3.4
Warm-moist	40	32.5	12.9	58	1601	868	54.2	623	38.9	110	6.8

[a]Average of values at beginning (5.8%) and end (5.8%) of treatment.
[b]Difference from succeeding treatment highly significant: $p < 0.001$. Values of $\chi^2_{df\ 1}$ = 31.7, 94.9, and 89.4, respectively.
[c]Average of values at beginning (5.8%) and end (4.0%) of treatment.

Diapause Termination

Unlike the eggs of certain other acridids, neither desiccation (as in *Locustana pardalina* [Matthee 1951]) nor the use of chemical reagents to modify the process of water uptake (as in *Melanoplus differentialis* [Slifer 1946]) appears to have any influence on the termination of diapause in *C. terminifera*. In this species, the only effective method of terminating diapause is exposure to temperatures close to the minimum for normal embryogenesis. Under laboratory conditions, maximum reactivation was achieved after 30–60 days at 10–15°C (Wardhaugh 1973). Temperatures above or below this range were progressively less effective for terminating diapause, as were exposures of less than 30 days.

Under field conditions, diapause is normally completed within 7–10 weeks of laying (Wardhaugh 1973; Hunter and Gregg, 1984). By then soil temperatures are normally too cold to allow more than slow or intermittent development, so that hatching is delayed until the return of warmer weather in spring. When soil moisture is not limiting, hatching may occur as early as mid to late August (in western and central Queensland [Figure 6.5]), whereas in southern New South Wales and Victoria, the spring hatch may be delayed until October-November (Wardhaugh 1973).

This sharp latitudinal gradient, coupled with a similar trend in the time of onset of diapause (Figure 6.5), has a marked effect on the potential number of generations. In the Southern Tablelands of New South Wales, the species is univoltine (Clark 1969), whereas in lower, or more northerly, breeding areas there is the potential for two to four generations per year (Figure 6.5). In the northernmost areas (i.e., in central and southern Queensland), winter-spring

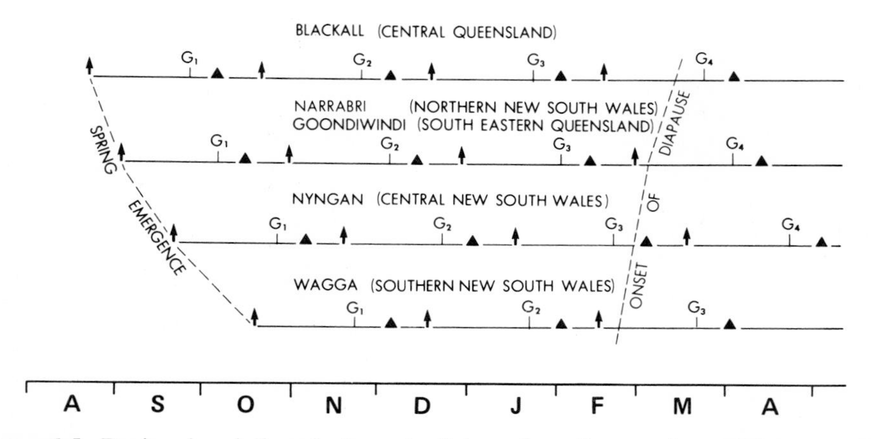

Figure 6.5. Regional variations in the potential number of generations of *C. terminifera* resulting from latitudinal differences in the time of spring emergence and the onset of diapause. (After Wardhaugh 1973.)

rainfall is usually low and often absent, so that an early spring hatch is likely to result in substantial nymphal mortality. Since postdiapause development is dependent on the moisture status of the egg, however, spring hatching, and hence the length of the reproductive season, may be determined more by the seasonal incidence of rainfall rather than by temperature in these northern areas.

Three or four generations per year are also possible in northern and central New South Wales (Figure 6.5). These areas receive more reliable winter-spring rains so that the early spring emergence is less likely to be affected by aridity. It seems probable, therefore, that multi-voltinism is realized more frequently in these areas than elsewhere.

Egg-Pod Morphology in Relation to Diapause

Although most acridids lay their eggs in the soil (Hunter-Jones 1972), the effects of environmental conditions on the shape of the eggpod and the depth at which it is laid have been little studied. Species size has been shown to influence the length of the oviposition hole, but the physical properties of the soil, such as hardness or moisture content (Uvarov 1966), are commonly regarded as being most important. Eggpods laid in autumn by *C. terminifera* are usually curved and confined to the top 5 cm of soil, whereas those laid in spring or summer are generally straight and reach depths of 8–10 cm (Wardhaugh 1972). This change of eggpod morphology in autumn appeared to be related to the production of diapausing eggs (Wardhaugh 1973). Accordingly, the experiments designed to examine the effects of temperature and photoperiod on the induction of diapause (see above) were also used to assess the extent to which these two factors were involved in modifying eggpod morphology.

Eggpods from each of the 36 temperature-photoperiod treatments described earlier were classified as either long, medium, or short on the basis of their length and/or shape. Long pods were laid more or less vertically and reached the base of the oviposition tube, which was 8 cm deep. Short pods were confined to the top 3 cm of soil and were usually curved or laid obliquely.

Eggpod morphology was found to be controlled by temperature and photoperiod experienced during both the nymphal and adult stages of development (Wardhaugh 1977). Among insects subjected to long day lengths or to an increase in day length during development (i.e., to conditions favoring the production of nondiapausing eggs), long pods accounted for 74.1% of the 928 pods examined. Only 2.8% of pods in these treatments were short. In contrast, insects subjected to short days or to a decrease in day length (i.e., to conditions favoring the production of diapausing eggs), produced 58.0% short pods and only 11.5% long pods ($n = 1003$). Temperature exerted no effect on eggpod morphology under long-day conditions, but at medium and short day lengths, either high temperature or an increase in temperature was associated with a predominance of long pods, whereas low temperature, or a decrease in temperature, resulted in the production of short pods (Wardhaugh 1977).

This unique adaptation, whereby temperature and photoperiod determine not only the type of egg that will be laid (diapause or nondiapause) but also the depth at which laying occurs, presumably has survival value. Eggs laid in spring and summer frequently encounter periods of hot, dry weather. Thus, the greater the depth of oviposition the greater the protection from desiccation and high soil temperatures. Moreover, eggs laid at greater depth are less likely to be affected by light rain, which might be enough to allow the completion of embryonic development but insufficient to promote plant growth adequate for nymphal development.

In autumn, because of cooler conditions, the reverse applies. Plant growth is often sustained by less rain (Biddiscombe et al. 1954; Slatyer 1962) and eggs laid close to the surface are better placed to use this moisture than those laid more deeply. Such an arrangement, coupled with higher soil temperatures in the surface layers, encourages rapid development in eggs conditioned to hatch before the onset of winter. In spring, similar advantages pertain with respect to eggs that spend the winter in diapause.

Development of Overwintering Nymphs

Eggs laid in the autumn, in which diapause is averted by exposure to high temperature, normally complete development and hatch before the onset of winter. The resulting nymphs rarely reach the adult stage before spring (Key 1938), spending most of the winter in the third instar (Farrow and Hunter, personal communication). Such an extended period of nymphal development was formerly thought to be explicable solely in terms of low seasonal temperatures (Key 1945; Clark 1947; Clark et al. 1969).

In the course of the diapause induction experiments described above, however, it was found that the duration of nymphal development varied with photoperiod and temperature (Wardhaugh 1979). Among insects reared at either long (15.5 hr) or medium (12.5 hr) day lengths, development was highly synchronized, but at a short photophase (11 hr) development in a proportion of nymphs was arrested at the third nymphal instar. This effect was more marked in insects held at low temperature than in those kept at high temperature. There was also an effect due to egg type. Insects hatching from diapausing eggs were less disposed to arrested development than those emerging from eggs that had not entered diapause. These latter individuals would normally comprise the overwintering nymphal population.

Among locusts in which development had been arrested at the third nymphal instar, development resumed sporadically at a photophase of 11 hr over a period of 57 days. On transfer to a longer photophase and/or to higher temperature, however, resumption of development was prompt and the molt to the fourth instar was generally completed within 7–14 days.

Over most of eastern Australia, day lengths of ca. 11 hr prevail for several weeks before and after the winter solstice. It would appear, therefore, that the slow rates of growth observed among overwintering nymphs may owe as much to day length as to low temperature. The net effect of these conditions is to synchronize the development of overwintering nymphal populations and to delay the production of adults until spring. As such, the winter arrest of nymphal development provides *C. terminifera* with an alternative survival strategy complementing that of diapause in bridging the gap between autumn and spring.

Summary of Developmental Options in *C. terminifera*

In response to a winter environment that is either too cool and/or too dry to support reproduction, *C. terminifera* has evolved a highly complex life system, which serves to maximize survival, synchronize seasonal development, and regulate voltinism. Photoperiod and temperature, acting during the nymphal and early adult stages, are the principal environmental cues regulating both the induction of diapause and the depth of oviposition. The latter determines the temperature and moisture conditions experienced by the eggs and thus moderates their survival, rates of development, and the proportion entering diapause.

Since the optimum conditions for diapause induction are associated with a photophase of 13.5 hr, inductive females are normally present well in advance of winter. Depending on the temperature and moisture content of the soil, a variable proportion of the eggs are conditioned to enter diapause, whereas the remainder retain the potential for continuous development. Those that hatch in early autumn, i.e., while temperature and photoperiod allow continuous nymphal development, may reach the adult stage and undergo a further cycle of breeding before the onset of winter. For later hatchlings, however, declining temperature and day length result in arrested development at the third nymphal instar, and so delay any further breeding until spring.

In an earlier model of the development of overwintering eggs of *C. terminifera,* the interaction of temperature and moisture during the period prior to blastokinesis was visualized as permitting nine alternative courses of embryonic development (Wardhaugh 1980b). With the recognition of an optimum soil temperature controlling the immediate expression of diapause (Hunter and Gregg 1984) the number of alternatives is increased to 18 (Figure 6.6). The simplest and most direct course of embryonic development is that associated with cool, moist conditions, which result in the suppression of diapause. This type of development is probably typical of eggs laid late in the season in areas of more reliable winter-spring rainfall, e.g., southern New South Wales and Victoria. At the opposite extreme are eggs laid in warm, dry conditions, which may cause arrested development at the prediapause stage of Q_1. Depending on seasonal temperatures and rainfall, subsequent development may be arrested at the diapause stage and again at Q_2. Such a complex sequence of events is most likely to occur in the warmer, more northerly parts of the species range, e.g., in central and western Queensland, where late autumn-early spring rainfall is often low and unreliable.

Matthee (1951) has proposed a similar though less complex scheme of embryonic development for the brown locust, *Locustana pardalina*. *L. pardalina* is endemic in areas of southern Africa, which are climatically similar to those occupied by *C. terminifera* in Australia.

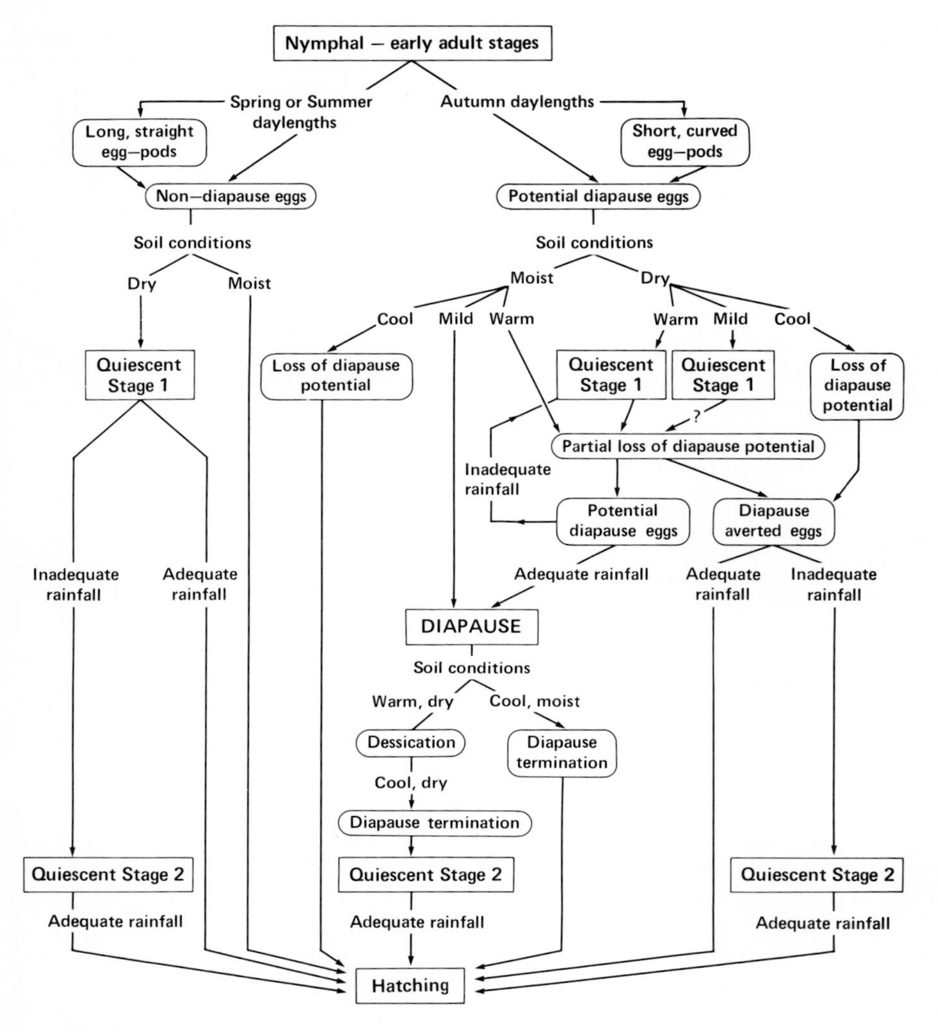

Figure 6.6. Developmental pathways of eggs of *C. terminifera* in relation to day length, temperature, and rainfall.

References

Andrewartha, H.G.: Diapause in relation to the ecology of insects. Biol. Rev. 27, 50–107 (1952).

Beck, S.D.: Insect Photoperiodism. New York: Academic Press, 1968.

Biddiscombe, E.F., Cuthbertson, E.G., Hutchings, R.J.: Autecology of some natural pasture species at Trangie, N.S.W. Aust. J. Bot. 2, 69–98 (1954).

Brown, V.K., Hodek, I., (eds.): Diapause and Life-Cycle Strategies in Insects. The Hague: Junk, 1983.

Casimir, M.: History of outbreaks of the Australian plague locust, *Chortoicetes terminifera* (Walk.), between 1933 and 1959 and analysis of the influence of rainfall on these outbreaks. Aust. J. Agric. Res. 13, 674–700 (1962).

Clark, D.P.: Night flights of the Australian plague locust, *Chortoice testerminifera* (Walk.), in relation to storms. Aust. J. Zool. 17, 329–352 (1969).

Clark, D.P., Ashall, C., Waloff, Z., Chinnick, L.: Field studies on the Australian plague locust (*Chortoicetes terminfera* Walk.) in the Channel country of Queensland. Anti-Loc. Bull. 44, 1–101 (1969).

Clark, L.R.: An ecological study of the Australian plague locust (*Chortoicetes terminifera* Walk.) in the Bogan-Macquarie Outbreak Area, N.S.W. Bull. Coun .Sci. Ind. Res. Melb. 236, 1–63 (1947).

Davidson, J.: On the ecology of the black-tipped locust (*Chortoicetes terminifera* Walk.) in South Australia. Trans. R. Soc. S. Aust. 60, 137–152 (1936).

Dingle, H. (ed.): Evolution of Insect Migration and Diapause. New York: Springer-Verlag, 1978.

Drake, V.A., Farrow, R.A.: The nocturnal migration of the Australian plague locust, *Chortoicetes terminifera* (Walker) (Orthoptera: Acrididae): quantitative radar observations of a series of northward flights. Bull. Entomol. Res. 73, 567–585 (1983).

Farrow, R.A.: Offshore migration and the collapse of outbreaks of the Australian plague locust (*Chortoicetes terminifera* Walk.) in southeast Australia. Aust. J. Zool. 23, 569–595 (1975).

Hunter, D.M., Gregg, P.C.: Variation in diapause potential and strength in eggs of the Australian plague locust, *Chortoicetes terminifera* (Walker)(Orthoptera: Acrididae). J. Insect Physiol. 30, 867–870 (1984).

Hunter-Jones, P.: Factors affecting egg survival in Acridoidea. In: Proceedings of the International Conference on the Current and Future Problems of Acridology. London, United Kingdom, 6-16 July 1970. Hemming,C.F., Taylor, T.H.C. (eds.). London: Centre for Overseas Pest Research, 1972, pp. 355–357.

Johnson, C.G.: Migration and Dispersal of Insects by Flight. London: Methuen, 1969.

Key, K.H.L.: The regional and seasonal incidence of grasshopper plagues in Australia. Bull. Coun. Sci. Ind. Res. Aust. 177, 1–87 (1938).

Key, K.H.L.: The general ecological characteristics of the outbreak areas of the Australian plague locust (*Chortoicetes terminifera* Walk.). Bull. Coun .Sci. Ind. Res. Melb. 186, 1–127 (1945).

Lees, A.D.: The Physiology of Diapause in Arthropods. Cambridge: University Press, 1955.

Magor, J.I.: Outbreaks of the Australian plague locust (*Chortoicetes terminifera* Walk.) in New South Wales during the period 1937–1962, particularly in relation to rainfall. Anti-Loc. Mem. No. 11, 1970, 68 pp.

Matthee, J.J.: The structure and physiology of the egg of *Locustana pardalina* (Walk.). Sci. Bull. Dept. Agric. Un. S. Afr., 316, 1951, 83 pp.

Slatyer, R.O.: Climate of the Alice Springs area. In: Lands of the Alice Springs Area, Northern Territory. Perry, R.A. (ed.). Melbourne: CSIRO Land Research Series No. 6., 1962, pp. 109–128.

Slifer, E.H.: The effects of xylol and other solvents on diapause in the grasshopper egg; together with a possible explanation of the action of these agents. J. Exp. Zool. 102, 333–356 (1946).

Swan, D.C.: Locusts and grasshoppers in South Australia. J. Agric. S. Aust. 59, 85–89 (1955).

Uvarov, B.P.: Grasshoppers and Locusts. Vol. 1. London: University Press, 1966.

Wardhaugh, K.G.: The development of eggs of the Australian plague locust, *Chortoicetes terminifera* (Walk.), in relation to temperature and moisture. In: Proceedings of the International Conference on the Current and Future Problems of Acridology. London, United Kingdom, 6–16 July 1970. Hemming, C.F., Taylor, T.H.C. (eds.). London: Centre for Overseas Pest Research, 1972, pp. 261–272.

Wardhaugh, K.G.: A study of some factors affecting egg development in *Chortoicetes terminifera* Walker (Orthoptera: Acrdidiae). Ph.D. Thesis, Australian Nat. Univ., Canberra (1973).

Wardhaugh, K.G.: The effects of temperature and photoperiod on the morphology of the egg-pod of the Australian plague locust (*Chortoicetes terminifera* Walker, Orthoptera: Acrididae. Aust. J. Ecol. 2, 81–88 (1977).

Wardhaugh, K.G.: A description of the embryonic stages of the Australian plauge locust, *Chortoicetes terminifera* (Walk.). Acrida 7, 1–9 (1978).

Wardhaugh, K.G.: Photoperiod as a factor in the development of nymphs of the Australian plague locust, *Chortoicetes terminifera* Walker (Orthoptera: Acrididae). J. Aust. Entomol. Soc. 18, 387–390 (1979).

Wardhaugh, K.G.: The effects of temperature and moisture on the inception of diapause in eggs of the Australian plague locust, *Chortoicetes terminifera* Walker (Orthoptera: Acrididae). Aust. J. Ecol. 5, 187–191 (1980a).

Wardhaugh, K.G.: The effects of temperature and photoperiod on the induction of diapause in eggs of the Australian plague locust, *Chortoicetes terminifera* (Walker) (Orthoptera: Acrididae). Bull. Ent. Res. 70, 635–647 (1980b).

Chapter 7
Life Cycle Strategies and Plant Succession

VALERIE K. BROWN

Variation in insect life cycle strategies is commonly encountered and has featured prominently in recent symposia (e.g., Dingle 1978; Brown and Hodek 1983). Such variability has been attributed to physical, seasonal, or geographical characteristics and has also been explained in terms of the r-K continuum (MacArthur and Wilson 1967) and bet hedging (Stearns 1976). Much more rarely have changes in insect life cycle strategies been related to habitat characteristics (e.g., Denno and Dingle 1981; Denno 1983) and seldom, if ever, have they been described in terms of the vegetational characteristics of the habitat. Indeed, studies of the insect fauna of specific habitats only rarely include consideration of the life cycle strategies of the species involved (e.g., Niemela et al. 1982). However, adaptive life cycle traits are of particular relevance in herbivorous species, since insect grazing is now considered by some to be a major selective force in the maintenance of plant diversity in time and space (e.g., Breedlove and Ehrlich 1972; Ehrlich 1970), in the evolution of anti-herbivore defense mechanisms such as secondary chemicals and morphological features of plants (Feeny 1976), and on the pattern and rate of plant succession (Brown 1982a, 1985, Stinson 1983).

In this chapter I aim to show how closely insect life cycle strategies are tuned to the vegetational characteristics of the habitat and furthermore to show how they change with the increasing maturity of the habitat. In so doing, I shall endeavor to test some of the theoretical models that predict changes in the structural and functional attributes of the communities as succession proceeds (Margalef 1968; Odum 1969). Finally, I hope to demonstrate the value of long-term field studies, in specific habitats, to the understanding of the ecological significance of insect life cycle strategies.

The natural succession from early colonization of bare earth to the development of a mature woodland provides an ideal opportunity to analyze the gradual change in specific life cycle traits. Such a study of secondary succession on sandy soil in southern England is being undertaken at Silwood Park, Berkshire, U.K. A series of experimental plots of known and different successional ages has been developed over the last 8 years. Recently harrowed fields, created annually since 1977 and left to recolonize naturally, have been used to represent early successional sites. An area of permanent pastureland with herbs represents a mid-successional site, whereas a predominantly birch woodland (*Betula pendula* and *B. pubescens*) represents a late successional site. Characteristics of the plant and insect communities in all these sites have now been monitored for an 8-year period. Details of the experimental sites and sampling procedures are given in Southwood et al. (1979) and Brown and Hyman (in press).

Successional Changes in Plant Community Organization

Any plant succession incorporates a number of seral stages each with its own vegetational characteristics. However, the rate of change varies considerably. In the experimental successional plots at Silwood Park there is a very rapid transition from an annual, herb-dominated plant community to a perennial herb and grass community before tree and shrub establishment begins (Figure 7.1). It is possible to recognize four major plant communities, namely ruderal, early, mid-, and late successional. In terms of the results discussed here, the ruderal community represents species colonizing during the first 5 months of succession, early successional up to 5 years, mid-successional from 7 to 11 years, and late successional, a woodland of estimated age over 60 years. The insect species found associated with each successional plant community can then be analyzed according to its life cycle strategies.

Vegetational Characteristics Relevant to Insect Life Cycle Strategies

The life cycle strategy of an insect is influenced by three characteristics of the habitat: habitat permanence, habitat complexity, and resource diversity and availability (Brown 1985).

These attributes have all been quantified and their degree of change along the successional gradient measured. In this way they provide a quantitative template on which to consider insect life cycles. Corresponding to later stages in plant community development is an increase in habitat permanence, habitat complexity, and resource diversity and availability (Table 7.1). Habitat permanence, or the durational complexity of the habitat (Southwood 1977), can be measured using the Sørensen's Index of Similarity *(Iw)* (see Southwood 1978a) based on a quantitative comparison of the vegetation in a site for consecutive years. Since the range of *Iw* is 0–1, it can be seen that the stability of the vegetation increases markedly as succession proceeds. Habitat complexity describes the spatial attributes of the vegetation and is based on the distribution of plant structures in space above ground level. This measure has been used in the study of bird diversity (MacArthur and MacArthur 1961; Karr 1968; Recher 1969; James 1971), but has only recently been related to insect species

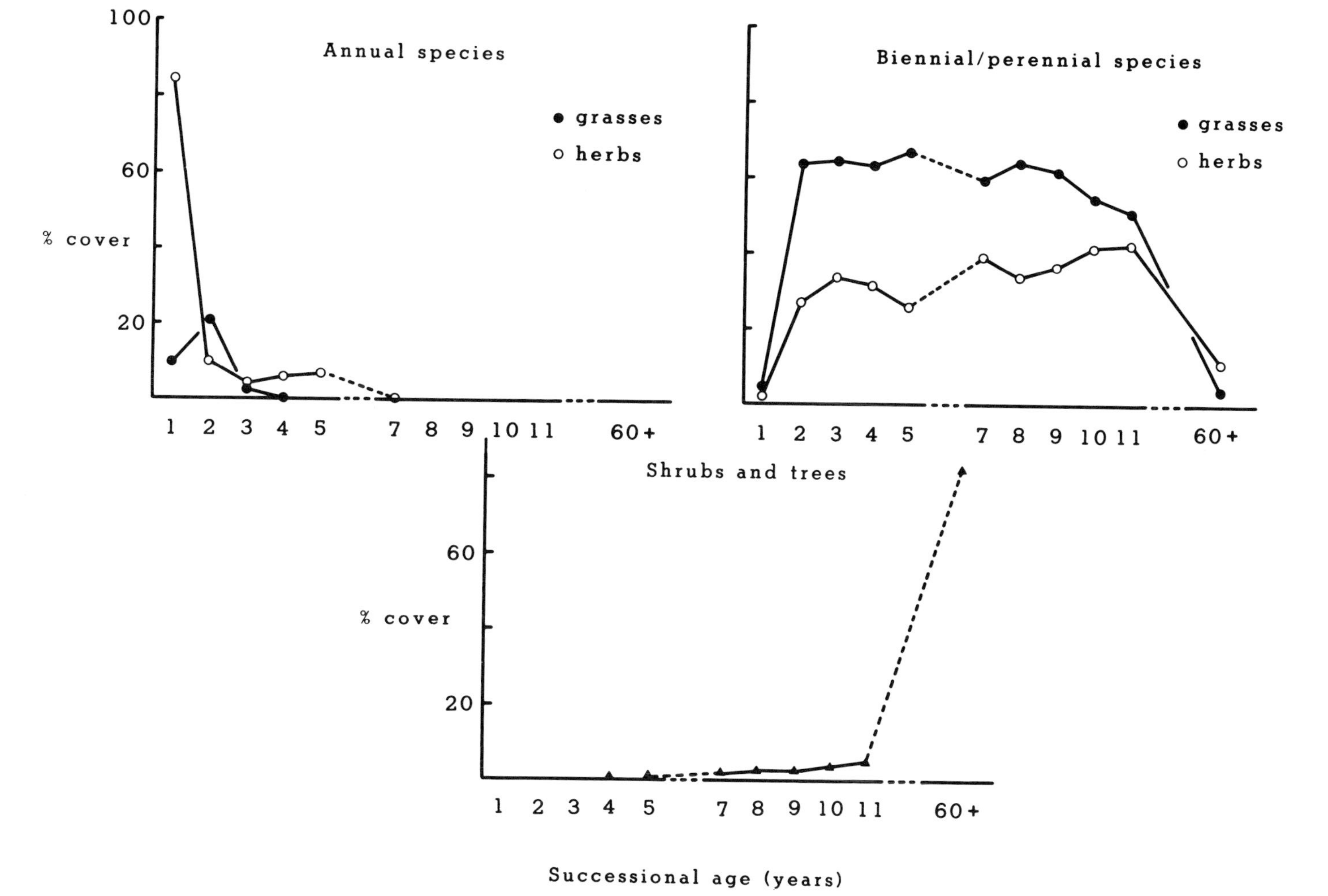

Figure 7.1. Composition of the vegetation during succession, in terms of annuals, biennials and perennials, and shrubs and trees. Each value is based on yearly totals of five samples each comprising records of at least 225 point quadrats.

Table 7.1. Successional changes in habitat characteristics influencing insect life cycle strategies

Characteristic	Successional community				Source of data
	Ruderal 0–5 months	Early 5 months–5 years	Mid 7–11 years	Late 60+ years	
Habitat permanence	0.23	0.45	0.79	0.87	Year–year comparison (Sørensen's Index of Similarity, *Iw*)
Habitat complexity	1.0	1.9	1.5	4.2	Spatial diversity (data from Southwood et al. 1979)
Resource diversity/availability	4.1	3.7	4.6	9.3	Architectural diversity (for method, see Southwood et al. 1979)

Modified from Brown (1985).

richness and diversity (Lawton 1978; Strong and Levin 1979; Southwood et al. 1979) and really reflects the size of the target for insects; larger plants offering greater potential for colonization, the ability to support larger populations, and therefore reduced chances of extinction through stochastic events. Resource diversity can be equated to architectural diversity (Lawton 1983) and is a measure of the variety of resources for insects. A more complex plant will provide a greater diversity of structures for feeding, resting, overwintering, and oviposition. It was first quantified at the community level by Southwood et al. (1979) and was subsequently refined by Stinson and Brown (1983). A new method of quantifying plant architecture has recently been developed by Morse et al. (1985), which makes use of the concept of fractals and involves the measurement of the fractal dimension of plant surfaces. In Table 7.1 both habitat complexity and resource diversity have been measured by the use of diversity indices; here William's α is used. Both measures increase as succession proceeds, although the greatest increase is seen in resource diversity and availability—a feature demonstrating the increase in available niches for insects.

Insect Life Cycle Strategies

The individual life cycle traits to be explored are categorized, as in Odum's model (1969), in terms of the nature of the life cycle, niche specialization, and the size of the organism. In the former category, in addition to generation time considered by Odum, other traits including two of those commonly associated with the r-K continuum, dispersal ability and reproductive potential, are also considered as is the overwintering stage—a trait so far not included in life cycle theory. The theoretical predictions for the traits seen in developing and mature ecosystems are given in Table 7.2.

In this chapter only herbivorous species are considered since these might be expected to be more closely tuned to the host plant (cf. tourists, which are nonpredatory species having no permanent association with the host plant, Moran and Southwood [1982]). To avoid destruction of the habitat, only external feeders on the aerial parts of the plant were sampled although the leafminers have been the subject of a special study (Godfray 1982, 1985). The major groups considered include Orthopera, Heteroptera, Homoptera excluding aphids, Thysanoptera (groups collected as immatures and adults), Curculionoidea and Chrysomeloidea (mostly collected as adults), and a relatively small number of larval Symphyta and Lepidoptera. The results for a 5-year period (1977–1981) have been analyzed and for each particular life cycle trait the taxonomic groups that display variation in respect of the trait were analyzed. Any deviation from the general pattern was noted and these groups discussed separately. The results are considered in terms of the proportion of species and also of the number of individuals displaying a particular life cycle trait. The latter gives a more realistic test of predictions and avoids the bias introduced by rare species. Differences between the successional communities were tested using the GLIM statistical package (Baker and Nelder 1978). GLIM allows one to fit a series of statistical

Table 7.2. Life cycle traits considered, including the predictions of Odum (1969), for developing and mature ecosystems and the observations from the successional sere at Silwood Park, Berkshire, U.K.

	Developing ecosystem		Mature ecosystem	
	Prediction	Observation	Prediction	Observation
Nature of life cycle				
Generation time	Short	Mostly multivoltine	Long	Mostly univoltine
	Selection pressure for rapid growth		Selection pressure for feedback control	
Dispersal ability	Not included by Odum	High	Not included by Odum	High in late succession, lower in mid-succession
	MacArthur and Wilson (1967) r-K theory: high		MacArthur and Wilson (1967) r-K theory: low	
Overwintering stage	Not included by Odum	Mostly adult	Not included by Odum	Mostly egg
Reproductive potential	Not included by Odum	High	Not included by Odum	Low
	MacArthur and Wilson (1967) r-K theory: high		MacArthur and Wilson (1967) r-K theory: low	
Niche specialization	Broad	Broad	Narrow	Mostly narrow—but broad in Curculionoidea
Size of organism	Small	Medium	Large	Mostly large—but in Heteroptera species display a wide range in size (and shape)

models, as in a stepwise regression, so that the statistical significance of "treatment" effects may be measured while controlling for differences in other variables.

Nature of Life Cycle

Generation Time

The number of generations a year shows considerable variation with the age of the habitat (Figure 7.2a). During the early years of succession a higher proportion of species is multivoltine, a trend that is more pronounced when the number of individuals is considered: 90% of the early colonizing insects are multivoltine, whereas only 5% of the late successional individuals have more than one generation a year. In later succession univoltinism is the norm, with a few species being semivoltine—requiring 2 years to complete a single generation. Even among the aphids, which are typically multivoltine, are some species with only a "few" generations a year and with a reproductive diapause during summer. Two of the common late successional species in this study fall into this category: *Euceraphis* spp. on *Betula* and *Sitobium ptericolens* on *Pteridium aquilinum* (see Brown and Southwood 1983). Multivoltine species tend to produce offspring of smaller body size. Selection for species in early successional habitats may not favor competitive ability for limited resources (mates, refugia from predators, etc.) but rather the ability to respond to environmental changes and opportunities. The occurrence of high levels of multivoltinism in early colonizing species enables them to exploit maximally the ephemeral nature of their host plants in terms of increase in population size. In addition, ruderal plant species, although short-lived, germinate throughout the growing season providing microsites are available (e.g., Brown, Leijn, and Meijer, unpublished). Thus young, nutritionally rich plants may be available throughout the growing season to support multivoltine species. On the otherhand, in mid- and late successional species, particularly trees and shrubs, the host plant may provide food in a suitable form for only a limited period each year (e.g., Feeny 1970), a feature that would favor selection for a univoltine life cycle.

Dispersal Ability

It has been shown by a number of workers that wing polymorphism and/or ability to fly is often associated with the degree of permanence of the habitat (e.g., Southwood 1962). This has been coupled with the ability to immigrate and exploit ephemeral, highly heterogeneous resources and to emigrate from such habitats. Indeed, the results obtained here for the early years of succession lend support to these ideas (Figure 7.2b). The level of macroptery is high in the first year of colonization and then declines through early to mid-succession, although the distributions (up to mid-succession) are significantly different only when the number of individuals is considered. In late succession there is a departure from this trend, with the majority of insects being fully winged. A similar picture was established by Waloff (1983) in an elegant appraisal of the

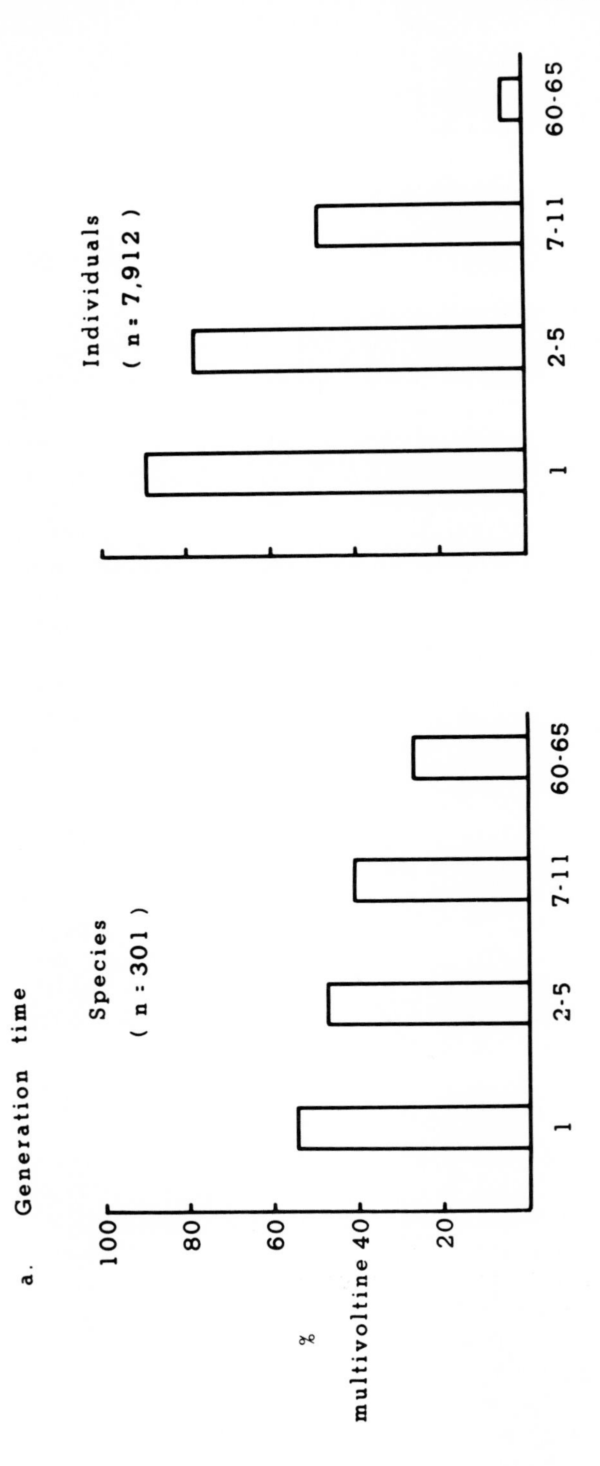
a. Generation time
Species
(n : 301)
Individuals
(n : 7,912)
% multivoltine
100
80
60
40
20
1
2-5
7-11
60-65
Successional age (years)

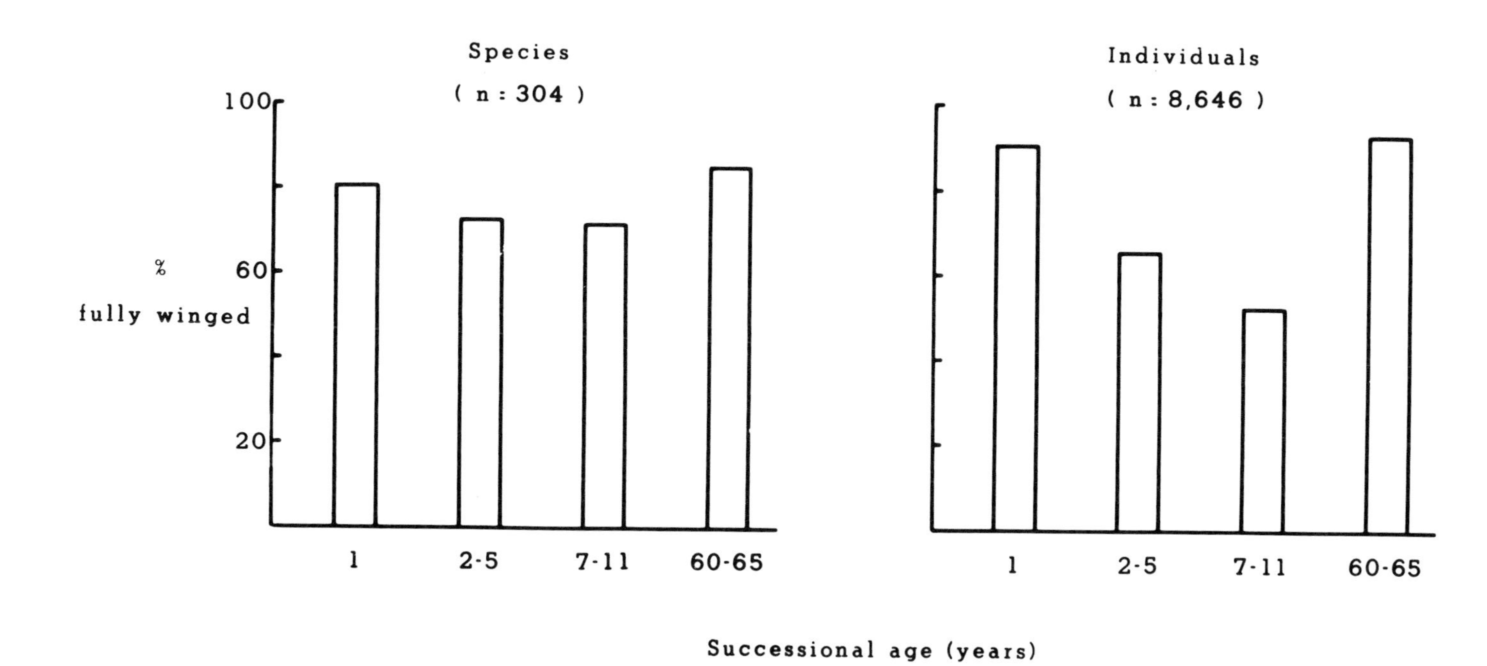

Figure 7.2. Differences in (a) generation time and (b) dispersal ability during plant succession, expressed in terms of proportion of species and individuals. Difference in distributions: generation time–species, $\chi^2_3 = 12.72$, $p < 0.01$; individuals $\chi^2_3 = 3842.0$, $p < 0.001$; dispersal ability—species, $\chi^2_2 = 1.84$, $p > 0.05$; individuals, $\chi^2_2 = 811.81$, $p < 0.001$.

incidence of wing polymorphism in British Hemiptera. She demonstrated that the occurrence of wing polymorphism was higher in species associated with low vegetation than in those associated with trees that were usually macropterous. Such findings cast some doubt on the frequently stated idea that more stable or permanent habitats select for aptery or brachyptery (e.g., Slater 1977; Vepsäläinen 1978). Waloff suggests that the architectural complexity of the habitat may be more important than its permanence in the evolution of flight ability. The high spatial and architectural complexity of a late successional habitat (e.g., Table 7.1), with the resources for insects being distributed in three dimensions, may select for macroptery. The advantages of aptery and brachyptery, in terms of increased resources for reproduction, would appear to be associated with mid-successional species, which can perhaps locate their more uniformly distributed resources by movement other than flight.

Overwintering Stage

This is a life cycle trait that is seldom considered in terms of the nature of habitat and has not been featured in theoretical models of successional attributes. Intuitively, it seems likely that overwintering as an adult might be an advantage to an early colonizing insect species, in so far as reproduction can take place immediately on colonization and the ability to be multivoltine enhanced. Indeed, it does appear that a higher proportion of early successional insect species, and particularly individuals, do overwinter as an adult (Figure 7.3). The only exception to this pattern is the auchenorrhynchan Homoptera—many species of which overwinter as eggs. However, this group is not among the very early colonizers, appearing towards the end of the first year of succession (Brown 1982a). Taking all herbivore groups together, the strategy of overwintering as an egg increases as succession proceeds with around 50% of both species and individuals overwintering in this way in late succession. Many of these species are known to have an egg diapause. In early and mid-succession, overwintering as an immature is common, a feature also seen in the Finnish Macrolepidoptera (Niemela et al. 1982) in which the proportion of species overwintering as larvae was higher on forbs than on trees. In the successional sere at Silwood Park, many of the immatures overwinter in the developing grass tussocks (e.g., *Dicranotropis hamata* (Delphacidae) and the cockroach, *Ectobius lapponicus*, commonly use tussocks of *Dactylis glomerata* [Brown 1973]).

The overwintering stage has important implications in terms of utilization of the host plant. Insect species associated with ruderal plants have to disperse in order to locate their food source and lay eggs as soon as foliage appears, making adult overwintering an advantageous strategy. With the exception of the grasses, many perennial plant species die back in winter and suitable foliage for egg laying and feeding is often not apparent until relatively late in the season, thereby reducing the time available for development. After a feeding period many species would then overwinter as immatures. In the case of insect species associated with late successional trees and shrubs, the "structure" of the plant

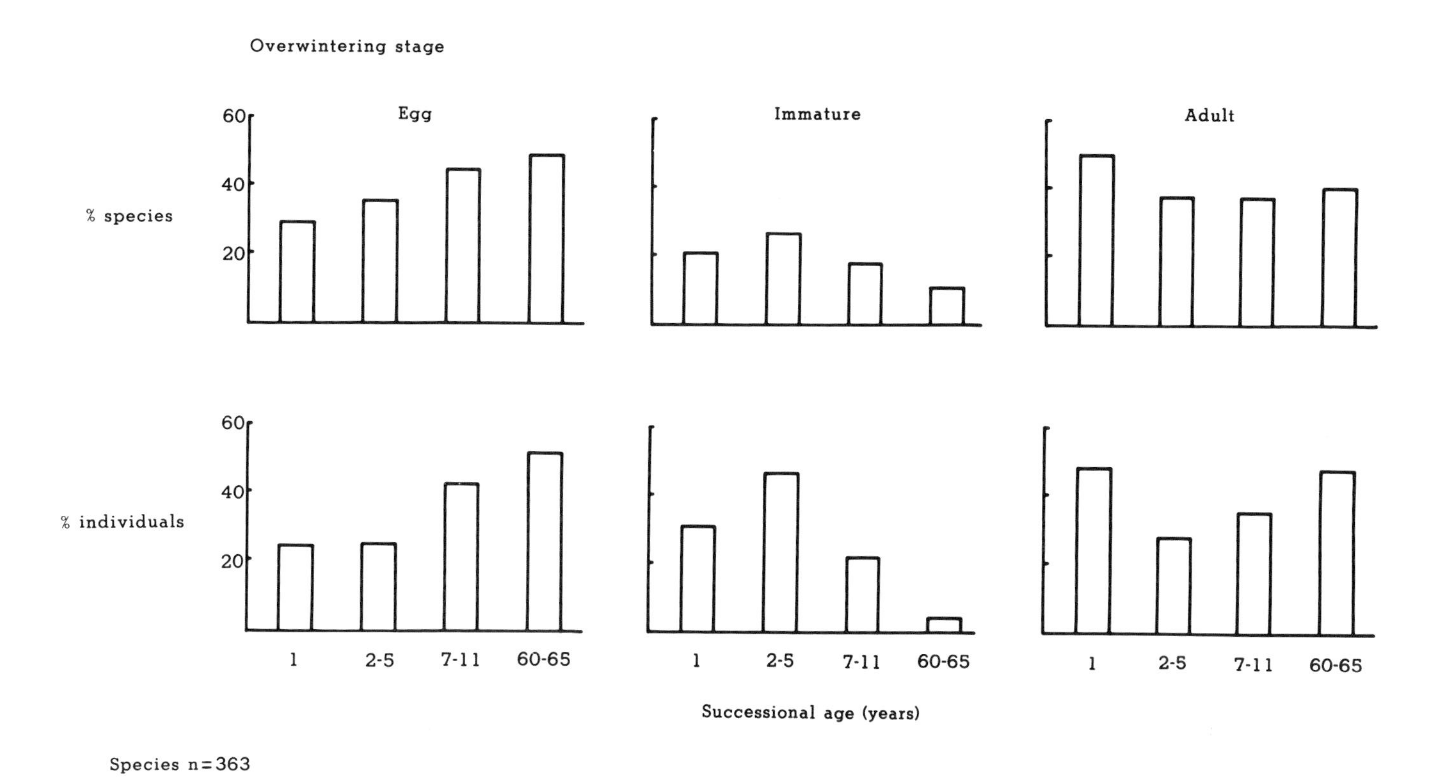

Figure 7.3. Proportion of species and individuals overwintering as eggs, immatures, and adults. Difference in distributions: species $\chi^2_6 = 14.02$, $p > 0.05$; individuals $\chi^2_6 = 1228.59$, $p < 0.001$.

is already present in spring and insects overwintering as eggs can exploit the early flush of spring foliage for larval feeding. In fact this explanation may also apply to the grassfeeding Auchennorhyncha, which commonly overwinter as eggs (Waloff 1980).

Reproductive Potential

Although reproductive potential has featured prominently in life cycle theory, tests of predictions on more than single species or small species groups are nonexistent. The reason for this lack of empirical data is undoubtedly the time-consuming nature of dissections of adult females or of egg counts in the field. Consequently, data are restricted here to one insect group, the aphids, and the results discussed fully in Brown and Llewellyn (1985) and Llewellyn and Brown (1985a,b). In many respects aphids are very suitable for this type of work since they are often numerous and can be readily collected in situ from their host plants. Using the successional sere at Silwood Park, more than 90 aphid species were collected from over 120 host plants representing different plant growth forms and phenologies (annuals, biennals and perennials, grasses, ferns, deciduous shrubs and trees, and evergreens). This enabled aphids associated with annual species, characteristic of ruderal plant communities, to be compared with those associated with the biennial and perennial species of early and mid-succession and with the late successional trees and shrubs. Wherever possible 10 specimens of the apterous and alate morph of a species were collected from each host plant every month and performance assessed by wet weight and number of embryos. The total number of embryos was established by dissection and provides a more accurate index of potential fecundity than the number of mature embryos (with pigmented eyes) used by many authors. In summary, aphids associated with herbaceous plants (annual, biennial, and perennial herbs) were more fecund than those on woody trees and shrubs (Figure 7.4). Within these categories, aphids on annuals generally have a higher reproductive potential than those on biennials and perennials whereas shrub-associated aphids are potentially more fecund than aphids on trees (Brown and Llewellyn 1985). However, there is also seasonal variation in fecundity, explicable in terms of plant chemistry (e.g., Dixon 1970). It is unfortunate that data are not yet available for other herbivorous groups to determine whether the phenomenon is widespread.

Niche Specialization

There are several ways of assessing niche specialization, although some of these rely on a degree of refinement of host records not normally available in the literature for large groups of insects. A relatively simple scheme for categorizing species into generalists and specialists was described by Brown (1985). The term generalist refers to species associated with more than one family of plants whereas the three grades of specialist (S_3, S_2, and S_1) refer to species feeding on more than one genera in a single family, more than one species in a single genus, and a single species, respectively. Figure 7.5 shows that the proportion

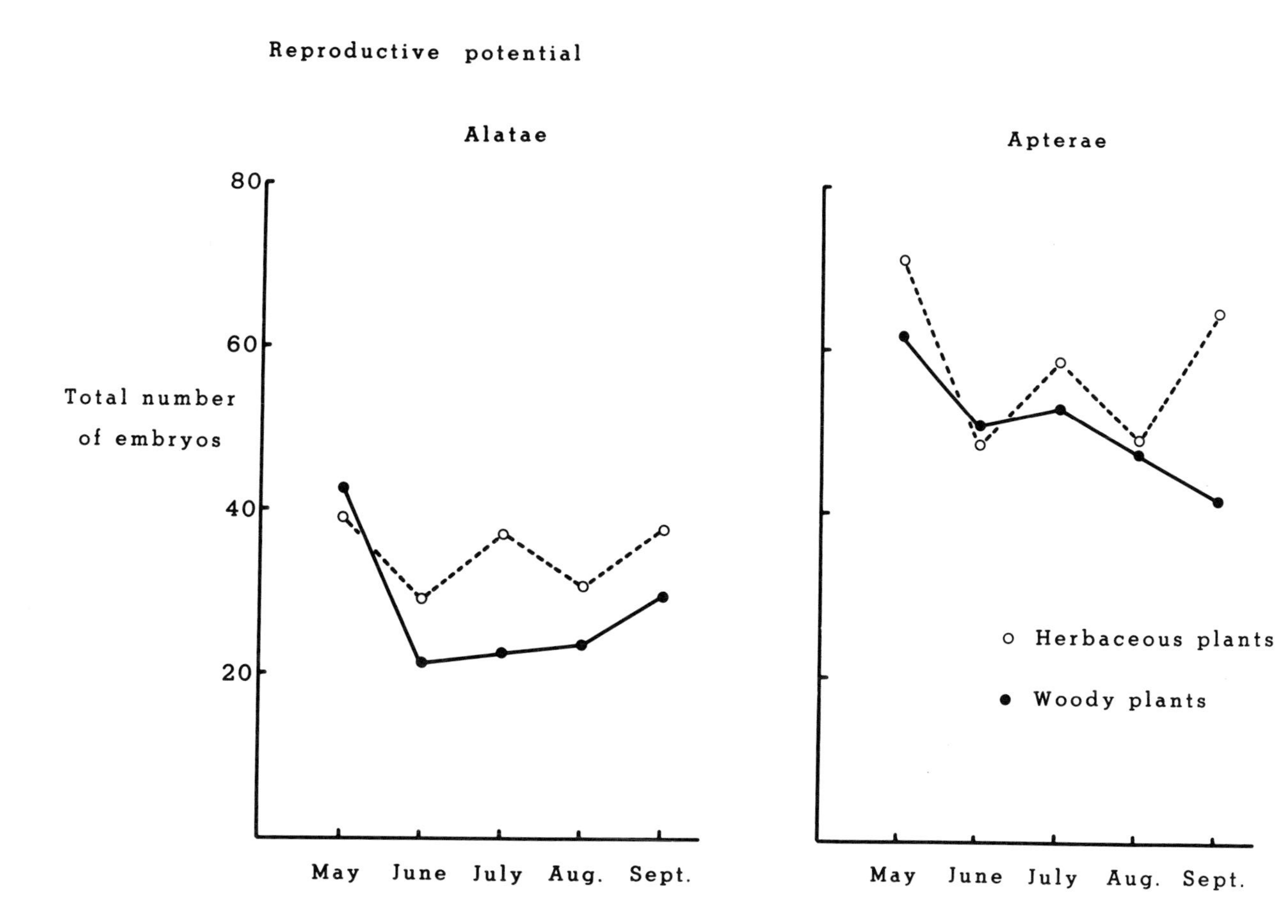

Figure 7.4. Reproductive potential, expressed in terms of number of embryos, of alate and apterous aphids associated with herbaceous and woody plants. Differences tested by two way ANOVA: alatae, $F = 60.43$, $p < 0.001$; apterae, $F = 6.9$, $p < 0.01$. (Modified from Brown and Llewellyn [1985]).

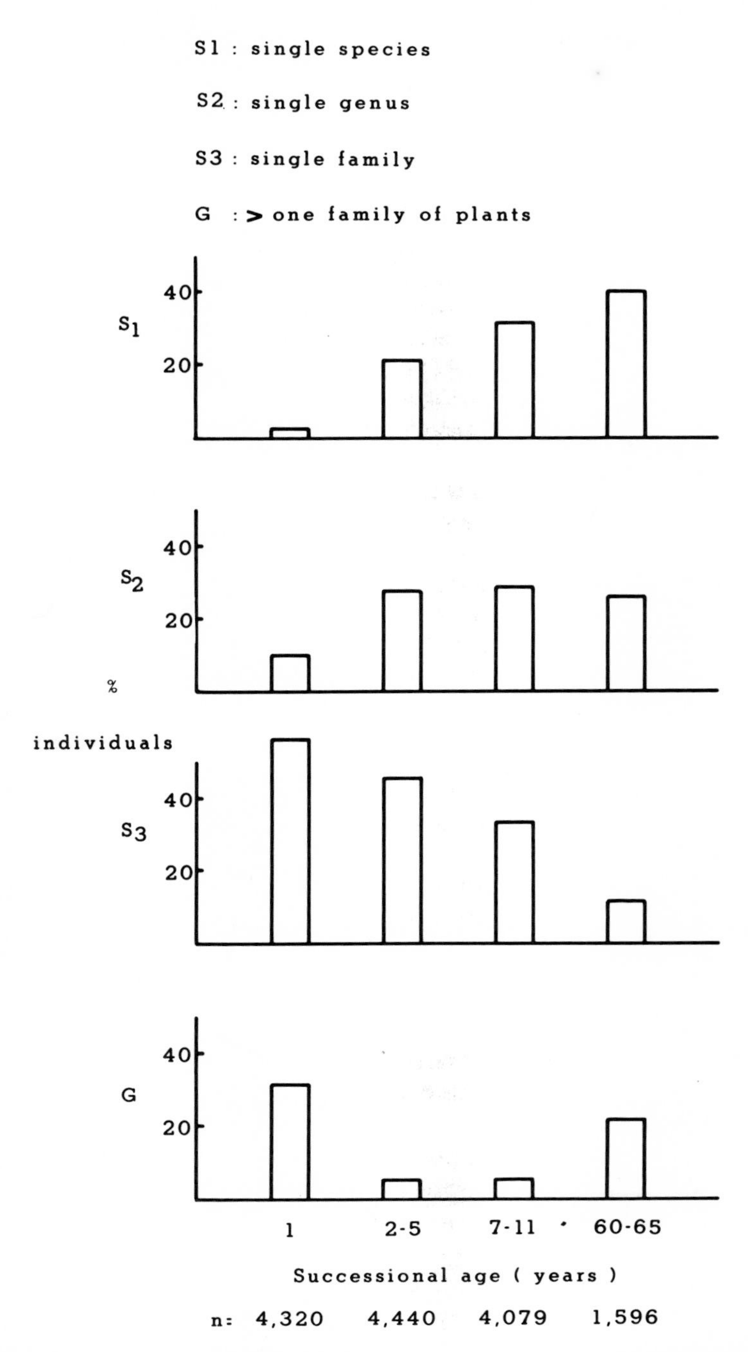

Figure 7.5. The occurrence of generalists (feeding on plants belonging to more than one family) and various grades of specialists (S_1, feeding on a single plant species; S_2, feeding on more than one species within a genus; S_3, feeding on more than one species within a family). Difference in distributions: species $\chi^2_9 = 32.77$, $p < 0.001$; individuals $\chi^2_9 = 3446.12$, $p < 0.001$.

of insects in the highly specialized categories, S_1 and S_2, increases as succession proceeds. On the other hand, the S_3 and generalist categories are numerically more abundant in early succession. This general pattern conforms to the predictions of Odum (1969) and Southwood (1977) and was found in an analysis of the sap-feeding insects in this particular successional sere by Brown and Southwood (1983). However, there is a discrepancy in this trend—the number of generalists in late succession is unexpectedly high. Examination of the individual taxonomic groups showed that this discrepancy was due to the weevils. This has been investigated in more detail by Brown and Hyman (in press) using a quantitative measure of niche breadth based on a numerical scale of 1–5 according to the range of host plants utilized (from extreme specialism to generalism; see Brown and Southwood 1983). For both species and individuals the median and mean niche breadth gradually declined through early and mid-succession (Table 7.3). However, there was a dramatic reversal of this trend in late succession.

There have been several attempts to relate the growth form and successional status of plants with the food selection strategies of insect herbivores, but the results have been contradictory. Rhoades and Cates (1976) proposed that ephemerality and predictability of food resources, based on different chemical defenses and apparencies, should select for generalism and specialism, respectively. Indeed the results displayed in Figure 7.5 do lend some support to these ideas. Later Rhoades (1979) proposed that although in predictable, late successional plants chemical defenses would be primarily against specialist herbivores, in certain situations there could be a convergence of chemicals, which might lead to the secondary selection towards generalism. This may be the case in the weevils and, according to Cates (personal communication), this is the first evidence, employing field observations from a single successional sere, to support these predictions. Studies, based on literature records of Lepidoptera, also demonstrated, however, that species feeding on herbs may be more specialized than those feeding on trees (e.g., Futuyma 1976; Niemela et al. 1982).

Size of Organism

The size of insect species has been considered in relation to theoretical successional models (Margalef 1968; Odum 1969), habitat differences in moisture and

Table 7.3. Niche specialization of weevils from sites of different successional age, expressed as means ± SEM and median (m) for samples taken in 1980

Successional age	Based on number of species			Based on number of individuals		
	n	$\bar{x}$ + SEM	m	n	$\bar{x}$ ± SEM	m
1	24	1.60 ± 0.16	1.75	966	1.86 ± 0.02	2.07
2	24	1.25 + 0.11	1.12	1001	1.31 ± 0.02	1.39
4	32	1.33 ± 0.15	1.05	459	1.11 ± 0.02	1.02
10	20	1.33 ± 0.18	1.10	1221	0.72 ± 0.01	0.60
60+	11	3.05 ± 0.48	3.50	912	3.88 ± 0.03	3.99

length of growing season (Schoener and Janzen 1968), habitat disturbance (Mason 1973), diet breadth (Wasserman and Mitter 1978), and resource pattern (Derr et al. 1981).

As yet, little attention has been given to insect size in terms of durational stability of the habitat (see Brown 1982b). Size is normally described by a measure of total body length, which, although easily measured and often available in taxonomic literature, does have limitations (see Brown 1982b). However, when a large number of species is to be considered, it is the only practical measure and for this reason has been employed here. Table 7.4 gives the mean body length for all insect species and individuals associated with each of the four plant communities. There is a general trend for insects to be larger in late succession, which follows predictions. However, when the number of individuals is considered there is a decline in mean size from species associated with the ruderal plant community to those associated with mid-succession. The size distributions of the taxonomic groups, treated separately, is to be presented elsewhere.

In addition to size, shape is also important (see Penrose 1953; Corruccini 1973). A multivariate, morphometric study of the size and shape of the heteropteran species associated with the successional gradient discussed in this paper was described by Brown (1982b). Contrary to theoretical predictions, in which small species were assumed characteristic of early successional habitats and large species of late successional habitats, it was found that the range in size and shape increased along the successional gradient. Species characteristic of early successional habitats were somewhat intermediate in size and tended to be fairly consistent both in size and shape, whereas late successional species were very variable in body form. These results were explained in terms of some other life history traits considered in this chapter, and the role of habitat permanence and complexity in determining these traits (see Brown 1982b).

General Discussion and Conclusions

In this chapter I have considered how insect life cycle strategies change as plant succession proceeds. Undoubtedly, certain characteristics of the habitat, such as structural complexity, resource availability, and durational stability

Table 7.4. Mean size of insects associated with four plant communities of different successional age

Successional age	n	Based on species ($\bar{x} \pm$ SEM)	n	Based on individuals ($\bar{x} \pm$ SEM)
1	159	3.68 ± 0.57	4494	3.33 ± 0.34
2–5	126	3.59 ± 0.63	3628	3.00 ± 0.38
7–11	129	3.86 ± 0.63	4752	2.53 ± 0.33
60+	114	4.14 ± 0.67	2625	4.27 ± 0.45

Likelihood of size distribution changing during succession: species χ^2_{117} = 69.4, p = 0.995; individuals χ^2_{117} = 7689, $p < 0.001$.

will affect the way in which insects exploit their immediate environment. The quantification of such habitat attributes is a useful means of describing relative change along a successional gradient, although it must be stressed that it is commonly the "microniche" of the insect that has the overriding effect (e.g., Slater 1977). Furthermore, it would be naive to believe that it is only the habitat that molds the life cycle traits of insects. For example, other biotic characteristics may be of importance, and for this reason the future study of insect–plant relationships during succession should include consideration of the contribution and role of guilds other than herbivores (e.g., Lawton and Strong 1981).

The pronounced changes in life cycle strategies along the successional gradient permit a categorization into ruderal, early, mid- and late successional species. Although the current work has revealed a distinct, but short-lived, ruderal insect community with specialized colonizing ability, other species probably lie along a continuum—a feature emphasized in Waloff (1983). In the weevils and leaf beetles, Brown and Hyman (in press) have defined annual successional communities by reciprocal averaging ordinations and have demonstrated such a continuum. It is, of course, possible that other taxa may behave differently under such analyses.

Any theoretical model requires empirical evidence to test it. The succession models of Margalef (1968) and Odum (1969) have been tested only in part (e.g., by Brown and Southwood 1983; Witkowski 1983). Here a number of discrepancies have been revealed, both in single groups of insects and in insects associated with a particular successional stage. The extent of those deviations can be seen in Table 7.1, and their ecological significance has been discussed in the appropriate section.

It is common for ecologists to explain ecological patterns based on collective literature records, sometimes from a range of habitat types, or geographical locations (e.g., Southwood 1978b; Lawton and Price 1979). Furthermore, little or no notice is taken of the relative abundance of the species concerned. Here, many of the trends in life cycle strategies have become clear only when the number of individuals belonging to species from a particular successional sere are considered. This perhaps provides the ideal support for Dingle's request at the conclusion of the last Symposium in this series in Kyoto "to do . . . more elegant and sophisticated, natural history. It is in nature . . . that both the questions and the answers lie" (Dingle 1983).

Acknowledgments A number of colleagues have contributed to this chapter: Mike Llewellyn collaborated in the work described on aphid reproductive potential, Nadia Waloff provided some information on the flight ability of Auchenorrhyncha, Matthew Cock and Brian Pitkin helped by providing information on some species of Chrysomelidae and Thysanoptera, respectively, Tony Ludlow gave statistical advice and Anne Storr drew the figures and ran some of the analyses. To them all I am extremely grateful. I should also like to thank Pekka Niemela and Seppo Neuvonen for some stimulating discussions during a visit to Turku, Finland. My thoughts on early colonization have been enhanced by a NATO-supported collaborative research project with Stephen Hendrix (Iowa) and Hugh Dingle (Davis, California). The effects of insect herbivores on the range of successional plant communities is currently being supported by NERC.

References

Baker, R.A., Nelder, J.A.: The GLIM System, Release 3, Generalized Linear Interactive Modelling. Oxford: Numerical Algorithms Group, 1978.

Breedlove, D.E., Ehrlich, P.R.: Coevolution: patterns of legume predation by a lycaenid butterfly. Oecologia 10, 99–104 (1972).

Brown, V.K.: The overwintering stages of *Ectobius lapponicus* (L.) (Dictyoptera: Blattidae). J. Entomol. Ser. A. Gen. Entomol. 48, 11–24 (1973).

Brown, V.K.: The phytophagous insect community and its impact on early successional habitats. In: Proceedings 5th International Symposium Insect-Plant Relations, Wageningen, 1982, Pudoc. Wageningen., pp. 205–213, 1982a.

Brown, V.K.: Size and shape as ecological discriminants in successional communities of Heteroptera. Biol. J. Linn. Soc. 18, 279–290 (1982b).

Brown, V.K.: Insect herbivores and plant succession. Oikos 44, 17–22 (1985).

Brown, V.K., Hodek, I. (eds.): Diapause and Life Cycle Strategies in Insects. The Hague: Junk, 1983.

Brown, V.K., Hyman, P.S.: Successional communities of plants and phytophagous Coleoptera. J. Ecol. (in press).

Brown, V.K., Llewellyn, M.: Variation in aphid weight and reproductive potential in relation to plant growth form. J. Anim. Ecol. 54, 651–661 (1985).

Brown, V.K., Southwood, T.R.E.: Trophic diversity, niche breadth and generation times of exopterygote insects in a secondary succession. Oecologia 56, 220–225 (1983).

Corruccini, R.S.: Size and shape in similarity coefficients based on metric characters. Am. J. Phys. Anthropol. 38, 743–753 (1973).

Denno, R.F.: Tracking variable host plants in space and time. In: Variable Plants and Herbivores in Natural and Managed Systems. Denno, R.F., McClure, M.S. (eds.). New York: Academic Press, pp. 291–341, 1983.

Denno, R.F., Dingle, M. (eds.): Insect Life History Patterns: Habitat and Geographic Variation. New York: Springer-Verlag, 1981.

Derr, J.A., Alden, B., Dingle, H.: Insect life-histories in relation to migration, body size, and host plant array: a comparative study of *Dysdercus*. J. Anim. Ecol. 50, 181–193 (1981).

Dingle, H.: Introduction Part 2: Diapause, development and phenology. In: Evolution of Insect Migration and Diapause. Dingle, H. (ed.). New York: Springer-Verlag, pp. 51–52, 1978.

Dingle, H.: Concluding remarks. In: Diapause and Life Cycle Strategies in Insects. Brown, V.K., Hodek, I. (eds.). The Hague: Junk, pp. 271–273, 1983.

Dixon, A.F.G.: Quality and availability of food for a sycamore aphid population. In: Animal Populations in Relation to Their Food Resources. Watson, A. (ed.). Oxford: Blackwell, pp. 271–287, 1970.

Ehrlich, P.R.: Coevolution and the biology of communities. In: Biochemical Coevolution. Chambers, K.L. (ed.). Corvallis, Oregon: University of Oregon Press, pp. 1–11, 1970.

Feeny, P.P.: Seasonal changes in oak leaf tannins and nutrients as a cause of spring feeding by winter moth caterpillars. Ecology 51, 565–581, 1970.

Feeny, P.P.: Plant apparency and chemical defense. In: Biochemical Interactions Between Plants and Insects. Wallace, J., Mansell, R. (eds.). Recent Adv. Phytochem. 10, 1–40, 1976.

Futuyma, D.J.: Food plant specialization and environmental predictability in Lepidoptera. Am. Natur. 110, 285–292, 1976.

Godfray, H.C.J.: Leaf-mining insects and their parasitoids in relation to plant succession. Ph.D. Thesis, University of London (1982).
Godfray, H.C.J.: The absolute abundance of leaf miners on plants of different successional stages. Oikos, 45, 17–25.
James, F.C.: Ordinations of habitat relationships among breeding birds. Wilson Bull. 83, 215–236 (1971).
Karr, J.R.: Habitat and avian diversity on strip-mined land in East Central Illinois. Condor 70, 348–357, 1968.
Lawton, J.H.: Host-plant influences on insect diversity: the effects of space and time. In: Diversity of insect faunas. Mound, L.A., Waloff, N. (eds.). Symp. R. Entomol. Soc. Lond. 9, 19–40, 1978.
Lawton, J.H.: Plant architecture and the diversity of phytophagous insects. Annu. Rev. Entomol. 28, 23–39 (1983).
Lawton, J.H., Price, P.W.: Species richness of parasites on hosts: agromyzid flies on the British Umbelliferae. J. Anim. Ecol. 48, 619–637 (1979).
Lawton, J.H., Strong, D.R. Jr.: Community patterns and competition in folivorous insects. Am. Natur. 118, 317–338 (1981).
Llewellyn, M., Brown, V.K.: Effect of host plant on aphid performance in polyphagous species. J. Anim. Ecol. 54, 639–650 (1985a).
Llewellyn, M., Brown, V.K.: Seasonal variation in aphid weight and reproductive potential. J. Anim. Ecol. 54, 663–673 (1985b).
MacArthur, R.H., MacArthur, J.W.: On bird species diversity. Ecology 42, 594–598 (1961).
MacArthur, R.H., Wilson, E.O.: The Theory of Island Biogeography. Princeton, New Jersey: Princeton University Press, 1967.
Margalef, R.: Perspectives in Ecological Theory. Chicago, Illinois: University of Chicago Press, 1968.
Mason, L.G.: The habitat and phenetic variation in *Phymata americana* Melin. Syst. Zool. 22, 271–279 (1973).
Moran, V.C., Southwood, T.R.E.: The guild composition of arthropod communities in trees. J. Anim. Ecol. 51, 289–306 (1982).
Morse, D.R., Lawton, J.H., Dodson, M.M., Williamson, M.H.: Fractal dimension of vegetation and the distribution of arthropod body lengths. Nature 314, 731–732 (1985).
Niemela, P., Tahvanainen, J., Sorjonen, J., Hokkanen, T., Neuvonen, S.: The influence of host plant growth form and phenology on the life strategies of Finnish macrolepidopterous larvae. Oikos 39, 164–170 (1982).
Odum, E.P.: The strategy of ecosystem development. Science (NY) 164, 262–270 (1969).
Penrose, L.S.: Distance, size and shape. Ann. Eugenics 18, 337–343 (1953).
Recher, H.F.: Bird species diversity and habitat diversity in Australia and North America. Am. Natur. 103, 75–80 (1969).
Rhoades, D.F.: Evolution of plant chemical defense against herbivores. In: Herbivores: Their Interaction with Secondary Plant Metabolites. Rosenthal, G.A., Janzen, D.H. (eds.). London: Academic Press, pp. 3–54, 1979.
Rhoades, D.F., Cates, R.G.: Towards a general theory of plant anti-herbivore chemistry. Recent Adv. Phytochem. 101, 168–213 (1976).
Schoener, T.W., Janzen, D.H.: Notes on the environmental determinates of tropical vs. temperate insect size patterns. Am. Natur. 102, 207–224 (1968).
Slater, J.A.: The incidence and evolutionary significance of wing polymorphism in Lygaeid Bugs with particular reference to those of South Africa. Biotropica 9, 217–229 (1977).

Southwood, T.R.E.: Migration of terrestrial arthropods in relation to habitat. Biol. Rev. 37, 171–214 (1962).
Southwood, T.R.E.: Habitat, the template for ecological strategies. J. Anim. Ecol. 46, 337–365 (1977).
Southwood, T.R.E.: Ecological Methods. (2nd ed.) London: Chapman and Hall, 1978a.
Southwood, T.R.E.: The components of diversity. In: Diversity of Insect Faunas. Mound, L.A., Waloff, N. (eds.). Symp. R. Entomol. Soc. Lond. 9, 19–40, 1978b.
Southwood, T.R.E., Brown, V.K., Reader, P.M.: The relationships of plant and insect diversities in succession. Biol. J. Linn. Soc. 12, 327–348 (1979).
Stearns, S.C.: Life history tactics: a review of the ideas. Q. Rev. Biol. 51, 3–47 (1976).
Stinson, C.S.A.: Effects of insect herbivores on early successional habitats. Ph.D. Thesis, University of London (1983).
Stinson, C.S.A., Brown, V.K.: Seasonal changes in the architecture of natural plant communities and its relevance to insect herbivores. Oecologia 56, 67–69 (1983).
Strong, D.R. Jr., Levin, D.A.: Species richness of plant parasites and growth form of their host. Am. Natur. 114, 1–22 (1979).
Vepsäläinen, K.: Wing dimorphism and diapause in *Gerris:* determination and adaptive significance. In: Evolution of Insect Migration and Diapause. Dingle, H. (ed.). New York: Springer-Verlag, pp. 218–253, 1978.
Waloff, N.: Studies of grassland leaf-hoppers and their natural enemies. Adv. Ecol. Res. 11, 82–215 (1980).
Waloff, N.: Absence of wing polymorphism in the arboreal, phytophagous species of some taxa of temperate Hemiptera: an hypothesis. Ecol. Entomol. 8, 229–232 (1983).
Wasserman, S.S., Mitter, C.: The relationship of body size to breadth of diet in some Lepidoptera. Ecol. Entomol. 3, 155–160 (1978).
Witkowski, Z.: Secondary succession of oak hornbeam biocenosis in the Niepolomice Forest against a background of the Margalef's and Odum's model of succession. Stud. Natur. A 27, 7–78 (1983).

Chapter 8

Polymorphism in the Larval Hibernation Strategy of the Burnet Moth, *Zygaena trifolii*

WOLFGANG WIPKING and DIETRICH NEUMANN

In temperate zones, organisms have evolved various adaptations for the timing of their developmental periods and dormancies to overcome annually changing weather conditions. For example, latitudinal gradients in the photoperiodic induction of a facultative diapause have been described in geographical populations of many insects from different latitudes (for reviews: see Beck 1980; Danilevskii 1965; Müller 1976; Saunders 1979; Tauber and Tauber 1976; Taylor and Spalding, this volume). Further attention has been focused on the occurrence of an obligatory diapause (Masaki 1961, 1978; Thiele 1977), a repetitive diapause (Matthes 1953; Pener and Orshan 1983; Tanaka 1983), and prolonged maintenance of diapause correlated to local climatic conditions (Halperin 1969; Philogene and Benjamin 1971; Waldbauer and Sternburg 1973). Although all these variations in seasonal control of insect life cycles have been mainly recognized between geographical populations, relatively few investigations have been carried out on the intrapopulational variation of diapause characters and its ecological significance (Bradshaw 1973; Lumme 1978; Sauer 1980, 1984).

The palaearctic Zygaenidae are known as burnet moths (Subfam. Zygaeninae) or foresters (Subfam. Chalcosiinae and Procridinae). All westpalaearctic Zygaeninae belong to the genus *Zygaena* F. At least 18 European species extend on the northwestern fringe of their distributions into West Germany and, of these, only four species have reached Scandinavia. Most of these species are ecologically classified as eurytopic, occupying various biotopes in the European Mediterranean area. The Zygaeninae, however, are generally thermo- or xerophilic moths specializing on particular food plants. Consequently, at the northern border of Germany, they prefer localized and discontinuously distributed dry grasslands and limestone habitats. Thus, the burnet moths exist only in small and mostly isolated habitats, but sometimes in relatively high population densities (Wipking 1983).

The first observations on the seasonal development and the generation time of European *Zygaena* species were made by Dorfmeister (1853, 1854, 1855). Further contributions have been published by Burgeff (1910, 1921, 1965), Holik

(1937, 1939), and Uebel (1974). The results can be summarized as follows. At temperate latitudes of Middle Europe, full-grown larvae of most species will be generally observed only in June. Pupation takes place in a cocoon attached to twigs, and the adults fly during the day from the end of June until the beginning of August. The young larvae of the succeeding, and hibernating, generation are active until the end of September when they enter diapause. The diapausing larvae are characterized by molting to a special decolorized morph of smaller size and high frost tolerance. The number of the larval dormancies, as well as their duration, may vary among the offspring of a single female, so that, in at least some of the species, the generation time varies markedly within a population, ranging from one to several years.

The aim of the following research program was to analyze intrapopulational variability in the zygaenid life cycle in more detail, especially with regard to the photoperiodic responses and the occurrence of repetitive diapauses.

Experimental Procedures

The burnet moth *Zygaena trifolii* Esp. has an Atlanto-Mediterranean dispersal center and ranges from Morocco and Tunesia to Denmark and Germany and east into Poland and the USSR (Naumann et al. 1984) (Figure 8.1).

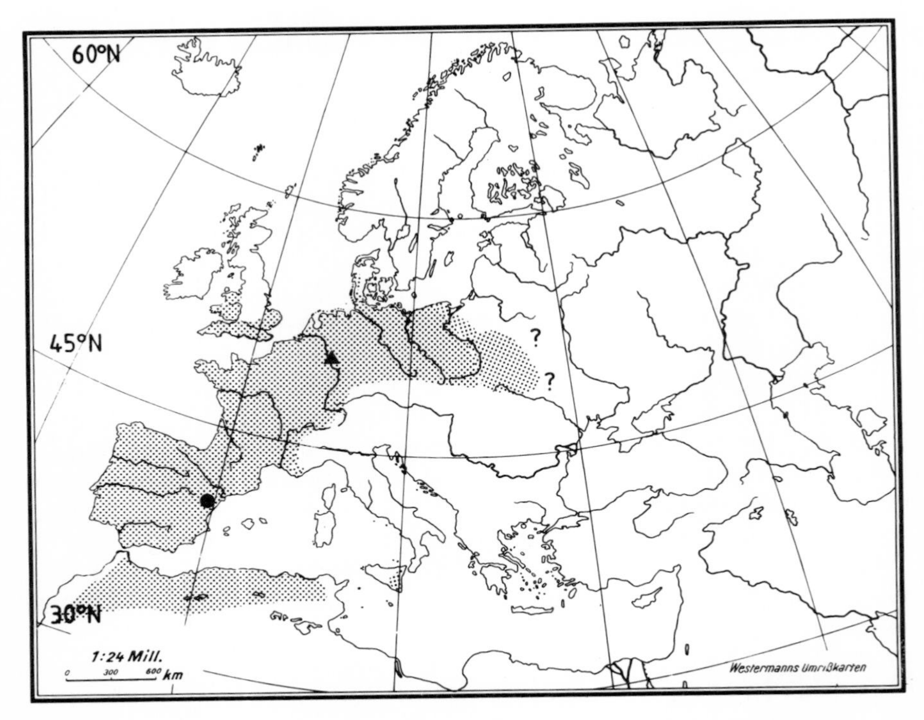

Figure 8.1. Distribution map of *Zygaena trifolii* Esp. Spain (circle): Valencia, El Saler (39°N), Western Germany (triangle): Cologne, Dellbrück (51°N).

Rearing experiments with stocks of two different European localities in Spain and West Germany were conducted under controlled laboratory conditions. The German stock was collected near Cologne (51°N) by W. Wipking, and the Spanish one by C. Naumann near Valencia (39°N) at about the same altitude. The larvae were reared at 20°C in small plastic cases and at different photoperiodic conditions in the range between LD 8:16 and 18:6. They were fed on a diet of fresh leaves of *Lotus corniculatus* L. (Fabaceae) growing throughout the year in a greenhouse. Every 2 days the larvae were monitored with respect to the date of molting and the stage of the diapausing larvae. In addition, the larvae of the Cologne stock were reared in the institute garden under natural light and temperature conditions.

Life Cycle Characteristics of the Spanish Population

The Valencia population is characterized by a facultative larval diapause and its critical photoperiod was about 14.5 hr (Figure 8.2a). More detailed observations (Figure 8.2b), however, revealed that diapause may occur in different

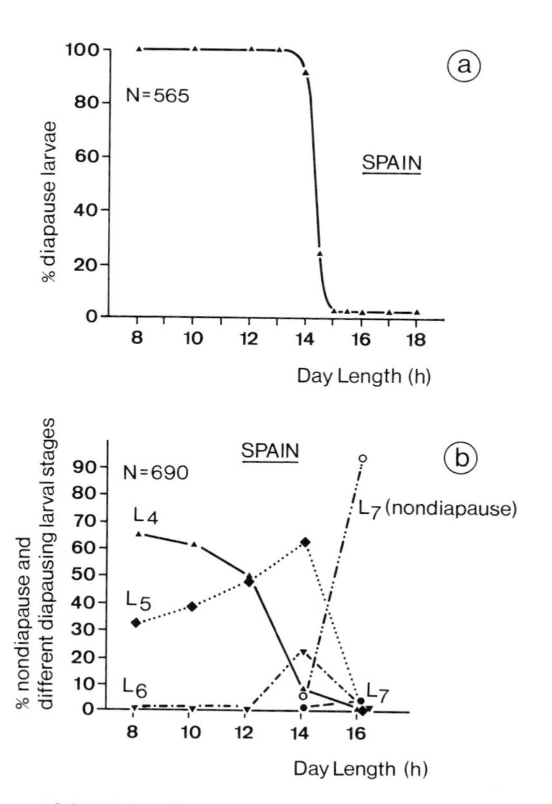

Figure 8.2. *Zygaena trifolii*, Valencia stock. (a) The photoperiodic responses at a constant temperature of 20°C. (b) Frequency of diapause larval instars and nondiapausing L_7 instar in relation to the increasing day length, according to an additional experiment, in which about 150 L_1 larvae were introduced to each of the photoperiods tested.

larval instars, mainly in the fourth and fifth instars, up to photoperiods of 14 hr. There was an obvious retardation of the physiological age of the diapausing stage with increasing day length such that at 14–16 hr even some individuals of the sixth and seventh instars entered into the developmental arrest with its characteristic morph. At day lengths of more than 14.5 hr, most of the larvae finished their ontogenesis without any dormancy.

These data suggest that the Valencia population has a potentially multivoltine life cycle with, generally, two generations per year. According to field observations of Naumann in Spain (Figure 8.3, lower part), however, at least three, and possibly four, different types of life cycles can be identified. Under favorable environmental conditions, the adults of the overwintering generation generally emerge and reproduce in May–June. Most of their progeny complete development under long-day conditions within six or seven larval instars by the middle of August so that the adults of this second generation fly during the end of August and early September. The succeeding generation consequently grows under short-day photoperiods. It enters diapause in the fourth larval instar and hibernates until the following spring when development is completed by May (Figure 8.3, line 9). Part of the summer generation, however, may enter an aestival diapause in stage 4, as detectable by the diapausing morph. As a con-

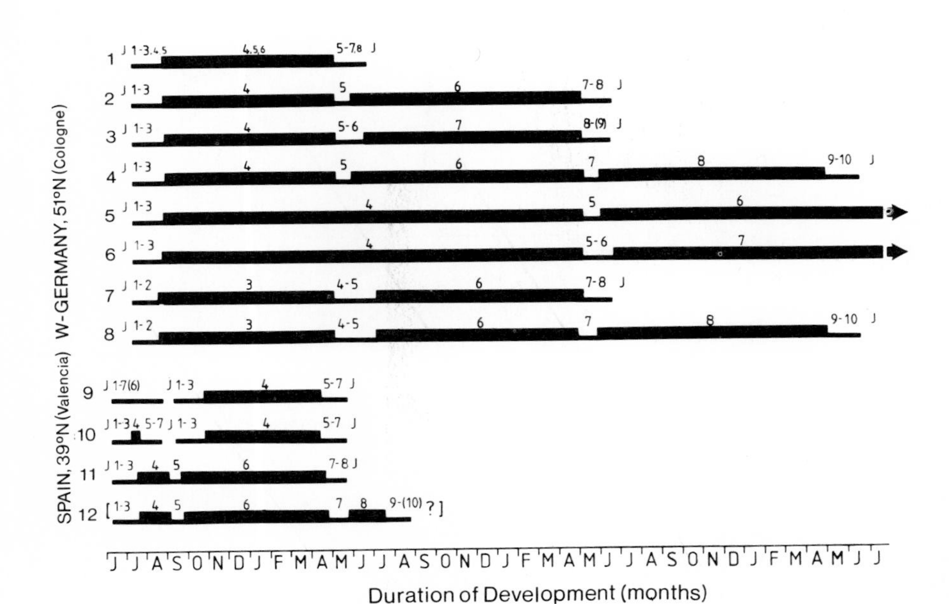

Figure 8.3. Variability in life cycles of *Zygaena trifolii* under natural conditions; in Cologne and Valencia showing the occurrence of diapause (wide bar) and the generation time from the first larval instar to the imago (J). The numbers indicate the specific nondiapausing and diapausing larval stages. For further explanations, see text. The Valencia data were obtained by C. Naumann and M. Venema (personal communication); the last line describes a hypothetical fourth type of life cycle.

sequence, long-day development is interrupted for some weeks or even months (Figure 8.3, lines 10–12). Individuals with a short aestivation more or less follow the bivoltine time schedule mentioned above (Figure 8.3, line 9). Those with a longer aestivation do not complete their development by autumn when, under short-day conditions, a second diapause is induced at about the sixth larval instar. The hibernal diapause of this third type of life cycle, characterized by repeated diapause, should be synchronously terminated with that of the other types (compare Figure 8.3, lines 9–12).

The laboratory experiments, in combination with the field obsevations, have demonstrated the range of potential photoperiod responses and their significance in generating quite different generation times and life cycle strategies. These results offer a first insight into the intrapopulation variability of the life cycles, which can be even more complex in northern populations of Zygaeninae as we show now.

Life Cycle Characteristics of the German Population

According to the photoperiodic response curve, all of the larvae of the Cologne stock entered an obligatory larval diapause, irrespective of the day length (Figure 8.4a). The detailed analysis again demonstrated a variable response of the larval instars (Figure 8.4b). Under short-day conditions of 8 or10 hrs, the third instar was the predominant dormant morph and, with longer day lengths, the diapausing stage was shifted to older larval instars (L_4–L_6). From these laboratory results, it may be concluded that the life cycle of the German population is univoltine provided that the generation time is not prolonged to 2 or even 3 years by an extremely long maintenance of diapause, or by repetitive diapause reactions, as discussed below.

The field experiment (Figure 8.3, lines 1–8) offers a survey of the quite different and numerous responses that have been observed up to now, 3 years after the start of the experiment with egg batches in June 1982. In very good agreement with the laboratory experiment (Figure 8.4), most of the larvae entered diapause in stage 4 during day lengths of about 14 hr (Figure 8.3, line 1). The second and third lines of Figure 8.3 represent the most common life cycles, with a repeated diapause in instars 4 (first winter) and 6 or 7, respectively, (second winter) resulting in a generation time of 2 years. In some individuals, yet a third diapause was induced during the third summer (in stage 8), with the consequence of an additional larval instar before pupation and a 3-year life cycle (Figure 8.3, lines 4 and 8). A prolonged maintenance of diapause over a period of two winters was also observed (Figure 8.3, lines 5 and 6).

Besides these different developmental pathways, with life cycles of one to three years and possibly more, temporal differentiation can be observed in the seasonal flight and reproductive period between local populations within the same geographical area. Locations exist where the moths fly during the middle of June (Figure 8.5, arrow a), and others where the moths appear first at the end of July (Figure 8.5, arrow b). The early larval stages, therefore, develop

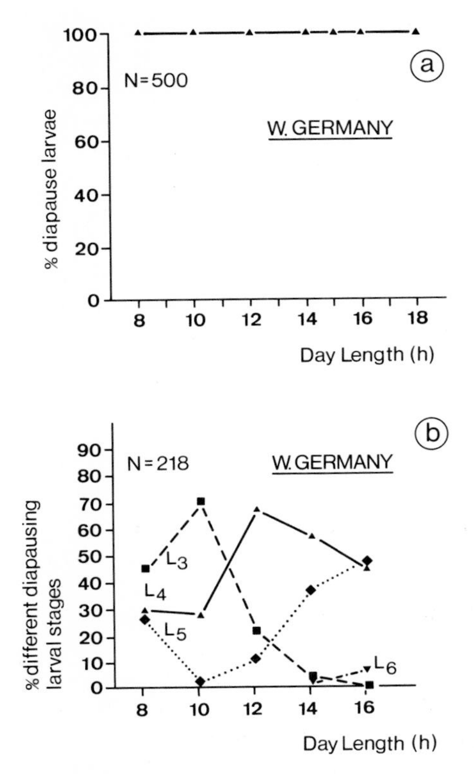

Figure 8.4. *Zygaena trifolii,* Cologne stock. (a) Photoperiodic response at 20°C. (b) Frequency of diapausing larval instars in relation to the increasing day length with about 50 L_1 introduced to each of the five conditions.

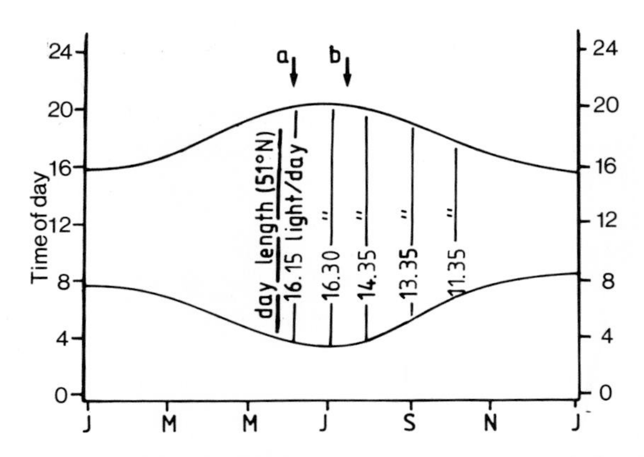

Figure 8.5. Natural photoperiodic amplitude between sunrise and sunset during 1 year in the area of Cologne, West Germany (51°N). The symbols a and b mark the different flight periods of two isolated local populations (a, Middle Rhine-Bornich, Fuchs, 1880; b, Cologne, Dellbrück).

at the specific locations under different photoperiod conditions as a consequence of the phase shifted flight periods. Thus it may be supposed that the early population, "a," will enter into diapause at a later larval instar of L_4 or L_5 associated with photoperiods of 16.30 hr during midsummer (Figure 8.4b), whereas the late population, "b," starts its development at photoperiods of < 15 hr with increasing frequencies of diapausing L_4 and even L_3 stages. These differences in flight time and hibernating larval stages may be transferred to the next year with the consequence of stabilizing early and later flight periods over a series of years, but some variability in the flight periods may be expected from year to year in response to variation in summer temperature conditions.

Concluding Remarks

Both populations analyzed above demonstrate a surprisingly high flexibility in the temporal programming of seasonal development. The life cycle of the Spanish population is characterized by a capacity for a photoperiodically induced, facultative diapause, in combination with a capacity for its prolonged maintenance and for a second diapause in a later instar. In comparison, the larvae of the Cologne population enter into an obligatory diapause, which might represent an adaptation to the more unstable and colder temperature conditions during late summer and autumn at the northern fringe of the distribution. The generation time in the Cologne population ranges from one to several years, either by repeated larval diapauses or by a prolonged maintenance of the dormancy of up to 1.5 years. This intrapopulation polymorphism in the physiological mechanism of larval dormancies was evident even within the progeny of a single female. The adaptive significance of this variability in the diapausing reactions and the resulting generation times might represent a strategy of "spreading of risk" (den Boer 1968) in succeeding years with regard to the fluctuating weather conditions during the growing season. The univoltine individuals of the northern populations may build up high-density populations in years with optimal temperature conditions, whereas those with longer generation times guarantee the survival of the population in unseasonable periods with nearly no reproductive success. Both types buffer this moth against elimination from its northern habitats, which sometimes exhibit detrimental weather conditions during the summer flying period.

The kinds of hibernal diapause strategies described here are found in numerous *Zygaena* species in Germany that have their distribution centers in southwest or southeast Europe. For instance, a repeated diapause, with a variation in generation time, was also established in laboratory cultures of *Zygaena transalpina* Esp., *Z. ephialtes* L., *Z. filipendulae* L., *Z. meliloti* Esp. (= *viciae* Den. and Schiff.), *Z. achilleae* Esp. (= *loti* Den. and Schiff.), *Z. carniolica* Scop., and *Z. purpuralis* Brün, however, taxas from related subfamilies as *Procris statices* L. (Procridinae) and *Aglaope infausta* L. (Chalcosiinae) developed without dormancy independently of the photoperiodic conditions at 20°C (Wipking, unpublished results). Rearing experiments at unfavorable temperature conditions and field observations are still lacking in this group. Thus, it cannot be excluded at the moment that potentially diapause responses have been overridden by the relatively high rearing temperatures; (Neumann and Krüger 1985).

It seems noteworthy that in temperate latitudes, many insect species are known to rely on similar adaptations to unpredictably varying, annually fluctuating climatic conditions. Thus, a prolonged maintenance of pupal diapause is common in sawflies (Philogene and Benjamin 1971; Waldbauer 1978) and lepidopteran species such as *Thaumetopoea wilkinsoni* Tams. (Halperin 1969) and *Attacus atlas* L. (Paukstadt and Paukstadt 1984). In these species the generation time of individuals that enter diapause at the same time may also extend over 2 or more years. With regard to the critical day length for diapause induction, the population-specific and gene-controlled value increases in several species by 1–1.5 hr per each 5° change of the geographical latitude (Beck 1980; Bradshaw 1976; Saunders 1979). Intrapopulation variation of the critical value also exists in relation to a clinal variation of the temperature factor in *Panorpa* flies (Sauer 1980 and this volume). A facultative diapause at southern locations of a species may turn into an obligatory diapause in the more northern population, as in a Japanese cricket (Masaki 1978).

Among insects, however, no other species have been found with such high intrapopulation variation with respect to the onset, prolonged maintenance, and repetition of diapause as in the family Zygaenidae. These moths appear to be excellent insect species for further evolutionary, ecological, and physiological studies on aspects of diapause control and the resulting life cycle strategies at different geographical latitudes.

Acknowledgments Our thanks are due to Professor C. Naumann (University of Bielefeld, FRG) for his kind support by offering egg batches and field data (Figure 8.3) of the Spanish population and to Mr. M. A. Brett for correcting the English text.

References

Beck, S.D.: Insect Photoperiodism. New York: Academic Press, 1980.

den Boer, P.J.: Spreading of risk and stabilization of animal numbers. Acta Biotheor. 18, 165–194 (1968).

Bradshaw, W.E.: Homeostasis and polymorphism in vernal development of *Chaoborus americanus*. Ecology 54, 1247–1259 (1973).

Bradshaw, W.E.: Geography of photoperiodic response in diapausing mosquito. Nature 262, 384–385 (1976).

Burgeff, H.: Beiträge zur Biologie der Gattung *Zygaena* F. II. Z. Wiss. Ins.—biol. 6, 39–44 and 97–98 (1910).

Burgeff, H.: Beiträge zur Biologie der Gattung *Zygaena* F. IV. Mitt. Münch. Entomol. Ges. 11, 50–64 (1921).

Burgeff, H.: Kreuzungsanalysen von Georassen der Gattung *Zygaena* Fab. (Lep.). Nachr. Akad. Wiss. Göttingen, II Math:—phys. Kl. 14, 187–205 (1965).

Danilevskii, A.S.: Photoperiodism and seasonal development of insects. Edinburgh: Oliver & Boyd, 1965.

Dorfmeister, G.: Beobachtungen über einige Zygaenen: angestellt in den Jahren 1851–1853 von Georg Dorfmeister in Bruck a.d.M. Verh. Zool. Bot. Ver. Wien 3, Sitz.—ber. 178–179 (1853).

Dorfmeister, G.: Abhandlungen über einige in der Steiermark vorkommende Zygaenen. Verh. Zool. Bot. Ver. Wien 4, 473–482 (1854).

Dorfmeister, G.: Über einige in der Steiermark vorkommende Zygaenen. Verh. Zool. Bot. Ver. Wien 5, 87–96 (1855).

Fuchs, A.: Lepidopterologische Mittheilungen aus dem unteren Rheingau. Stett. Entomol. Z. 41, 115–128 (1880).

Halperin, J.: Prolonged pupal diapause in *Thaumetopoea wilkinsoni* Tams. Z. Angew. Entomol. 64, 62–64 (1969).

Holik, O.: Quelques problèmes au sujet du genre *Zygaena* Fabr. III. Hibernation des chenilles. Lambillionea 37, 80–91 (1937).

Holik, O.: Quelques problèmes au sujet du genre *Zygaena* Fabr. V. La signification biologique du développement fractioné des chenilles de *Zygaena*. Lambillionea 39, 82–89, 104–111, 123–127 (1939).

Lumme, J.: Phenology and photoperiodic diapause in northern populations of *Drosophila*. In: Evolution of Insect Migration and Diapause. Dingle, H. (ed.). New York: Springer-Verlag, 1978, pp. 145-170.

Masaki, S.: Geographic variation of diapause in insects. Bull. Fac. Agric. Hirosaki Univ. 7, 66–98 (1961).

Masaki, S.: Seasonal and latitudinal adaptations in the life cycles of crickets. In: Evolution of Insect Migration and Diapause. Dingle, H. (ed.). New York: Springer-Verlag, 1978, pp. 72–100.

Matthes, E.: Diapause, Bivoltinismus und zweimalige Überwinterung bei *Fumea crassiorella* (Lep., Psychidae). Mem. Est. Mus. Zool., Univ. Coimbra 220,1–16 (1953).

Müller, H.J.: Formen der Dormanz bei Insekten als Mechanismen ökologischer Anpassung. Verh. Dtsch. Zool. Ges. 1976, 46–58 (1976).

Naumann, C., Feist, R., Richter, G., Weber, U.: Verbreitungsatlas der Gattung *Zygaena* Fabricius 1775. Theses Zoologicae 5, Braunschweig: Cramer-Verlag, 1984.

Neumann, D., Krüger, M.: Combined effects of photoperiod and temperature on the diapause of an intertidal chironomid. Oecologia - (Berlin) 67 154–156 (1985).

Paukstadt, U., Paukstadt, L.H.: Beitrag zur Kenntnis der Ökologie und Biologie von *Attacus atlas* Linnaeus 1758 aus westjavanischen und balinesischen Populationen. Entomol. Z. 94, 225–240 (1984).

Pener, M.P., Orshan, L.: The reversibility and flexibility of the reproductive diapause in males of a 'short day' grasshopper, *Oedipoda miniata*. In: Diapause and Life Cycle Strategies in Insects. Brown, V.K., Hodek, I. (eds.). Ser. Entomol. 23, The Hague: Junk, 1983, pp. 67–85.

Philogene, B.J.R., Benjamin, D.M.: Diapause in the swaine jackpine sawfly *Neodiprion swainei* as influenced by temperature and photoperiod. J. Insect. Physiol. 17, 1711–1716 (1971).

Sauer, K.P.: Klinale Unterschiede in der genetischen Variabilität der photoperiodischen Reaktion von *Panorpa vulgaris* (Mecoptera). Zool. Jb, Syst. 107, 113–147 (1980).

Sauer, K.P.: The evolution of reproductive strategies as an adaptation to fluctuating environments. In: Advances in Invertebrate Reproduction 3. Engels, W. et al. (eds.). Amsterdam: Elsevier North Holland, 1984, pp. 317–326.

Saunders, D.S.: Insect Clocks. Oxford: Pergamon Press, 1979.

Tanaka, S.: Seasonal control of nymphal diapause in the spring ground cricket, *Pteronemobius nitidus* (Orthopt., Gryllidae). In: Diapause and Life CycleStrategies in

Insects. Brown, V.K., Hodek, I. (eds.). Ser. Entomol. 23, The Hague: Junk, 1983, pp. 35–53.
Tauber, M.J., Tauber, C.A.: Insect seasonality: diapause maintenance, termination, and postdiapause development. Annu. Rev. Entomol. 21, 81–107 (1976).
Thiele, H.U.: Carabid Beetles in Their Environments. Zoophysiol. Ecol. 10, Berlin: Springer-Verlag, 1977.
Uebel, W.: Zygaenenzuchten. Mitt. Entomol. Ver. Stuttgart 9, 43–61 (1974).
Waldbauer, G.P.: Phenological adaptation and the polymodal emergence patterns of insects. In: Evolution of Insect Migration and Diapause. Dingle, H. (ed.). New York: Springer-Verlag, 1978, pp. 127–144.
Waldbauer, G.P., Sternburg, J.G.: Polymorphic termination of diapause by *Cecropia:* genetic and geographical aspects. Biol. Bull. 145, 627–641(1973).
Wipking, W.: Ökologische Untersuchungen über die Gründe der Habitatbindung der rheinischen Zygaenidae (Insecta, Lepidoptera)—unter besonderer Berücksichtigung der Fauna der Trockenrasen. Staatsexamensarbeit Univ. Köln, 127 pp., 1983.

Chapter 9

r-K Selection at Various Taxonomic Levels in the Pierine Butterflies of North and South America

ARTHUR M. SHAPIRO

The theory of r and K selection (MacArthur and Wilson 1967) attempts to make very sweeping predictions about the life history traits of organisms evolving under different forms of population control; it has generated a large body of literature in which it has been tested observationally, experimentally, and mathematically. Many of the empirical tests have consisted of looking for evidence of r- and K-selected traits in populations of species along various gradients, of which the most common have been latitude, climatic predictability, and habitat stability (Cody 1966, 1971; Gadgil and Solbrig 1972; Abrahamson and Gadgil 1973; Solbrig and Simpson 1974; Gaines et al. 1974; Abrahamson 1975; McNaughton 1975; Derickson 1976). The gradients chosen were initially very simplistic (Pianka 1970): thus we find Krebs (1972, p. 579) saying: "The simplest illustration of these two extremes might be organisms in tropical (K selection) vs. polar environments (r selection)." Although we have become less naive, the initial suggestions of the 1970s persist in the minds of many ecologists. As a result, it is often *assumed* that populations in high latitudes are under density-independent regulation, whereas those in low latitudes are density-dependent. Yet most ecologists will readily admit that evidence bearing on this proposition is scarce at best, and even classic studies of population regulation are frequently subjected to reinterpretation. Needless to say, "validations" of r-K theory are potentially spurious when not supported by convincing studies of population regulation.

In the heady, early days of r-K theory, comprehensive adaptive syndromes were recognized as the likely outcomes of such selection. r-Selection was supposed to generate high reproductive capacity, reduced body size, high productivity, short generation time, low efficiency, low competitive ability, and high dispersability; K-selection, the reverse. It was already apparent to Downes (1962, 1964, 1965) that these generalizations did not apply to arctic insects—but, being in the entomological literature, these papers were overlooked. Yet the old notions persist, though an increasing number of papers address the complications (Boyce 1984; Livdahl 1984).

It has never been stated explicitly at what taxonomic level(s) r and K selection should operate, or its effects be sought. In a sense, this is a subset of the general problem of levels of selection (Lewontin 1970); it also touches on the questions of "punctuated equilibrium" and the causes of "stasis"(Van Valen 1982). If r and K selections are visualized as modes of natural selection, the alleged latitudinal gradients can then be seen as clines in genetically determined life history phenomena; many such have now been documented. On the other hand, MacArthur and Wilson, Pianka, and Krebs all imply that comparisons may be made fruitfully not only among conspecific populations but among species and higher categories, up to classes at least. Pianka (1970) goes so far as to suggest that terrestrial vertebrates overall are K-selected, insects r-selected. Is there any inherent inconsistency here? Can selection operate on life history traits enough to produce recognizable clines within species, yet be so constrained within lineages that they can be classified as r- or K-selected as wholes? This rather peculiar problem—hitherto noted only by Stearns (1984)—seems unexplored, largely because there are few reasonably speciose and diverse lineages within which suitable reproductive data are available both within and among taxa.

This chapter examines the reproductive biology of one such lineage, the pierine butterflies (Lepidoptera: Pieridae: Pierinae: Pierini), asking at what taxonomic level(s) r- and K-selected traits can be recognized, and whether they are disposed along the gradients usually invoked. The pierines include the familiar "cabbage butterflies." They are sexual, outcrossing animals, essentially cosmopolitan but with greater diversity in cool and temperate climates than in the lowland tropics. Although many tropical and subtropical genera feed on Loranthaceae, most cool-temperate ones (including all considered here) are associated with Cruciferae and related families. All treated here lay singly or in very small batches, though there are large-batch layers scattered in various genera. The last comprehensive revision at the generic level was by Klots (1933). The phylogeny of the Pierini is being studied in my laboratory using both morphological and biochemical approaches. Comments on phylogeny and historical biogeography in this chapter are based on the latest (generally unpublished) information but should be taken as highly provisional.

I have reared many pierine taxa in mass culture over the past 20 years, resulting in a substantial accumulation of reproductive data. The availability of these data permits inter-taxon comparisons at the levels of conspecific populations, subspecies, congeneric species and species groups, genera and groups of genera, as described below. Representatives of all these groups are shown in Figure 9.1.

Two Holarctic Species Groups

"Lumpers" put all the Crucifer-associated Holarctic pierines in one genus, *Pieris,* with several subgenera or species groups. "Splitters" divide it into four or five genera corresponding to these. Two of these—the *Pieris napi* species

Figure 9.1. Representative Pierini of the various groups discussed in the text, drawn to scale. Upper left: *Phulia nymphula* female, Aguadas Blancas, Depto. Arequipa, Peru. Center left: *Pierphulia rosea* female, same data. Lower left: *Pieris napi mogollon* female, Sandia Mts., New Mexico. Upper right: *Tatochila sterodice macrodice* male, Caquiaviri, Bolivia. Lower right: *Pieris protodice* female, Arapahoe County, Colorado.

group (subgenus or genus *Artogeia,* in part) and the *P. callidice* group (subgenus or genus *Pontia,* in part, or *Pontieuchloia* or *Synchloe,* which are more restrictive)—have very broad latitudinal distributions and have been cultured for studies of photoperiodic adaptation for as long as 36 consecutive generations. I have laboratory rearing data for populations of the *napi* group from Dawson City, Yukon Territory to Monterey County, California and central New Mexico, and for the *callidice* group from Fairbanks, Alaska to Xochimilco, Mexico. Data on egg numbers and sizes produced in the laboratory are given in Table 9.1. These were obtained under our usual rearing conditions, described in various publications (Shapiro 1975a,b, 1977). There seems to be very little variation in egg size among individuals within populations of pierines. Jones et al. (1982) report a decrease in egg weight during the lifetimes of individual females of *P. rapae,* which belongs to another species group. This may be a fairly common occurrence in butterflies (Wiklund and Karlsson 1984). The egg size data in Table 9.1 and subsequent tables are based on dimensions rather than weight,

Table 9.1. Egg size and egg production in various nearctic populations of two species groups of the holarctic genus *Pieris*

Taxon	Number of females[a]	Source locality and latitude	Egg size	Days lived[b] $\bar{x} \pm$ SD	Eggs laid[b] $\bar{x} \pm$ SD	Eggs/day[b] $\bar{x} \pm$ SD
		callidice species group (*Synchloe*)				
nelsoni	5	Fairbanks, Alaska 64°51′N (subarctic, disturbed unforested)	1.1 × 0.3 mm	13.20 ± 2.39	260.00 ± 81.89	20.27 ± 8.17
occidentalis	5	Colorado Rockies 40°00′ N (montane disturbed and alpine grassland)	1.1 × 0.3 mm	13.40 ± 2.97	284.80 ± 25.56	22.31 ± 6.38
occidentalis	12	Donner Pass area, California 39°20′N (montane, subalpine, unforested)	1.1 × 0.3 mm	11.75 ± 4.16	312.75 ± 113.66	27.49 ± 6.46
protodice	17	Philadelphia, Pennsylvania 39°57′N (urban vacant lots)	1.1 × 0.3 mm	14.53 ± 2.67	330.59 ± 107.16	23.24 ± 7.92
protodice	10	Sierra County, California 39°32′N (Great Basin rangeland-steppe)	1.1 × 0.3 mm	14.30 ± 3.74	310.50 ± 113.65	22.20 ± 7.40
protodice	10	Los Angeles,California 34°00′N (urban vacant lots, Mediterranean climate)	1.1 × 0.3 mm	13.60 ± 2.17	309.20 ± 88.00	22.65 ± 4.98
protodice	6	Houston, Texas 29°46′N (humid subtropical, disturbed)	1.1 × 0.3 mm	12.67 ± 3.08	309.17 ± 165.48	23.92 ± 9.64

protodice	6	Xochimilco, D.F., Mexico 19°16′N (vegetable gardens, dry winter climate)	1.0 × 0.27 mm	11.50 ± 3.62	298.83 ± 80.00	28.51 ± 11.87
		napi species group (*Artogeia* in part)				
? subspecies	5	Dawson, Yukon Territory 64°04′N (boreal coniferous forest)	1.2 × 0.4 mm	16.20 ± 4.66	126.60 ± 45.41	7.76 ± 1.73
virginiensis	3	Tioga County, Pennsylvania 41°50′N (beech-maple-hemlock forest)	1.2 × 0.4 mm	12.33 ± 3.21	91.67 ± 35.73	7.32 ± 1.31
macdunnoughi	5	Colorado Rockies 40°00′N (edges of montane bog meadows)	1.2 × 0.4 mm	14.60 ± 3.36	166.80 ± 41.33	11.40 ± 0.42
microstriata	11	Solano County, California 38°28′N (foothill riparian woodland in canyons)	1.2 × 0.4 mm	13.45 ± 3.30	130.91 ± 39.34	10.12 ± 3.54
venosa	7	San Mateo County, California 37°45′N (coastal fog-belt Douglas Fir forest)	1.2 × 0.4 mm	13.43 ± 5.09	108.14 ± 57.57	7.80 ± 2.63
mogollon	5	Sandia Mts., New Mexico 35°20′N (riparian forest in canyons)	1.2 × 0.4 mm	13.40 ± 2.07	109.40 ± 37.22	8.38 ± 1.76

Egg size = maximum height × width at base. Only fully fertile females with normal lifespans were included.
[a]All lab-reared and -mated females.
[b]Counted from day of mating onward.

and were derived from eggs preserved for morphological study. These eggs were selected haphazardly with no particular attention to the age of the female. The patterns to be described appear robust to any lifetime egg size variation, especially insofar as egg samples from many females were examined for most populations. Egg production is not constant throughout the lives of most pierines; commonly many eggs are matured quickly after mating and egg production then decreases unless a second mating occurs. Occasionally egg production increases spontaneously again without remating. Virgin females mature and lay few or no eggs, as do also females mated to males of incompatible taxa (e.g., the hybrid of British *P. n. napi* female × Californian *P. n. microstriata* male produces males that are fully fertile when backcrossed to California *microstriata* but fully or nearly sterile with pure Colorado *P. n. macdunnoughi*).

It may be seen that both egg size and egg number are essentially invariant with latitude within each group. *Callidice* group populations—regardless of species or locality—lay many small (orange or red) eggs; *napi* group populations lay fewer, large (white) eggs. Although the southernmost *callidice* group population studied (*protodice* from Xochimilco, Mexico) lays smaller eggs than the others, no latitudinal trend is apparent.

These data can be interpreted in either phylogentic or environmental terms. This is because each species group has its characteristic habitat preference, such that although taxa of both are sympatric on a coarse scale, they rarely can be found flying at the same spot. All members of the *callidice* group occur in unforested habitats—grassland, steppe, arctic and alpine tundra, and desert. Most *napi* occur in boreal or mesic forest, but some, such as *macdunnoughi* of the Rockies and Great Basin, occur in swales, marshes, and bogs. *Macdunnoughi* has a significantly higher egg production than other *napi* in the Nearctic, but it is not clear whether this is related to its habitat preference; it is *not* a function of size, as Colorado *macdunnoughi* are indistinguishable in size or weight from second-brood Californian *microstriata*. We need to sample more marsh populations.

There is also a major eco-behavioral difference between the two groups. *Callidice* group populations are inflorescence feeders by preference, whereas those in the *napi* group are normally foliage feeders. The former engage in egg load assessment, the latter do not (Shapiro 1981); this implies that food is more likely to be limiting to *callidice* than to *napi* populations, but the natural control of their numbers remains essentially unstudied. *Callidice* group females appear much more dispersive than *napi* group ones; both *P. protodice* and *P. occidentalis* are "colonizing," "weedy," or "fugitive" species (Shapiro 1976a). Morphology, biochemistry, and biogeography all agree in suggesting that the taxa in these groups represent quite recent, almost certainly Quaternary, differentiation (Geiger 1981; Geiger and Shapiro unpublished; Varga and Toth 1978). The nearctic *napi* taxa are, however, older and better differentiated genetically than the European ones. It might be argued that insufficient time has elapsed for life history traits of the sort discussed here to differentiate along latitudinal gradients. There is, however, no a priori reason why such traits should be more conservative than such things as diapause and polyphenism, which we know to be so differentiated (cf. Shapiro 1975a,b, 1976a,b, 1977). Moreover, oviposition behavior and egg-production pattern are different in Australian and Canadian *Pieris rapae*, although both populations represent introductions from Europe within the past 130 years (a maximum of 780 generations) (Jones 1977; Jones et al. 1982). Furthermore, egg production appears to be closely tied physiologically to diapause (Shapiro, unpublished) in a variety of pierines, including *P. rapae, P. napi,* and *P. protodice*. For this reason, all the reported data are for nondiapausing animals.

The Tatochila Sterodice Species Group

This is a group of endemic South American Pierini ranging in the Andes from central Colombia to Tierra del Fuego, and in the temperate low lands of Argentina and Chile. The latest taxonomic revision (Herrera and Field 1959) recognizes three species—*mercedis, vanvolxemii,* and *sterodice* itself (then called *microdice*), the last with four allopatric subspecies. *Mercedis* differs from the others in its genital morphology. *Vanvolxemii* is genitally identical to *sterodice* (except for the statistical distribution of edeagal torsion), but differs in wing pattern and averages about 20% larger. Shapiro (1984a, 1985a,b) has demonstrated that all these entities (except *sterodice* ssp. *fueguensis,* which was not available for breeding) are vigorously interfertile, showing no consistent pre- or postzygotic reproductive barriers. Because they occupy different biomes, the opportunity to hybridize is largely restricted to ecotones and disturbed areas. Electrophoretically they are about as diverse as the nearctic *napi* taxa, which seem to show much more reproductive isolation. About 13,000 individuals have been reared over 7 years, and a very complete set of crosses is available for analysis.

Table 9.2 demonstrates considerable differentiation in life history traits among these entities—about as much as occurs between the two holarctic species—groups previously discussed. (Preliminary analyses of reciprocal complete crossing programs suggest that egg size and number are under polygenic control. These data will be reported elsewhere.) This is perhaps not entirely surprising, since the *sterodice* group has radiated into all the climates occupied by holarctic pierines, plus a few not represented in the northern hemisphere.

Tatochila vanvolxemii, which lives in the subhumid to subarid *pampa* and in the *monte* desert, and *T. mercedis* from Mediterranean grassland, between them occupy the same range of habitats and climates as *Pieris protodice* (*callidice*group) of the nearctic, and like it lay many small (orange, rarely yellow) eggs. *T. sterodice macrodice,* from the subarid to arid *puna* and *altiplano,* has the same size eggs as *T. s. arctodice,* from the northern Andean high-altitude fog belt (*páramo*), but lays more of them. *T. s. sterodice,* from the humid to subhumid Patagonian Andes and steppe forest ecotone, lays shorter, squatter yellow eggs in smaller numbers. Climate seems to correlate much better than latitude with reproductive "strategies" in this group. This is true of phenotype as well: the subspecies of *sterodice* from humid climates, at opposite ends of

Table 9.2. Egg size and egg production in various populations of the *Tatochila sterodice* species group

Taxon	Number of females[a]	Source locality and latitude	Egg size	Days lived[b] $\bar{x} \pm SD$	Eggs laid[b] $\bar{x} \pm SD$	Eggs/day[b] $\bar{x} \pm SD$
vanvolxemii	19	Southern Buenos Aries province, Argentina 38°50′S (subhumid grassland)	1.2 × 0.33 mm	14.37 ± 3.79	374.58 ± 144.39	26.26 ± 8.44
mercedis	14	vic. Santiago, Chile 33°30′S (Mediterranean grassland)	1.05 × 0.27 mm	10.21 ± 3.17	260.79 ± 91.43	27.02 ± 11.65
macrodice	6	Tucumán province, Argentina ca. 27°S (*puna*)	1.15 × 0.40 mm	11.50 ± 3.56	207.00 ± 42.34	19.19 ± 5.46
arctodice	6	Ecuadorian Andes, ca. 0° (*páramo*)	1.15 × 0.4 mm	12.17 ± 2.86	187.17 ± 43.08	15.66 ± 3.34
sterodice	12	NW Patagonia, Argentina ca. 41°S (steppe-forest ecotone)	1.12 × 0.47 mm	12.42 ± 4.40	159.00 ± 61.46	14.03 ± 7.04

Egg size = maximum height × width at base. Only fully fertile females with normal lifespans were included.
[a]All lab-reared and -mated noninbred females.
[b]Counted from day of mating onward.

the range, have both evolved yellow, heavily melanized females, making them much more sexually dimorphic than the intervening populations of drier climates. The same phenomenon occurs in the *Pieris napi* group in the Holarctic. Convergent evolution is very common in Pieridae.

The High-Andean Genera

From Peru to northeastern Chile and northwestern Argentina in the *puna* and *altiplano* occur numerous species of diminutive, poorly known pierines, formerly all placed in the genus *Phulia* (and here referred to collectively as such) but now scattered among several genera (Field and Herrera 1977). These animals average less than half the size of normal Pierini (Figure 9.1). They often have bizarre morphospecializations associated with a life spent entirely within the "boundary layer" of air near the ground. They are assumed to be derived from more normal pierines. *Phulia nymphula* is simultaneously the most widespread and normal of the group, but it shows great interpopulation differentiation in a variety of characters (Shapiro and Geiger, unpublished). Other, more local *Phulia* display more derivative character states, and the various taxa in *Infraphulia* and *Pierphulia* are still further from normality. *Infraphulia madeleinea* has the most reduced venation of any Pierid (and perhaps of any butterfly).

Until very recently the biology of these animals was completely unknown. We now know that most, if not all, of them are tied to the large, flat peat bogs (*turberas altoandinas*) of the *altiplano* and *puna,* which lie interspersed among screes, sparse bunchgrass, and snow fields; often three to five species may coexist, showing fine-scale habitat differences. What is known of their ecology and behavior in Peru may be found in Shapiro (1985a) and in various papers by Courtney and Shapiro (in preparation).

All of these tiny animals studied to date (*Phulia nymphula,* which is probably two or three sibling species: *P. paranympha ernesta; P. garleppi; Pierphulia rosea,* two subspecies; *Infraphulia madeleinea; I. ilyodes*) lay relatively enormous eggs. The egg of *Phulia nymphula,* which is yellow, is as large as that of *Pieris protodice,* a butterfly twice its size; that of *Pierphulia rosea,* an even smaller butterfly, is orange and larger still. Indeed, the *P. rosea* egg is almost the size of a *P. napi* egg! We have not yet bred these animals in continuous culture and have data only from wild-collected Peruvian females (Table 9.3), but they are very long-lived and it seems certain that they do not normally lay more than three or four (*P. rosea*) or five or six (*P. nymphula*) eggs per day when they are able to lay every day. This makes their egg production comparable to the long-lived tropical lowland *Heliconius* often used as a paradigm of K selection (Turner 1971; Ehrlich and Gilbert 1973; Benson et al. 1976).

Electrophoretically, the *Phulia* group of genera looks very old—older than the split between the *napi* and *callidice* groups of *Pieris,* i.e., probably Tertiary, but speciation looks definitely Quaternary. It seems increasingly unlikely that *Phulia* and its relatives are derived from *Tatochila* or its sister genus *Hypsochila,* though they may share a common ancestor with them further back in time.

Table 9.3. Egg production by caged, wild-collected *Phulia* from the Department of Arequipa, Peru (ca. 16°30′S, 4000+ m)

Day	Number of females alive	Eggs laid	Eggs/female/day
		Phulia nymphula	
1	8	5	0.625
2	8	10	1.250
3	8	13	1.625
4	7	12	1.714
5	7	16	2.285
6	7	9	1.222
7	7	8	1.143
8	6	12	2.000
9	6	14	2.333
10	5	10	2.000
11	5	8	1.600
12	5	9	1.800
13	4	*a*	*a*
14	4	10	2.500
15	4	13	3.250
16	4	*a*	*a*
17	3	11	3.667
18	2	8	4.000
19	1	2	2.000
20	0	—	—
		Pierphulia rosea	
1	3	5	1.667
2	3	8	2.667
3	2	5	2.500
4	2	0	0
5	2	4	2.000
6	2	3	1.500
7	2	5	2.500
8	2	4	2.000
9	2	4	2.000
10	2	5	2.500
11	2	6	3.000
12	2	1	0.500
13	2	*a*	*a*
14	2	5	2.500
15	2	6	3.000
16	2	*a*	*a*
17	2	6	3.000
18	2	5	2.500
19	2	4	2.000
20	1	1	1.000
21	0	—	—

	Egg size	Mean eggs/female
P. nymphula	1.1 × 0.3 mm	35.1
P. rosea	1.2 × 0.33 mm	38.5

All specimens were from bogs in *altiplano*.
[a]Females kept refrigerated all day; no laying.

Discussion

Any hope that the pierines would help to define "the" level at which r-K phenomena should be sought has been dashed: apparently species groups (or subgenera, or splitters' genera) are r- or K-selected in the holarctic, whereas in *Tatochila* the same amount of differentiation is found among populations in a complex in which the process of speciation has not yet been completed. If we consider the *Phulia* group of genera as K-selected relative to other pierines, then such differentiation can be demonstrated at the subtribe level—which is to say it antedates the separation of the various genera in which it is now maintained.

In both *Pieris* and *Tatochila* the environmental correlates of r-K traits are consistent and intuitively reasonable (fewer, larger eggs in humid, cool temperate, forested environments; more, smaller eggs in dry, mild to warm temperate, unforested ones). Admittedly, this is not a causal or mechanistic "explanation" of anything, but at least there is a pattern. The *Phulia* story, however, seems paradoxical: why should animals that live in very harsh climates above 4000 m near the equator lay few, very large eggs?

As noted earlier, Downes (1962, 1964, 1965) had refuted the notion that arctic insects were r-selected even before it was published. Chernov (1978) and Danks (1981) reiterated the point, adding that very low fecundity is sometimes partially compensated for by prolonged or repeated breeding seasons. In this sense, the *Phulia* are not so surprising; what is surprising is that "theory" failed to explain the attributes of arctic bumblebees or aphids *or* of pierines from the *puna*.

Dunbar (1968), arguing from a perspective of genetic feedback at the level of whole communities, countered the conventional wisdom with a set of unorthodox predictions. He felt that long-term adaptation to extreme environments such as the arctic would entail reduction in fecundity as buffers were built into population regulation. He saw this happening either through small body size with few small eggs but fast growth, or large body size with few large eggs and slow growth. These predictions, too, fail with *Phulia,* which has small body size, few large eggs, and very slow growth. (Actually, the *P. nymphula* "race" from the *puna* of Jujuy, Argentina, grows much faster than that from the *altiplano* of Arequipa, Peru, under identical conditions. It should also be remembered that *Phulia* larvae exercise partial control over their microenvironment by living in individual silken nests, a unique situation in the tribe; they also pupate in loose cocoons.) There is no evidence that any *Phulia* is capable of diapause, which is used as a hedge against the adverse season by pierines in middle and high latitudes. But all seasons are more or less adverse in *Phulia* habitat, albeit for different reasons—less so on the bogs than elsewhere—and no diapause has yet been found in any circumequatorial high-altitude butterfly.

No one knows what regulates populations of any of these organisms, so calling their life history traits r- or K-selected is clearly premature. Still, all the observed patterns make sense in the light of what is now known of the population dynamics of some other butterflies. The remainder of this discussion is basically an evolutionary scenario which, while highly speculative, is at least coherent.

Dempster (1983) and Shapiro (1984b) independently seized on two studies as crucial to the understanding of butterfly population dynamics. These papers, by Courtney (1980) and Hayes (1981), working independently on the Pierids *Anthocharis cardamines* in England and *Colias alexandra* in Colorado, respectively, identify failure of the females to lay a full complement of eggs as the "key factor" in the dynamics of their populations. Since then a veritable cascade of similar work has appeared, implicating weather as the density-independent factor underlying such failure to lay. Pierines are obligate heliotherms and are inactive in cloudy or cold weather. Kingsolver (1983a,b) showed that flight time can be a limiting resource for *Colias,* and that high-altitude females in the Rockies may have a lifetime egg production less than half that of lowland conspecifics (one needs complete life tables, however, to assess the contribution of this to net selection and population growth). *Pieris napi* and *Tatochila s. sterodice* habitats average cloudier and colder than *P. protodice* and *T. vanvolxemii* ones. We would then predict that the first pair of species should either be more efficient at finding oviposition sites, or lay fewer eggs, or both. In fact, *napi* commonly oviposit multiply on single plants (Shapiro 1980), a fact suggesting they may indeed be pressed for time; their larvae are noncannibalistic, unlike red egg species.

Pieris virginiensis, a spring univoltine member of the *napi* group, is very subject to flight interference by bad weather (Cappuccino and Kareiva 1985); in 1982 at one site it enjoyed favorable flight weather during only 22% of daylight hours. Given their sunnier habitats, members of the *Pieris callidice* group should have less of a problem getting and staying airborne, in spite of the need to search out many hosts due to egg load assessment and cannibalism. They are, however, probably wind-limited more often than *napi,* and in deserts *P. protodice* must seek shade in mid-late afternoon to avoid overheating. Throughout the Holarctic, *callidice* group populations go both higher and further north than *napi* ones, except in Fennoscandia where *callidice* was apparently eradicated in the Pleistocene and failed to recolonize, and in the eastern Canadian Arctic.

The early discussion of r and K selection confounded environmental severity and unpredictability. Colwell (1974) subsequently pointed out that "predictability" was itself a mixed concept, and helped to clear up some of the muddles in the literature. *Altiplano* and *puna* pierines live in a predictably severe environment, in which in the best weather, flight activity is possible only for a few hours per day, and cannot be long sustained. Figure 9.2 represents the situation at Morococha (Dept. Junín, Peru). Temperatures at dawn are below freezing, and may rise only to 15°C by midday. The butterflies must spend considerable time thermoregulating in order to be able to fly at all. Globally, pierines rest with the wings closed over the back like most butterflies. *Phulia* and allied genera, however, keep the wings fully open, thus exposing the heavily melanized wing bases and the body at all times; they are basking any time the sun shines. On the very rare warm days with little wind—I have seen two—this posture can lead rapidly to overheating, and the butterflies revert to the normal pierid posture for heat avoidance. Activity is usually terminated by high winds beginning in early afternoon, even if the sky is clear (Shapiro 1985a;

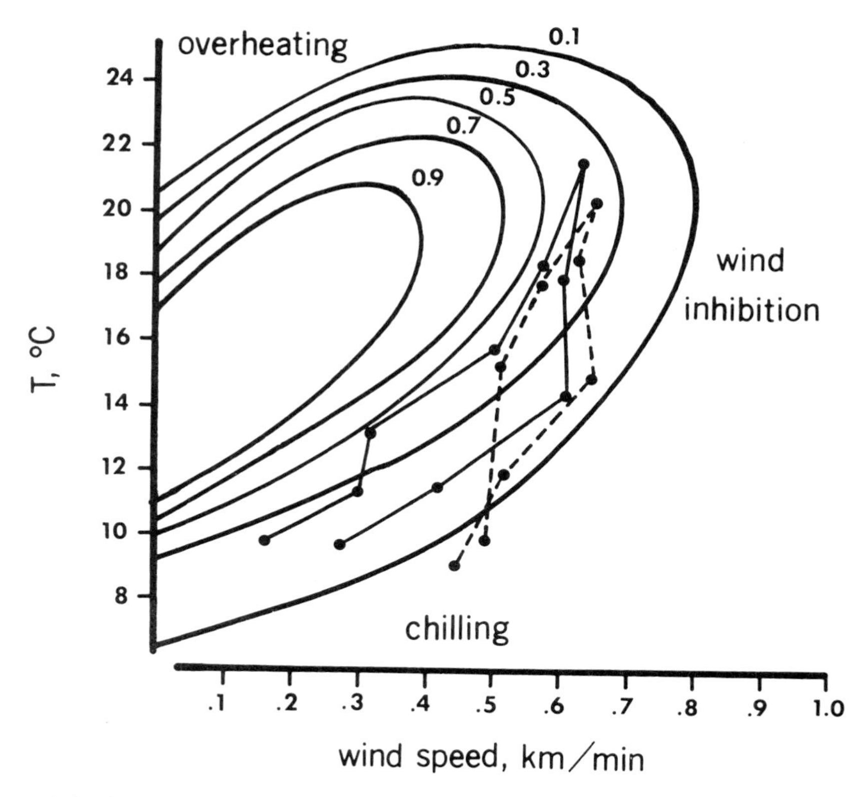

Figure 9.2. Generalized flight probabilities under full sun for *Infraphulia madeleina*, based on field observations at Morococha, Department of, Junín, Peru, with some extrapolation. Two typical days are superimposed (observations at 1-hr intervals beginning at 0900) to ilustrate the limited flight conditions available. Normally there are high, thin cirrus clouds in early morning, little or no cloudiness in late morning, and variable cumulus tending to overcast in late afternoon. Probabilities may be thought of as the proportion of the adult population in the air at a given time.

Courtney and Shapiro in preparation). There is thus very little time for oviposition, so few eggs could be laid in any case. Field and Herrera (1977) report finding larvae in the oviducts of female *Infraphulia ilyodes*, and speculate that this species might be larviparous. We have not seen this phenomenon. Larviparity may occur rarely, but *I. ilyodes* normally lays eggs. I suspect larviparity reflects excessively long egg retention due to poor weather; I have found ready-to-hatch eggs in the oviducts of female *Colias eurytheme* from upper New York State in October and November.)

Small wing size may have been selected as an adaptation to living in the boundary layer—the long, narrow wing shapes certainly are—and carried body size along with it. Or body size may have been selected down, carrying along wing size, as a consequence of the minuteness of most of the Cruciferous hosts available. (The *Phulias* are so specialized to using tiny rosettes that it is next to impossible to obtain eggs using larger, leafy Crucifers, even of the same species; the female stands "on tiptoe" on the rosette and probes extensively within it using a very long ovipositor.) A small animal with little flight time is unlikely to reduce its egg size proportionately to body size in order to maintain a fixed quantity. On the other hand, it is easy to envision pressure to increase egg size. With laying time so limited, females often lay rather indiscriminately within the correct habitat; this is especially true of the bog obligates, which often lay on emergent Juncaceae which are completely inedible to the larvae. Captive specimens will often lay on inappropriate objects (dandelion flowers, the mesh of cages) if Crucifers are present nearby and it is possible to probe with the ovipositor. Female *Phulia nymphula* seem able to detect Crucifers at a distance of a few centimeters by olfaction, and will walk a beeline to them on the ground. The newly hatched larva of *Pierphulia rosea* is very large, and can live for 3–4 days at 20°C while wandering in search of food. During this time it must be living on reserves carried over from the egg. First-instar food foraging is not unknown in other butterflies; it seems to be true of the nearctic fritillaries (*Speyeria*) in general, is probably common in the genus *Parnassius* (which also has large eggs and reduced egg number), and was first recorded for the skipper *Hesperia lindseyi* by MacNeill (1964). I hasten to add that Karlsson and Wiklund (1984) found no correlation between starvation resistance of the first-instar larva and egg size in the European Satyrid *Lasiommata megera,* though it is not clear how often such larvae of *Pierphulia rosea* are as large as a second-instar larva of *Pieris protodice,* a much greater difference than exists among first instars from early and late eggs of *L. megera.*

This narrative could be extended to incorporate the other bizarre aspects of *Phulia* biology, such as mating behavior, but the message should be clear: the oddities can be seen as an internally consistent adaptive syndrome fitting them to a predictably severe environment—exactly the sort of thing envisioned in r-K theory, yet in its specifics quite different from what the theoreticians had in mind. It is summarized as a schematic drawing in Figure 9.3.

Wiklund and Karlsson (1984) find that egg size in Satyrids is positively correlated with overall body size, but also with phylogeny, and invoke phylogenetic constraints as a partial explanation of the variation. In the pierines body size is not a good predictor of egg size and phylogenetic constraints seem to vary, depending on the lineage. The antiquity of the lineage and of speciation may be relevant, but clearly there is no fixed pattern.

The biology of *Pieris* and *Tatochila* fits fairly comfortably within r-K theory. The difference in taxonomic levels of adaptation is actually useful in underscoring the importance of environment, even while debunking "latitude" as a meaningful gradient in itself. One of the reasons why latitudinal gradients and the like persist in our folklore is that not enough *Phulia* type stories have been published. Taylor and Condra (1980), Constanz (1979), Barclay and Gregory (1981), Luckinbill (1984), and Livdahl (1984) are among the authors who have described circumventions of the r-K dichotomy. Their message, like that from *Phulia,* can be paraphrased: there are plenty of ways to skin a cat.

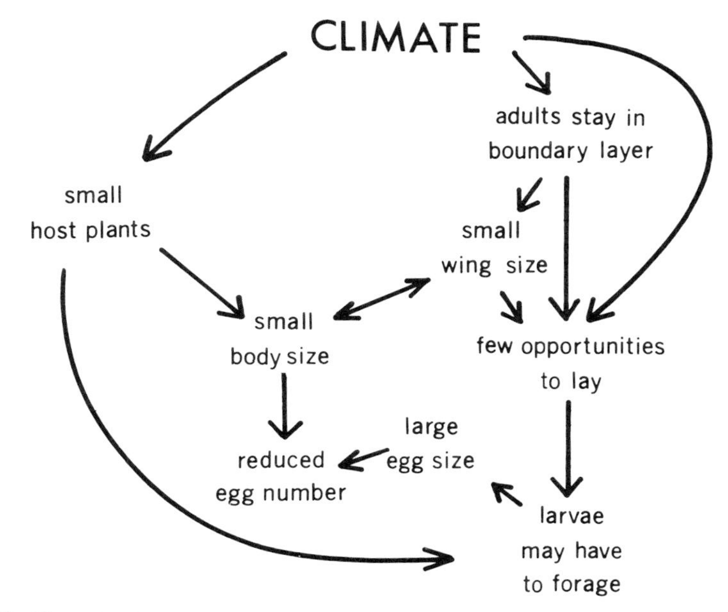

Figure 9.3. Hypothetical selective regimen leading to the unusual characteristics of *Phulia* and its relatives.

Acknowledgments I thank Steve Courtney, Jane Hayes, Peter Kareiva, Joel Kingsolver, Rick Karban, P. Lounibos, and S. Masaki for discussion; Steve Courtney and Peggy Stern for data and specimens; Henri Descimon, Gerardo Lamas M., Estela Neder de Roman, and Marta Arce de Hamity for facilitating collections in the Andes and for field companionship; and NSF Grant BSR-8306922 (Systematic Biology) for supporting this work. Figures 9.1–9.3 were drawn by Karen English-Loeb.

References

Abrahamson, W.G.: Reproductive strategies in dewberries. Ecology 56, 721–726 (1975).

Abrahamson, W.G., Gadgil, M.: Growth form and reproductive effort in goldenrods (*Solidago*, Compositae). Am. Natur. 107, 651–661 (1973).

Barclay, J.J., Gregory, P.T.: An experimental test of models predicting life-history characteristics. Am. Natur. 117, 944–961 (1981).

Benson, W.W., Brown, K.S. Jr., Gilbert, L.E.: Coevolution of plants and herbivores: passion-flower butterflies. Evolution 29, 659–680 (1976).

Boyce, M.S.: Restitution of r- and K-selection as a model of density-dependent natural selection. Annu. Rev. Ecol. Syst. 15, 427–447 (1984).

Cappuccino, N., Kareiva, P.: Coping with a capricious environment: a population study of a rare pierid butterfly. Ecology 66, 152–162 (1985).

Chernov, Y.I.: Adaptive features of the life cycles of insects of the tundra zone (in Russian). Zh. Obshch. Biol. 39, 394–402 (1978).

Cody, M.L.: A general theory of clutch size. Evolution 20, 174–184 (1966).
Cody, M.L.: Ecological aspects of reproduction. In: Avian Biology. Farner, D.S., King, J.R. (eds.). New York: Academic Press, 1971, v. 1, pp. 461–512.
Colwell, R.K.: Predictability, constancy, and contingency of periodic phenomena. Ecology 55, 1148–1153 (1974).
Constanz, G.D.: Life history patterns of a live-bearing fish in contrasting environments. Oecologia 40, 189–201 (1979).
Courtney, S.P.: Studies on the biology of the butterflies *Anthocharis cardamines* (L.) and *Pieris napi* (L.) in relation to speciation in Pierinae. Ph.D. Thesis, University of Durham (1980).
Danks, H.V.: Arctic Arthropods: A Review of Systematics and Ecology with Particular Reference to the North American Fauna. Ottawa: Entomol. Soc. Canada, 1981, 608 pp.
Dempster, J.P.: The natural control of populations of butterflies and moths. Biol. Rev. 58, 461–481 (1983).
Derickson, W.K.: Ecological and physiological aspects of reproductive strategies in two lizards. Ecology 57, 445–459 (1976).
Downes, J.A.: What is an arctic insect? Can. Entomol. 94, 143–162 (1962).
Downes, J.A.: Arctic insects and their environment. Can Entomol. 96, 280–307 (1964).
Downes, J.A.: Adaptations of insects in the arctic. Annu. Rev. Entomol. 14, 271–298 (1965).
Dunbar, M.J.: Ecological Development in Polar Regions: A Study in Evolution. Englewood Cliffs, N.J.: Prentice-Hall, 1968, 119 pp.
Ehrlich, P.R., Gilbert, L.E.: Population structure and dynamics of the tropical butterfly *Heliconius ethilla*. Biotropica 5, 69–82 (1973).
Field, W.D., Herrera, J.: The Pierid butterflies of the genera *Hypsochila* Ureta, *Phulia* Herrich-Schaeffer, *Infraphulia* Field, *Pierphulia* Field, and *Piercolias* Staudinger. Smithson. Contrib. Zool. 232, 1–64 (1977).
Gadgil, M., Solbrig, O.: The concept of r and K selection: evidence from wildflowers and some theoretical considerations. Am. Natur. 106, 14–31(1972).
Gaines, M., Vogt, K., Hamrick, J., Calwell, J.: Reproductive strategies and growth patterns in sunflowers (*Helianthus*.) Am. Natur. 108, 889–894 (1974).
Geiger, H.J.: Enzyme electrophoretic studies on the genetic relationships of Pierid butterflies (Lepidoptera: Pieridae). I. European taxa. J. Res. Lepidpot. 19, 181–195 (1981).
Hayes, J.L.: The population ecology of a natural population of the Pierid butterfly *Colias alexandra*. Oecologia 49, 188–200 (1981).
Herrera, J., Field, W.D.: A revision of the butterfly genera *Theochila* and*Tatochila* (Lepidoptera: Pieridae). Proc. US Nat. Mus. 108, 467–514(1959).
Jones, R.E.: Movement patterns and egg distribution in cabbage butterflies. J. Anim. Ecol. 46, 195–212 (1977).
Jones, R.E., Hart, J.R., Bull, G.D.: Temperature, size and egg production in the cabbage butterfly, *Pieris rapae* L. Aust. J. Zool. 30, 223–232 (1982).
Karlsson, B., Wiklund, C.: Egg weight variation and lack of correlation between egg weight and offspring fitness in the brown wall butterfly, *Lasiommata megera*. Oikos 43, 376–385 (1984).
Kingsolver, J.G.: Thermoregulatory strategies in *Colias* butterflies. I. The determinants of body temperature and flight activity: elevational patterns and mechanistic limitations. Ecology 64, 534–546 (1983a).

Kingsolver, J.G.: Thermoregulatory strategies in *Colias* butterflies. II. Thermoregulation and the ecological significance of flight activity: implications for reproductive strategy and population structure. Ecology 64, 546–551 (1983b).
Klots, A.B.: A generic revision of the Pieridae (Lepidoptera), together with astudy of the male genitalia. Entomol. Am. 12, 139–242 (1933).
Krebs, C.J.: Ecology: The Experimental Analysis of Distribution and Abundance. New York: Harper & Row, 1972, 694 pp.
Lewontin, R.C.: The units of selection. Annu. Rev. Ecol. Syst. 1, 1–18 (1970).
Livdahl, T.P.: Interspecific interactions and the r-K continuum: laboratory comparisons of geographic strains of *Aedes triseriatus*. Oikos 42, 193–202 (1984).
Luckinbill, L.S.: An experimental analysis of a life history theory. Ecology 65, 1170–1184 (1984).
MacArthur, R.H., Wilson, E.O.: The Theory of Island Biogeography. Princeton, New Jersey: Princeton University Press, 1967, 203 pp.
MacNeill, C.D.: The skippers of the genus *Hesperia* in western North America, with special reference to California. Univ. Calif. Pub. in Entomol. 35, 1–230 (1964).
McNaughton, S.J.: r- and K-selection in *Typha*. Am. Natur. 109, 251–261 (1975).
Pianka, E.R.: On r- and K-selection. Am. Natur. 104, 592–597 (1970).
Shapiro, A.M.: Developmental and phenotypic responses to photoperiod in uni- and bivoltine *Pieris napi* in California. Trans. Roy. Entomol. Soc. Lond. 127, 65–71 (1975a).
Shapiro, A.M.: Photoperiodic control of development and phenotype in a subarctic population of *Pieris occidentalis* (Lepidoptera: Pieridae). Can. Entomol. 107, 775–779 (1975b).
Shapiro, A.M.: The biological status of Nearctic taxa in the *Pieris protodice occidentalis* group (Pieridae). J. Lepidopt. Soc. 30, 289–300 (1976a).
Shapiro, A.M.: Seasonal polyphenism. Evol. Biol. 9, 259–333 (1976b).
Shapiro, A.M.: Phenotypic induction in *Pieris napi* L.: role of temperature and photoperiod in a coastal California population. Ecol. Entomol. 2, 217–224 (1977).
Shapiro, A.M.: Egg-load assessment and carryover diapause in *Anthocharis* (Pieridae). J. Lepidopt. Soc. 34, 307–315 (1980).
Shapiro, A.M.: The pierid red-egg syndrome. Am. Natur. 117, 276–294 (1981).
Shapiro, A.M.: Experimental studies on the evolution of seasonal polyphenism. In: The Biology of Butterflies. Vane-Wright, R., Ackery, P. (eds.). New York: Academic Press, 1984a, pp. 297–307.
Shapiro, A.M.: Lo que sabemos y lo que ignoramos de la regulación poblacional de mariposas. Medio Ambiente (Valdivia) 6, 19–22 (1984b).
Shapiro, A.M.: Behavioral and ecological observations of Peruvian high-Andean Pierid butterflies. Stud. Neotrop. Fauna Envt., 20, 1–14 (1985a).
Shapiro, A.M.: The genetics of polyphenism and its role in the phylogenetic interpretation of the *Tatochila sterodice* species-group in the Andean-Neantarctic region. J. Res. Lepidopt., 24, suppl., 24–31 (1985b).
Solbrig, O., Simpson, B.: Components of regulation of a population of dandelions in Michigan. J. Ecol. 62, 473–486 (1974).
Stearns, S.C.: Models in evolutionary ecology. In: Population Biology and Evolution. Woehrmann, K., Loeschke, V. (eds.). Berlin: Springer-Verlag, 1984, pp. 261–265.
Taylor, C.E., Condra, C.: r- and K-selection in *Drosophila pseudoobscura*. Evolution 34, 1183–1193 (1980).
Turner, J.R.G.: Experiments in the demography of tropical butterflies. II. Longevity and home-range behavior of *Heliconius erato*. Biotropica 3, 21–31 (1971).

Van Valen, L.M.: Integration of species: stasis and biogeography. Evol. Theor. 6, 99–112 (1982).

Varga, Z., Toth, I.: Ubersicht der taxonomischen und chorologischen Verhaeltnisse in der *Pieris (Artogeia) napi-bryoniae*-Gruppe (Lep., Pieridae) mit Ruecksicht auf die Anwendung des Superspezies-Begriffe in der Lepidopteren-Taxonomie. Act. Biol. Debrecina 15, 297–322 (1978).

Wiklund, C., Karlsson, B.: Egg size variation in Satyrid butterflies: adaptive vs. historical, ''Bauplan,'' and mechanistic explanations. Oikos 43, 391–400 (1984).

Chapter 10

Adaptive Significance of Genetic Variability of Photoperiodism in Mecoptera and Lepidoptera

KLAUS PETER SAUER, HUBERT SPIETH, and CORNELIA GRÜNER

Successful reproduction is significant for the evolutionary success of any population. Therefore, the mode of reproduction is under strong selective control. A single optimal reproductive strategy does not exist because the reproductive strategy depends on the environmental conditions. Here we define a strategy as a set of coadapted traits molded by natural selection to solve particular ecological problems. Looking at reproductive strategies as a part of life history strategies, we must distinguish between two aspects. First, with varying predictability of favorable environmental conditions, selection results in differential investment in the offspring. Second, natural selection favors maximal exploitation of the duration of favorable environmental conditions. Here we consider only the second aspect.

What Are the Causes of Periodism in the Life Cycles of Insects?

Competition and Periodism

Natural selection favors maximal exploitation of the duration of favorable environmental conditions, i.e., for a short-lived insect in an expanding population to produce as many generations as possible. An increase in number of generations is best achieved by synchronization of the individuals with the favorable environmental conditions to which a given population is exposed.

In general, we can say that, with increasing latitude, the temporal variation of food resources increases as a result of seasonal changes in productivity. In the warm season, food resources of quite different qualities are readily available. In contrast, in the cold season availability of food resources is very limited. During the course of the year, insect species compete for these unequally distributed resources. A strategy for lessening competition is temporal partitioning of the exploitation of these limited food resources. Only a few insect species are found that share the very limited food resources during the cold season, e.g., *Chionea* (Diptera) and *Boreus* (Mecoptera). The majority of insects are active during the warm season. Temporal resource partitioning generally results in ecological separation of competing species. Ecological separation results in different life cycles and, therefore, in different adaptations to climate. Differing life cycles and adaptations to climate consequently result in different synchronization not only with the limited food resources but also with the climatic conditions. In this chapter, we emphasize the observation that during its life cycle every individual of a species experiences a number of critical points in time.

Critical Points in Time

The time of reproduction, the time of starting activity, the time of terminating activity, the time of reactivating development after a dormant stage, and the time for inducing a diapause to survive adverse conditions are only some examples of critical points in time within an individual life cycle. We hypothesize that competition for temporally unequally distributed food resources is the evolutionary, ultimate cause of periodism in insects.

Here we consider only some special aspects of this problem: The determination of a critical point in time at the level of individuals, populations and, species. We are sure many strategies for determining a critical point in time have evolved. Here we consider only the significance of genetic variability in the innate recognition of the critical day length and the critical number of days for the determination of the critical point in time at the end of the growing season.

Significance of the Critical Day Length for the Determination of the Critical Point in Time

The End of the Growing Season: A Critical Point in Time

Multivoltine species have to solve the ecological problem of determining an unpredictable critical point in time at the end of the favorable conditions to minimize the likelihood of catastrophic weather-induced mortality. After the critical point in time, only diapause makes survival possible. Multivoltine species exposed to changing conditions display eudiapause (in the sense of Müller 1970) in which the mode of development (diapause or nondiapause) is exogenously determined by the proximate factor (*Signalfaktor*) day length. As the day length is correlated with the seasonal change of environmental conditions, it is a proximate factor of considerable significance.

We examined the induction of hibernation in the scorpion-fly *Panorpa vulgaris,* the pierid butterfly, *Pieris brassicae,* and hibernation and aestivation in the noctuid moth, *Mamestra brassicae,* by exposing the larvae to a constant regimen of different day lengths and measuring the percentage of individuals developing without diapause. When the percentage of incidence of nondiapause is plotted against day length, we obtain a photoperiodic response curve. This sigmoid curve is the cumulative distribution of day length thresholds for nondiapause development. As Grüner and Sauer (1984), Sauer (1976, 1977, 1984), and Spieth (1985) have shown through a large number of inbreeding (selection) experiments, each individual shows an innate critical day length (Figure 10.1). The results of these investigations confirm the hypothesis that the photoperiodic response curve represents largely the genetic variability in a population with respect to the recognition of a critical day length inducing a photoperiodic response (diapause or nondiapause).

Beginning of Reproduction: A Critical Point in Time

In univoltine species that display parapause (in the sense of Müller 1970), an obligatory dormancy, the end of reproduction is genetically fixed. As an example, consider the carabid beetle, *Pterostichus nigrita*. In carabid beetles, reproduction is a "critical state" within the life history of a female as shown by Müller (1983). We define "critical state" to be an age particularly sensitive

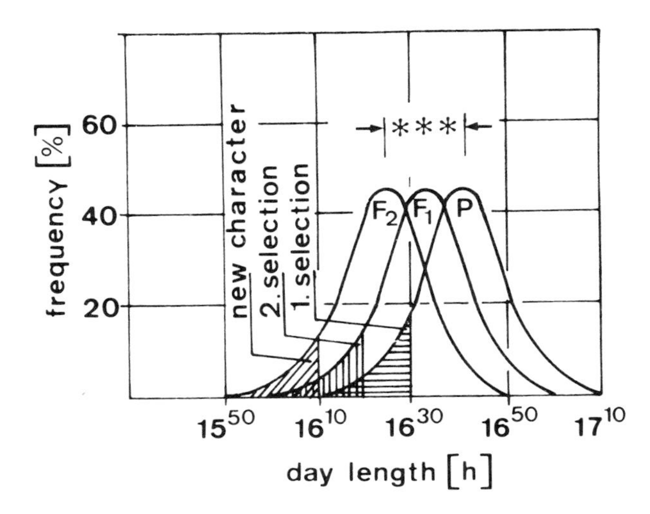

Figure 10.1. Frequency of individuals of *Panorpa vulgaris* responding with different day length thresholds. P, frequency within the parental generation (Freiburg population); F_1, frequency within the F_1 generation after selection of P individuals responding to low day length thresholds (hatched horizontally); F_2, frequency within the F_2 generation after selection within the F_1 generation (selected portion hatched vertically). The diagonally hatched portion indicates the frequency of individuals that expresses thresholds not normally found in the parental generation. The symbol *** indicates that the difference between the means of the parental generation and the F_2 generation obtained by transforming selection is significant ($p \leq 0.001$, F test)

to competitive effects that reduce reproductive success. Synchronization of the beginning of reproduction with the optimal availability of the limited food resource is advantageous and represents a critical point in time. Vitellogenesis for *Pterostichus nigrita,* and with that the beginning of reproduction of the females, is determined by day length thresholds (Thiele 1971; Ferenz 1975) (Figure 10.2). The phenotypic variation of critical day lengths must have a ge-

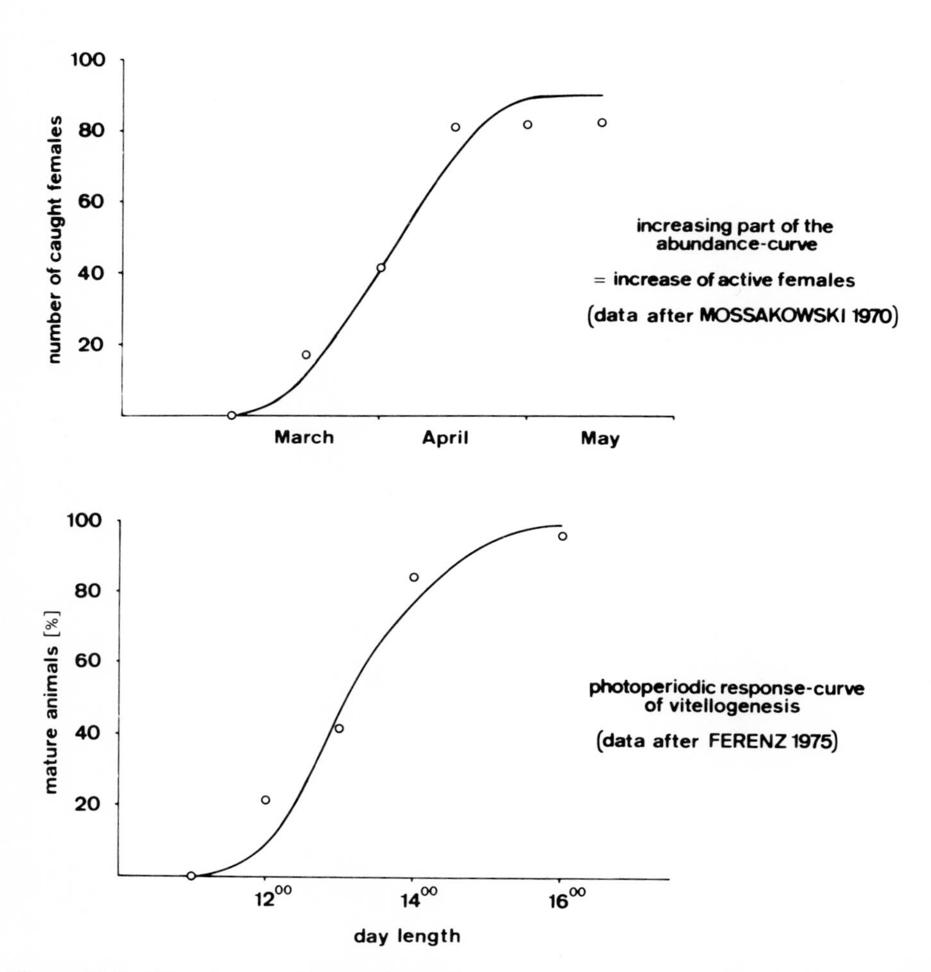

Figure 10.2. Abundance and maturity of females of *Pterostichus nigrita* as a function of season and day length. Upper panel indicates the absolute number of females caught from March to May. The number of females caught is a criterion for the number of active females and activity in turn correlates with achievement of maturity (vitellogenesis) of females. In the study area, day length increases from 12.0 hr to 16.5 hr in the time span mentioned above. Lower panel indicates the photoperiodic response curve with respect to vitellogenesis. Above a day length of 11.5 hr, the proportion of females maturing under constant day length conditions increases with increasing day length.

netic basis as shown by raising samples from geographically different populations under identical conditions (Ferenz 1975).

What Are the Causes of Genetic Variability in Critical Day Length Determining a Critical Point in Time?

Proximate and Ultimate Causes

Competition for limited food resources, the ecological, ultimate cause, has several effects, e.g., a physiological effect in which exceeding an inborn threshold of day length causes a special program (nondiapause or vitellogenesis)to be realized. During evolution, different species have evolved different programs depending on the selective environmental conditions. The program is a mechanism to transform causes into effects. Such effects we often interpret as proximate causes of the adaptations.

The proximate factor day length is only approximately correlated with the course of environmental conditions (the ultimate selective factors). The seasonal change of day length is the same every year. But the end of the growing season, in particular the availability of food for the beginning of reproduction, changes unpredictably from year to year. This is why critical points in time change unpredictably. In every year a different day length correlates with the critical point of time. Consequently the relative fitness of the innate ability to recognize a specific critical day length changes from generation to generation and from year to year. Let us consider the following *"Gedankenexperiment":* If all individuals of a given population used the same critical day length to synchronize themselves with the critical point in time, the population would not be optimally fitted to the unpredictably varying critical point in time. In the case of *Panorpa vulgaris,* the population would not be optimally fitted to the end of the growing season or, similarly, the population of *Pterostichus nigrita* would not be optimally fitted to the annually varying beginning of optimal food availability for successful reproduction.

The cause of variation is as follows: The inexact correlation of proximate (day length) and ultimate (food supply) factors results in variation in the thresholds for the photoperiodic response. As environmental conditions favor other genotypes with different critical day lengths in successive generations and years, the frequency of these genotypes changes permanently (Sauer 1977, 1984).

How Much Variation Is Adaptive?

At a latitude of 48°N, the end of the growing season may fluctuate between 3 and 4 weeks and in this period the day length changes between 50 and 80 min. For populations of *Panorpa vulgaris* from this latitude, the range of day length thresholds is close to 70 min (Figure 10.1). We interpret this variation in the innate recognition of the critical day length in *Panorpa vulgaris* as reflecting the frequency of correlation of a distinct day length with a critical point in time in the past; in other words it represents a sort of fitness curve for the different genotypes, which respond photoperiodically to distinct critical day lengths. (Note that our term "fitness curve" differs from the fitness functions used by Taylor in this volume.)

We hypothesize that in the case of the induction of vitellogenesis, and with that the triggering of the beginning of reproduction in females of *Pterostichus nigrita,* the photoperiodic response curve is the same sort of fitness curve mentioned above. As Figure 10.2 shows, the proportion of active (i.e., food searching) females increases from March to May (data drawn from Mossakowski 1970). In this period the day length increases from 12.5 to 16.5 hr. For populations of *Pterostichus nigrita* at a latitude of 51°N the range of day length thresholds is approximately 3.5–5 hr. The variation in the innate recognition of the critical day length with respect to the triggering of vitellogenesis (beginning of reproduction) reflects the frequency of correlation of a distinct day length with the time of the optimal beginning of reproduction in the past.

Significance of the Critical Number of Days for the Determination of the Critical Point in Time

The experience of a particular number of days with definite day lengths provides an additional cue for the determination of the critical point in time (e.g., the end of the growing season). The programming of the mode of development involves not only measurement of a day length threshold but also the summation of successive cycles during the sensitive period to a point at which induction of nondiapause development occurs; in most species a certain number of days is required, usually between 3 and 30 (Saunders 1982). A transfer of the larvae of *Panorpa vulgaris* during a definite period from short- to long-day cycles was used to map out the number of days required to induce nondiapause development. Plotting the increase in percentage of nondiapausing individuals versus the number of days experienced with a definite length, we obtain a sigmoid curve like the photoperiodic response curve (Figure 10.3). The tested populations of *Panorpa vulgaris* show phenotypic variation with respect to the recognition of a critical number of days inducing a photoperiodic response. This phenotypic variation must be at least partly genetically determined because it varies geographically as shown by raising populations under identical conditions (Figure 10.3). We tested this hypothesis by the following experiment: Individuals that had responded to 8 days at a day length of 16.5 hr with nondiapause were selected to be parents and their offspring were tested again under the same conditions. In the wild population, 17% responded to these conditions after exposure throughout larval development. After inbreeding and testing again this percentage increased to 35% ($p \leq 5\%$, Weir Test) (Sauer 1984; Sauer and Spieth 1986.) We conclude that each individual responds to its innate critical number of days.

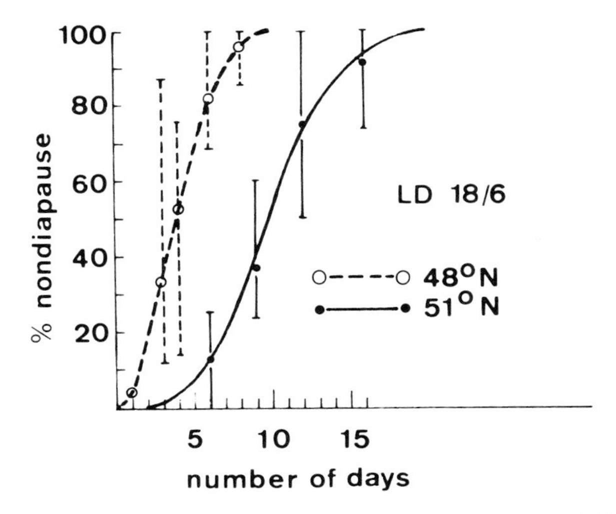

Figure 10.3. Clinal variation of number of days inducing photoperiodic response in *Panorpa vulgaris*. Points give mean values; bars indicate the range. The individuals of the two populations from Freiburg (48°N) and Gießen (51°N) were raised under a constant temperature regimen of 20°C. A light regimen of LD 18:6 is sufficient to induce nondiapause in all individuals.

Correlation of the Two Characters: Critical Day Length and Critical Number of Days

The photoperiodic programming of the mode of development is thus a two-stage process. An individual of *Panorpa vulgaris* is able to recognize not only its innate day length threshold but also its innate critical number of days. Is there a correlation between the recognition of a critical day length and a critical number of days? As explained above, every year another day length with another number of days is correlated with the critical point in time at the end of the growing season after which only diapause makes survival possible.

It may be possible that different individuals respond to the same innate day length threshold but with variation in the innate critical number of days. To test this hypothesis, we performed the following experiment: Groups of sibling larvae were kept at three different day lengths (16.33 hr, 16.66 hr, 17 hr) and five different numbers of days (3, 4, 6, 8, and as control 16, encompassing the whole of larval development). The results are shown in Figure 10.4. The shorter the day length to which an individual is exposed, the more days are required on the average to induce nondiapause development. These results can be explained in two different ways: (1) The effects on nondiapause development of the two factors, day length and length of exposure (number of days), are inversely correlated, once an innate threshold in day length is exceeded. (2) A

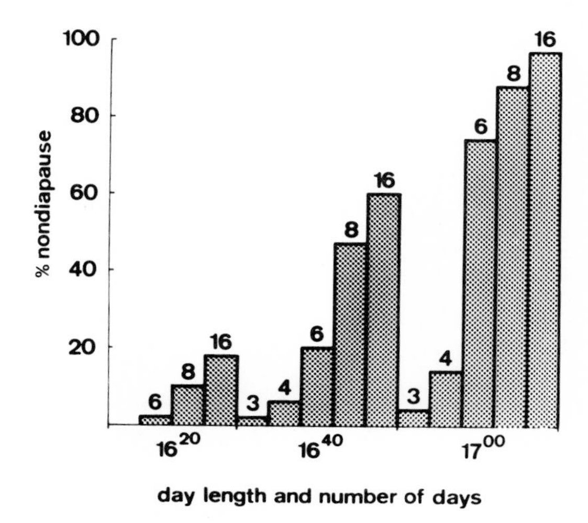

Figure 10.4. Cumulative distribution of nondiapause in *Panorpa vulgaris* as a function of day lengths and number of days. Three day lengths were tested (16:20, 16:40, 17:00). The numbers above the bars give the number of days experienced at the respective day lengths.

definite threshold in day length in connection with a definite number of days must be exceeded for nondiapause development; in this case no "compensation" occurs in the effects of the two factors, day length and number of days. The two explanations are demonstrated hypothetically in Figure 10.5a.

Explanation 1: If "compensation" exists, then an increase in day length above the day length of the saturation level of the photoperiodic response curve

→

Figure 10.5. Differential effects of day lengths exceeding the uppermost day length of the photoperiodic response curve. (a) Two hypothetical explanations. Hypothesis 1: Exposure for an insufficient number of days may be compensated for by increasing day length. For a constant number of days insufficient to induce nondiapause in all individuals, the portion of nondiapausing individuals will increase with increasing day length exceeding the uppermost critical day length. Hypothesis 2: Each "day length type" is composed of various "numbers of days" phenotypes. According to this model the portion of nondiapausing individuals should remain constant. (b) Experimental evidence for hypothesis 1. In *Pieris brassicae* compensation of an insufficient number of days (10 days) by longer day lengths exists. Points give mean values; bars indicate the range. The U test indicates significant ($p \leq 5\%$) differences in the proportion of nondiapausing individuals between day lengths of 15.5 hr and 18 hr, 16 hr and 18 hr, and 17 hr and 18 hr. (c) Experimental evidence for hypothesis 2. In *Panorpa vulgaris* an insufficient number of days (10 days) cannot be compensated by longer day length. This indicates a fixed linkage of day length and number of days. Points give mean values; bars indicate the range. The U test indicates no differences in the proportion of nondiapausing individuals between any pair of day lengths.

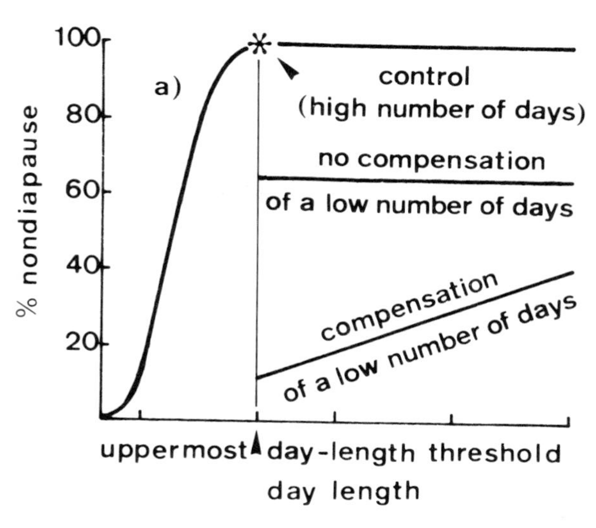

a)
100
80
60
40
20
% nondiapause
control
(high number of days)
no compensation
of a low number of days
compensation
of a low number of days
uppermost day-length threshold
day length

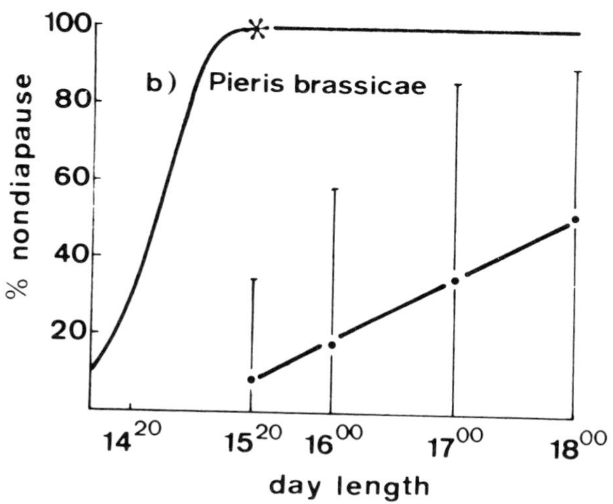

b)
Pieris brassicae
100
80
60
40
20
% nondiapause
14^20
15^20
16^00
17^00
18^00
day length

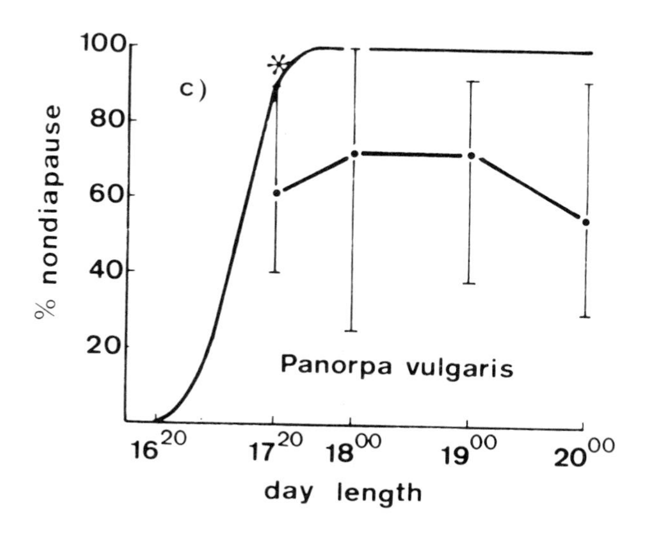

c)
Panorpa vulgaris
100
80
60
40
20
% nondiapause
16^20
17^20
18^00
19^00
20^00
day length

together with a constant number of days must cause an increase in the proportion of nondiapause development. Above the uppermost day length threshold, all individuals of a given population will respond with nondiapause development when exposed for a definite number of days (a long enough period). With the same day length but a smaller number of days (too short of a period) fewer individuals respond in this way. If the day length is now increased above the threshold of saturation, the proportion of nondiapause development should rise with increasing day length in the case of "compensation." We did not find evidence for this hypothetical explanation in *Panorpa vulgaris* but did in the great white cabbage butterfly, *Pieris brassicae* (Spieth 1982; Sauer 1984; Sauer and Spieth in preparation). In this species the principle just outlined is realized (Figure 10.5b). Apparently *Pieris brassicae* must build up a definite "light sum" (*Lichtsummenwert*) to display nondiapause development. If the innate day length threshold, which first enables an individual to show a photoperiodic response, is exceeded, it is possible for each individual to sum the experienced number of days with different day lengths to a "light sum," which induces nondiapause development. As shown by our experiments, the individuals of *Pieris brassicae* are able to measure the absolute length of a day, since an increase of day length alone above the saturation threshold causes a higher rate of nondiapausing individuals while the number of days remains constant. Further experiments (Sauer and Spieth in preparation) have demonstrated the ability of the individuals of *Pieris brassicae* to discriminate the absolute day length not only above the innate threshold but also below this innate level (short day), once the innate day length threshold has been exceeded for a definite time during larval development. These results suggest that we must revise the orthodox conception that only the absolute day length of a short day (day length shorter than the innate critical day length) causes any alteration in the expression of a phenotypic trait (diapause or nondiapause). So far we know little about the mechanism. But we hypothesize that *Pieris brassicae* is able by this mechanism to discriminate between increasing and decreasing day lengths. The adaptive significance of this principle may be based on the migratory behavior of *Pieris brassicae*.

Explanation 2: The results of the experiments with *Panorpa vulgaris* in Figure 10.5c show that a lower number of days cannot be compensated by a higher day length. The percentage of incidence of nondiapause remains nearly constant. In *Panorpa vulgaris* a definite day length together with a definite number of days must be exceeded for nondiapause development. In this case, a definite critical day length is strictly correlated with a definite critical number of days. Each "day length threshold type" is composed of various types, which are characterized by different innate numbers of days. This correlation between the two components of the signal perhaps reflects the frequency of correlation of a definite day length and a definite number of days with the critical point of time at the end of the growing season in the past. Owing to this correlation a population becomes highly variable in determining the predictable critical point in time.

Innate Recognition of Day Length for Aestivation in *Mamestra brassicae*

Appearance and Adaptive Significance of Aestivation

Whereas the hibernal diapause of *Mamestra brassicae* (Grüner and Sauer 1984) is a true eudiapause (in the sense of Müller 1970) comparable to that of *Panorpa vulgaris* and *Pieris brassicae*, aestival diapause is only a modified, nondiapause development (an obligopause in the sense of Müller 1970). Different geographical populations of *Mamestra brassicae* are exposed to diverse climatic conditions. In southwest Europe, periods of favorable environmental conditions (availability of food plants of good quality) are interrupted by periods of drought in summer and by periods of frost in winter. Competition for limited food resources causes differentiation in food plant utilization, but geographically diverse climatic conditions cause temporally different distributions of food plants. Therefore, in general, we can say that this temporally and spatially varying distribution of food resources selects for locally specialized life cycle adaptations in insect species that utilize plants. Masaki (1968) and Masaki and Sakai (1965) were the first to interpret aestival diapause as an adaptive strategy for synchronization of the life cycle with the cycle of temporally changing environmental conditions. Poitout and Bues (1977) concluded the same for populations of southwest Europe. Under natural conditions, the southern populations (44°N) of *Mamestra* display both aestival and hibernal diapause in the pupae. As the breeding season progresses, the larvae of *Mamestra brassicae* are exposed to relatively long day lengths and high temperatures. Larvae exposed to such environmental conditions display aestival diapause. Under natural conditions, the northern populations (48°N) normally display only hibernal diapause. Depending on the day length and temperature regimen during the larval development, however, the larvae may produce nondiapausing, aestivating, *and* hibernating pupae. The development of nondiapausing pupae at a temperature range between 20 and 25°C requires 18–32 days. But, if larvae experience relatively long day lengths and high temperatures, the developmental time of a portion of nonhibernating pupae is increased to 40–60 (or more) days. Duration of pupal development exceeding 40 days is defined as aestival diapause.

Three questions must be answered: (1) Does the incidence of aestival diapause vary geographically? (2) Does an innate critical day length trigger the aestival diapause, if this threshold is exceeded? (3) Do northern populations possess a cryptic ability to aestivate, which is never expressed phenotypically in nature?

Clinal Variation

To answer the first question, we examined three geographically different populations from Argelès (42.5°N), Avignon (44°N), and Freiburg (48°N). Figure 10.6 shows that the incidence of aestivation diapause varies geographically. As

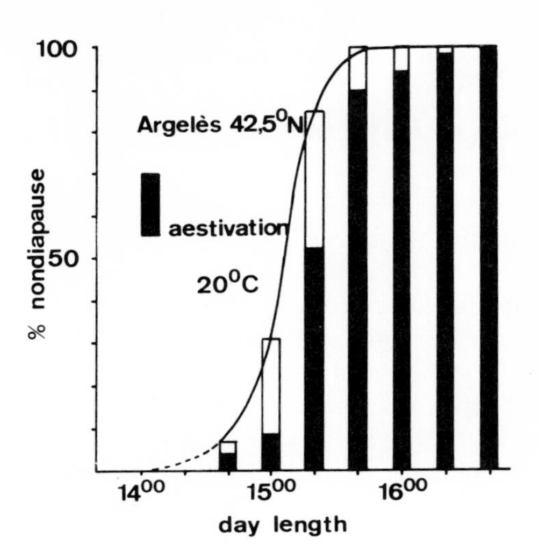
100
50
% nondiapause
Argelès 42,5°N
aestivation
20°C
14⁰⁰
15⁰⁰
16⁰⁰
day length

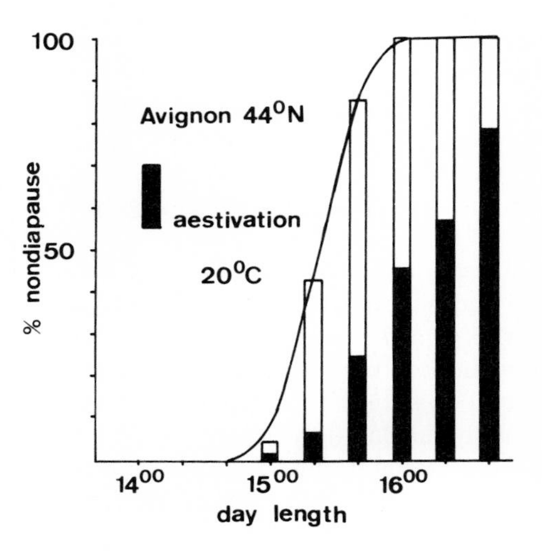
100
50
% nondiapause
Avignon 44°N
aestivation
20°C
14⁰⁰
15⁰⁰
16⁰⁰
day length

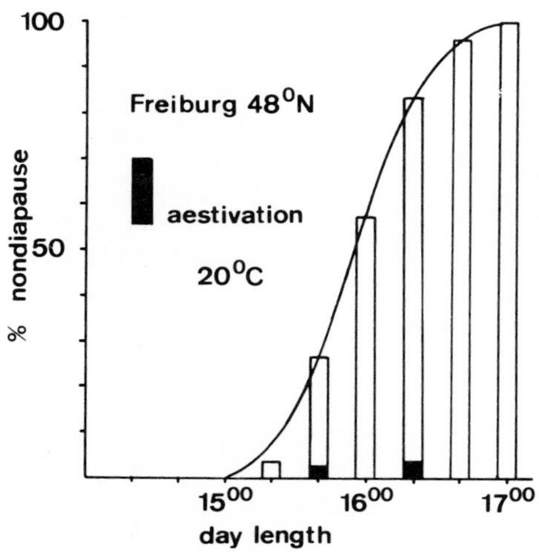
100
50
% nondiapause
Freiburg 48°N
aestivation
20°C
15⁰⁰
16⁰⁰
17⁰⁰
day length

shown in the populations of Argelès and Avignon, at a constant temperature regimen of 20°C, the portion of aestivating individuals increases with increasing day length. In the population of Argelès (42.5°N) a day length of 16 hr causes aestival diapause in nearly 100%, in the population of Avignon (44°N) in 50%, and in the population of Freiburg in 0% of the individuals. In the southern populations the portion of aestivating individuals increases continuously with increasing day length but remains negligible for the northern population under this temperature regimen.

Genetic Basis of the Day-Length Recognition

To answer the second question, we carried out the following experiments: Individuals that responded under a constant temperature regimen of 20°C to a relatively short day length (15 hr) with aestival diapause were sib-mated (inbred) and the offspring of these selected animals were tested again under the same experimental conditions as the offspring of the wild population. We also selected for differential duration of aestival diapause. In the wild population reared as described above, the portion of aestivating individuals was 18.9%; this portion increases significantly to 48% in the population after inbreeding (selection). The mean duration of aestivation after inbreeding (selection) changes, too, depending on the mean duration that was selected for. By selecting individuals with a mean pupal duration not very different from that of the wild (parental) population, the amount of change was conspicuous but not significant. Only by selecting individuals with a very long aestival diapause does the change in this trait become significant. By inbreeding of nonaestivating individuals, the mean duration of pupal development remains nearly constant.

These experiments confirm our hypothesis that each individual responds to an innate day length threshold for aestival diapause. Plotting the increase in percentage of aestivating individuals against day length, we obtain a sigmoid curve (filled columns in the figures) similar to the photoperiodic response curve described above. This sigmoid curve is the cumulative distribution of day length thresholds for aestivation in the population. Each individual of *Mamestra brassicae* of southern populations must be able to distinguish between two day length thresholds, inducing hibernation diapause (eudiapause) or, alternatively, aestival diapause (oligopause). We call this phenomenon a "mixed strategy" of determining a critical point in time within the life cycle of an insect species.

Figure 10.6. Photoperiodic response curves of three geographical populations of *Mamestra brassicae*. Open columns indicate nondiapausing individuals (duration of nondiapausing pupae at a temperature of 20°C–25°C ranges from 18 to 32 days). Filled columns indicate the aestivating portion of nondiapausing individuals (duration of aestivating pupae at a temperature of 20°C–25°C ranges from 40 to 60 days).

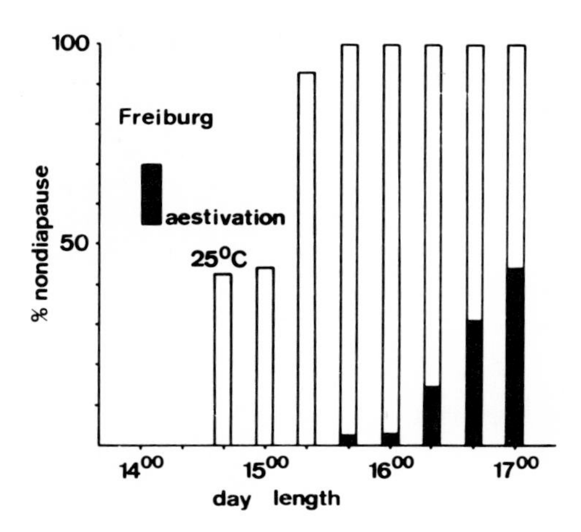

Figure 10.7. Effect of high temperature on the incidence of aestivation (filled columns) in a northern population of *Mamestra brassicae*. For further explanation see Figure 10.6.

Cryptic Variation

To answer the third question, we tested larvae of the northern population (Freiburg, 48°N) under a regimen of a very high temperature (25°C). Temperatures between 20°C and 25°C alter the value of the day length threshold that induces aestival diapause. For southern populations this temperature range is ecologically relevant. A temperature of 25°C is very high and rare for the Freiburg area. Therefore, this high temperature is ecologically not very relevant. But, as Figure 10.7 shows, aestivation behavior will be phenotypically expressed in northern populations at such a high temperature. These results demonstrate that what is inherited is not a fixed day length threshold for aestivation, but rather a temperature-dependent "reaction norm." Therefore a cryptic ability for aestivation behavior in northern populations exists, but it is not normally expressed phenotypically under natural conditions.

The Strategy of Flexible Response

Temperature Dependence of the Critical Day Length

A great many species exist under environmental conditions that do not permit a high genetic load. Genotypes with a wide "reaction norm," which produce different phenotypes under different environmental conditions, are favored under such circumstances. Adaptation by modification or a flexible response "protects" individuals exposed to unpredictable conditions.

How does an individual become more flexible in synchronizing its developmental mode with the unpredictably fluctuating critical point in time? If the innate recognition of the day length threshold or critical number of days that induces the photoperiodic response is to a certain degree temperature-dependent, then an individual is able to use the actual weather conditions to determine the unpredictable critical point of time.

Figure 10.8 indicates that in *Panorpa vulgaris*, experiencing constant temperatures between 11°C and 15°C alters the value of the day length threshold. This temperature range is ecologically relevant. Furthermore, this experiment shows that the inherited factor is not a definite day length threshold but rather a "reaction norm." Within the "reaction norm," the actual temperature effective during a sensible phase alters the position of the day length threshold. This "reaction norm" makes it possible for the individuals of a population to modify their response depending on the actual temperature. More individuals enter diapause before the critical point in time and therefore are not eliminated by selection. The genetic load of the population is reduced by a temperature-dependent flexible response to day length. As Sauer (1984; Sauer and Spieth 1986) has shown, the critical number of days is temperature-dependent, too, and not temperature-compensated as Saunders (1982) generally hypothesized. This temperature dependence of the critical number of days fortifies the behavioral flexibility of an individual of *Panorpa vulgaris* in the sense mentioned above. Possibly depending on the predictability of the critical point in time at the end of the growing season, the temperature dependence of the critical day length

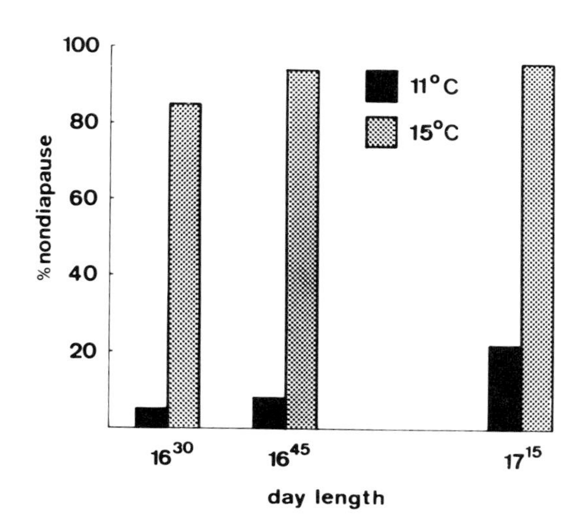

Figure 10.8. Temperature dependence of the innate day length threshold of *Panorpa vulgaris*.

and critical number of days in *Panorpa vulgaris* varies geographically (Riebel 1984; Riebel and Sauer 1984; Sauer and Riebel in preparation).

Clinal Variation in Temperature Dependence of the Critical Day Length

So far we have good evidence for this phenonomen of clinal variation of temperature dependence of the critical day length in the great white cabbage butterfly *Pieris brassicae* (Spieth 1985). Figure 10.9 indicates the diverse photoperiodic response curves of three geographical populations (Horsens 56°N, Bodensee 47.5°N, and Banyuls 42.5°N) obtained under two constant temperature regimens of 15°C and 20°C. The temperature-dependent "reaction norm," as measured by the difference in critical day lengths at the two temperatures, is greatest (68 min) in the most southern population and decreases continuously with increasing latitude (46 min in the Bodensee population and 10 min in the Horsens population). Other experiments (Spieth 1985) have shown that the temperature threshold above which the effective day length threshold will be altered varies geographically, too. This threshold is shifted from 12°C in Banyuls to 20°C in Horsens (Figure 10.9).

Adaptive Significance of Temperature-Dependent Critical Day Lengths

With increasing latitude in eastern Europe the predictability of the critical point in time at the end of the growing season increases. In the Horsens population of *Pieris brassicae,* the genetic variability with regard to the critical day length is relatively small in comparison with those in more southern populations. The genetic variability with regard to the critical day length is inversely correlated with latitude; it increases with decreasing latitude. Together with the genetic variability, the temperature-dependent "reaction norm" increases with decreasing latitude.

The length of the growing season is highly variable in Banyuls (42.5°N). There the critical point in time may occur in the middle of September or at the end of October or any time between. The long growing season allows numerous generations. Maximal exploitation of this long duration of suitable environmental conditions is favored by natural selection. But with decreasing predictability of the length of the growing season, the risk increases that individuals will fail to enter diapause before the critical point in time is exceeded. In contrast, the length of the growing season between years is nearly constant in Horsens (56°N). There it amounts to 110 days with mean daily temperatures between 13°C and 16°C. This time permits only two generations per growing season.

The Banyuls population has responded to its risky situation with increased genetic variability with respect to the critical day length and by magnification of the temperature-dependent reaction norm of the innate critical day length. The Horsens population, in contrast, is adapted to its particular situation by a relatively small genetic variability and a nearly temperature-compensated critical day length.

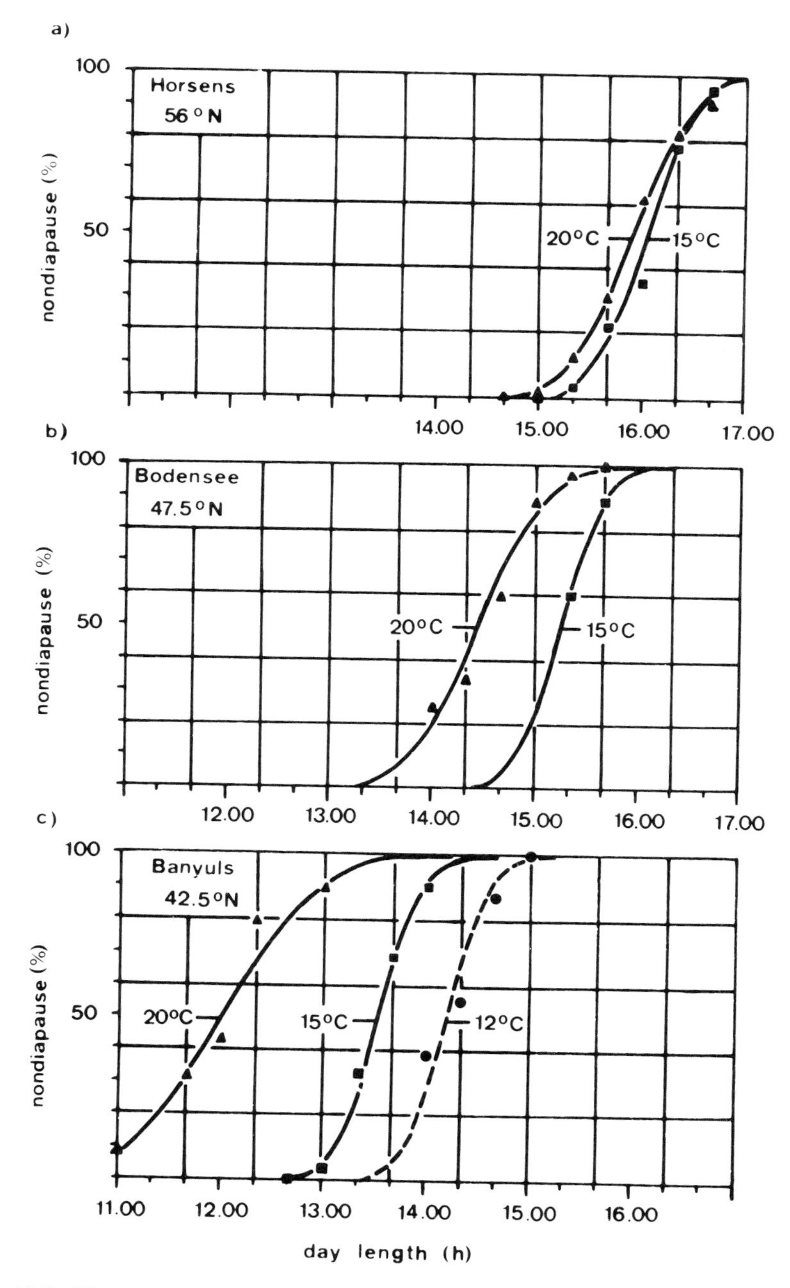

Figure 10.9. Photoperiodic response curves of three geographical populations of *Pieris brassicae* under different temperature regimens.

The temperature-dependent critical day length is a trait of adaptive significance with respect to the migration behavior of *Pieris brassicae,* too. Individuals migrating northward are exposed to environmental conditions that are characterized by lower temperatures and longer days compared to their place of origin. In contrast, individuals migrating southward change to shorter day lengths and higher temperatures. By the temperature-dependent "reaction norm" the offspring of migrating females compensate for the changed conditions and, therefore, may be able to synchronize their life cycle better to the new conditions.

Furthermore, we hypothesize that a relationship between temperature compensation of the critical day length and nonmigratory populations may exist. Investigations of the migration behavior of populations from Denmark have demonstrated that these populations, characterized by a temperature-compensated critical day length, display very little migration behavior (Roer 1959).

Adaptive Significance of the Genetic Variation in Recognition of the "Day Length Threshold" and the "Critical Number of Days"

Genotypes within a population are often characterized by different relative fitnesses. The relative fitness of a given genotype is expressed by its change in frequency in succeeding generations. In stable environments, phenotypic similarity is the result of stabilizing or canalizing selection. The success of canalizing selection depends on the constancy of the environments. A day length-controlled, facultative diapause (eudiapause) cannot be evolved under constant environmental conditions without seasonal change. As the length of the growing season fluctuates unpredictably, the relative fitness of the innate recognition of a definite day length threshold and a definite number of days also fluctuates. This causes a change in the frequency of different genotypes, thereby maintaining a genetic polymorphism. In any year, the selection resulting from the weather conditions in the current year may actually favor the "wrong" genotypes for the environmental conditions in the next year. The evolutionary changes within the population lag behind the actual situation.

The synchronization of the number of generations to the unpredictably fluctuating length of the growing season by means of an innate day length threshold connected with a critical number of days is a strategy that is maintained only if the innate recognition of these two components of the signal is determined by multiple-factor inheritance. Only in this case can genotypes that were completely eliminated by selection in one generation be reconstituted by recombination in the following generations. Figure 10.1 shows the results of inbreeding of the individuals that display nondiapause development at relatively short day lengths; in the following generations, response types appear that were not detectable in the parental generation. Accordingly, the second annual generation of a wild population contains individuals that display a photoperiodic response to a very short day length, which could not be detected in the first annual generation (Sauer 1977). The first generation had already passed through the *filter of selection,* but the second annual generation had not. Only the offspring of this second annual generation and perhaps a small portion of the offspring of the first generation, which build up the first annual generation of the next year, must still pass this filter. In the temperate zone, fluctuating conditions select insect populations with a facultative diapause to be genetically polymorphic with respect to the determination of the unstable critical point in time.

The adaptive significance of the genetic variation and the genetic correlation of the recognition of different photoperiodic thresholds and critical number of days depends on a polygenic system. Individuals genetically programmed to respond on or after the critical point in time with nondiapause development are "incorrectly programmed," and will be eliminated by selection. Thus, a genetic load is generated that the population has to bear. As long as the polymorphism persists, a seasonal load exists because of the presence of currently less favored response types.

Acknowledgments Ruth Baumann has contributed excellent technical help to this study. This work was supported by the Deutsche Forschungsgemeinschaft (Sa 259/3–1).

References

Ferenz, H.J.: Anpassungen von *Pterostichus nigrita* F. (Coleopt., Carabid.) an sub arktische Bedingungen. Oecologia (Berl.) 19, 49–57 (1975).

Grüner, C., Sauer, K.P.: Aestivation of *Mamestra brassicae:* an adaptive strategy. Abstract in: Advances in Invertebrate Reproduction 3, 588 (1984).

Masaki, S.: Geographic adaptation in the seasonal life cycle of *Mamestra brassicae* L. (Lepidopt. Noct.). Bull. Fac. Agri. Hirosaki Univ. 14, 16–26 (1968).

Masaki, S., Sakei, F.: Summer diapause in the seasonal life cycle of *Mamestra brassicae* C. (Lepidopt., Noct.). Jpn. J. Appl. Entomol. Zool. 9, 191–205 (1965).

Mossakowski, D.: Ökologische Untersuchungen an epigäischen Coleopteren atlantischer Moor- und Heidestandorte. Z. Wiss. Zool. 181, 233–316 (1970).

Müller, H.J.: Formen der Dormanz. Nova Acta Leopoldina 35, 7–25 (1970).

Müller, J.K.: Konkurrenzverminderung durch ökologische Sonderung bei Laufkäfern (Coleopt., Carabidae). Dissertation, Freiburg (1983).

Poitout, S., Bues, R.: Études compareès des diapauses nymphales estivales existant dans les populations de basse valleè du Rhône de deux Noctuidae (*Mamestra brassicae* L., *M. oleracea* L.) Ann. Zool. Ecol. Anim. 9, 225–234 (1977).

Riebel, W.: Klinale Variation im Erkennen von Zeitgebern in marginalen Populationen von *Panorpa vulgaris*. Diplomarbeit Bielefeld (1984).

Riebel, W., Sauer, K.P.: Clinal variation in photoperiodic reaction in central and marginal population of *Panorpa vulgaris*. Abstract in: Advances in Invertebrate Reproduction 3, 627 (1984).

Roer, H.: Über Flug- und Wandergewohnheiten von *Pieris brassicae*. Z. Ang. Entomol. 44, 272–309 (1959).

Sauer, K.P.: Untersuchungen zur klinalen Variation des Diapauseverhaltens von *Panorpa vulgaris* unter besonderer Berücksichtigung der Unterschiede zwischen Berg- und Flachlandpopulationen. Verhand. d. Ges. f. Ökologie, Wien 1975, 77–88 (1976).

Sauer, K.P.: Die Adaptive Bedeutung der genetischen Variabilität der photoperiodischen Reaktion von *Panorpa vulgaris* (Mecopt., Panorpidae). Zool. Jb. Syst. 104, 489–538 (1977).
Sauer, K.P.: The evolution of reproductive strategies as an adaptation to fluctuating environments. In: Advances in Invertebrate Reproduction 3, 317–326 (1984).
Sauer, U.D., Spieth, H.: Zeitliche Orientierung in fluktuierender Umwelt durch Messen einer Anzahl von Laugtagen. Zool. Jb. Syst. 113, 373–388 (1986).
Saunders, D.S.: Insect Clocks. Oxford and London: Pergamon Press (1982).
Spieth, H.: Die Anpassung des jährlichen Entwicklungszyklus an unterschiedlich lange Vegetationsperioden geographisch getrennter Lebensräume beim Wanderfalter *Pieris brassicae* L. (Lepidopt. Pieridae). Dissertation Freiburg (1982).
Spieth, H.R.: Die Anpassung des Entwicklungszyklus an unterschiedlich lange Vegetationsperioden beim Wanderfalter *Pieris brassicae*. Zool. Jb. Syst .112, 35–69 (1985).
Thiele, H.U.: Die Steuerung der Jahresrhythmik von Carabiden durch exogene und endogene Faktoren. Zool. Jb. Syst. 98, 341–371 (1971).

Chapter 11

Mosquito Maternity: Egg Brooding in the Life Cycle of *Trichoprosopon digitatum*

L. P. Lounibos and C. E. Machado-Allison

Subsocial or presocial behavior, as distinguished from the more sophisticated eusociality of ants, bees, or termites, has evolved independently many times among otherwise solitary insects. Perhaps the simplest of subsocial behaviors is egg attendance, a form of parental care, which has been documented in virtually every major insect order (Hinton 1981). Presocial behaviors appear to have evolved in the course of exploitation of either particularly favorable or unusually severe environments (Wilson 1971). For example, when resources are abundant yet ephemeral, scarabaeid parents may protect offspring from intense competition by provisioning dung balls to nourish their larvae. In the exploitation of harsh environments, parental attendance is known to reduce lace bug mortality due to predation (Tallamy and Denno 1981), fungal invasion of earwig nests (Lamb 1976) and oxygen depletion in staphylinid burrows (Bro Larsen 1952). Although parental care is one evolutionary solution to environmental pressures, Tallamy (1984) cautions that it is a costly trait, predicted to evolve only in the presence of intense selective pressure and specific behavioral precursors.

The purpose of this chapter is to describe parental care in a neotropical mosquito, *Trichoprosopon digitatum* (Rondani), to identify through experimental evidence a key environmental pressure that probably influenced the evolution of egg brooding in this species, and to emphasize the general importance of this factor, rainfall, as a source of egg mortality among mosquitoes. A brief account of the relationship between rainfall and brooding behavior has been presented elsewhere (Lounibos and Machado-Allison 1986). Lastly, we consider a possible case of evolutionary convergence of brooding behavior in an unrelated mosquito from tropical Asia. *Trichoprosopon digitatum* is at present the only confirmed example (Lounibos and Machado-Allison 1983) of presocial behavior among the more than 3000 recognized species of Culicidae (Knight and Stone 1977).

Natural History of *Trichoprosopon Digitatum*

Habits and Habitats

The neotropical genus *Trichoprosopon* consists of 21–22 species, some undescribed, in four complexes (Zavortink 1981). The immature stages of all species are restricted to phytotelmata (Maguire 1971), especially bamboo internodes, fallen leaves and spathes, and fruit husks. *Trichoprosopon digitatum* is the most commonly collected and ecologically diverse species. It represented more than 95% of all specimens examined by Zavortink (1981) for a generic revision, and is the only species common in phytotelm habitats created by man, such as

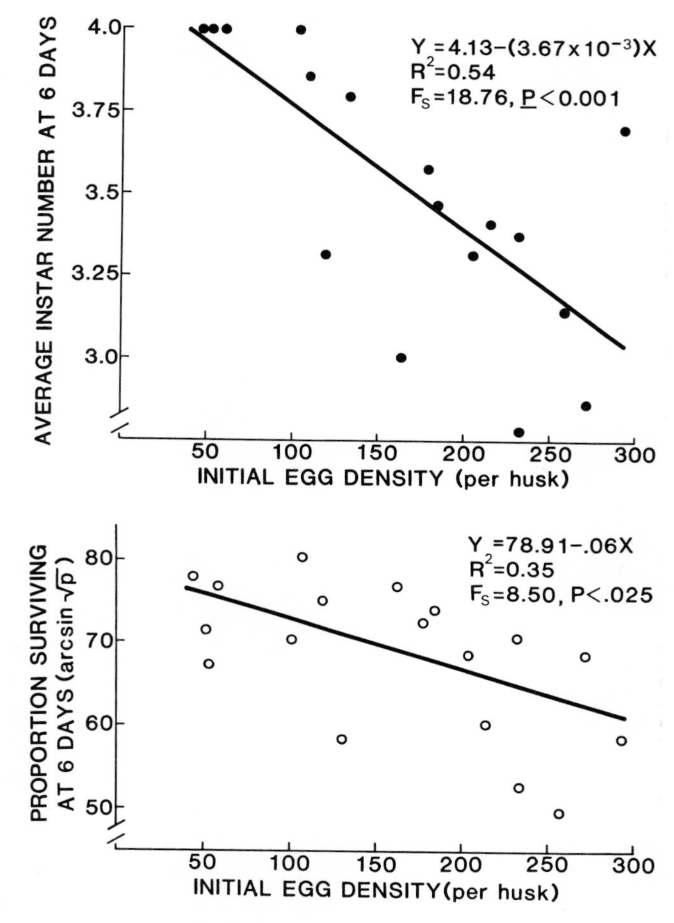

Figure 11.1 Relationship between initial egg density and larval development time and survivorship in 18 experimental cacao husks sheltered from rainfall but otherwise exposed to natural conditions at Panaquire, Venezuela.

cacao and coconut husks (Zavortink et al. 1983). Because of its abundance in association with man, the propensity of females to bite humans, and the isolation of numerous arboviruses from this species, *T. digitatum* is regarded as a potential vector of human diseases (Zavortink et al. 1983).

Our observations on this species were made at a cacao (*Theobroma cacao* L.) plantation in lowland tropical rain forest near Panaquire (10°13′N, 66°14′W), Miranda State, Venezuela. During the harvest season from December through July, discarded cacao husks collect rain water and provide rich, if transitory, larval habitats preferred by ovipositing females. Fresh husks artificially filled with rain water were colonized by *T. digitatum* within one-half day after setting, and gravid females were shown to be attracted to fluid extracts from cacao (Lounibos and Machado-Allison 1983).

Cacao husks, strewn in large discard piles, are of two shapes, depending on whether cut by male or female plantation workers at Panaquire. Men cut the cacao pod lengthwise with a machete, producing an open, boat-shaped husk after extraction of the seeds. Women typically use a hand knife to make a transverse cut and this leaves a deeper, less exposed cavity. Female *T. digitatum* strongly prefer female-cut "crown" husks to male-cut "boats" for oviposition (Lounibos and Machado-Allison 1983).

Because the 60–85 eggs in a *T. digitatum* raft (Aitken et al. 1968) are laid in less than 2 hr, larvae hatch synchronously and maintain cohorts during development (Seifert and Barrera-R. 1981). In bracts of *Heliconia,* a habitat infrequently used by *T. digitatum* at our study site, larval development required 14–17 days and larval survivorship was not density-dependent within the range of commonly observed densities of 20–60 individuals per bract (Seifert 1980). The cacao habitat is a richer yet more ephemeral resource than *Heliconia*. The pupal stage is reached in less than 10 days, and both developmental time and survivorship are density-dependent within the range of commonly observed egg inputs of 50–300 per husk (Figure 11.1). Thus, intraspecific competition may be more important for population regulation in cacao husks than in *Heliconia* bracts.

Egg Brooding

Although egg raft formation by *T. digitatum* was observed some decades ago (Howard et al. 1915; Pawan 1922), early investigators apparently did not notice that females continued to hold egg rafts between their mesothoracic legs long after oviposition was complete. Aitken et al. (1968) noted egg guarding by females in a laboratory colony of *T. digitatum* derived from cacao in Trinidad but did not comment on the potential ecological significance of brooding. Egg guarding is maintained in nature for the entire period from oviposition until larvae hatch, ca 26–30 hr at 26–29°C (Lounibos and Machado-Allison 1983). Our preliminary speculation that brooding might protect eggs from predation or desiccation in cacao husks has not been substantiated.

Eggs of *T. digitatum* float vertically, the hydrofuge lower third submerged (Figure 11.2). The upper two-thirds of the egg are hydrophobic, the surface

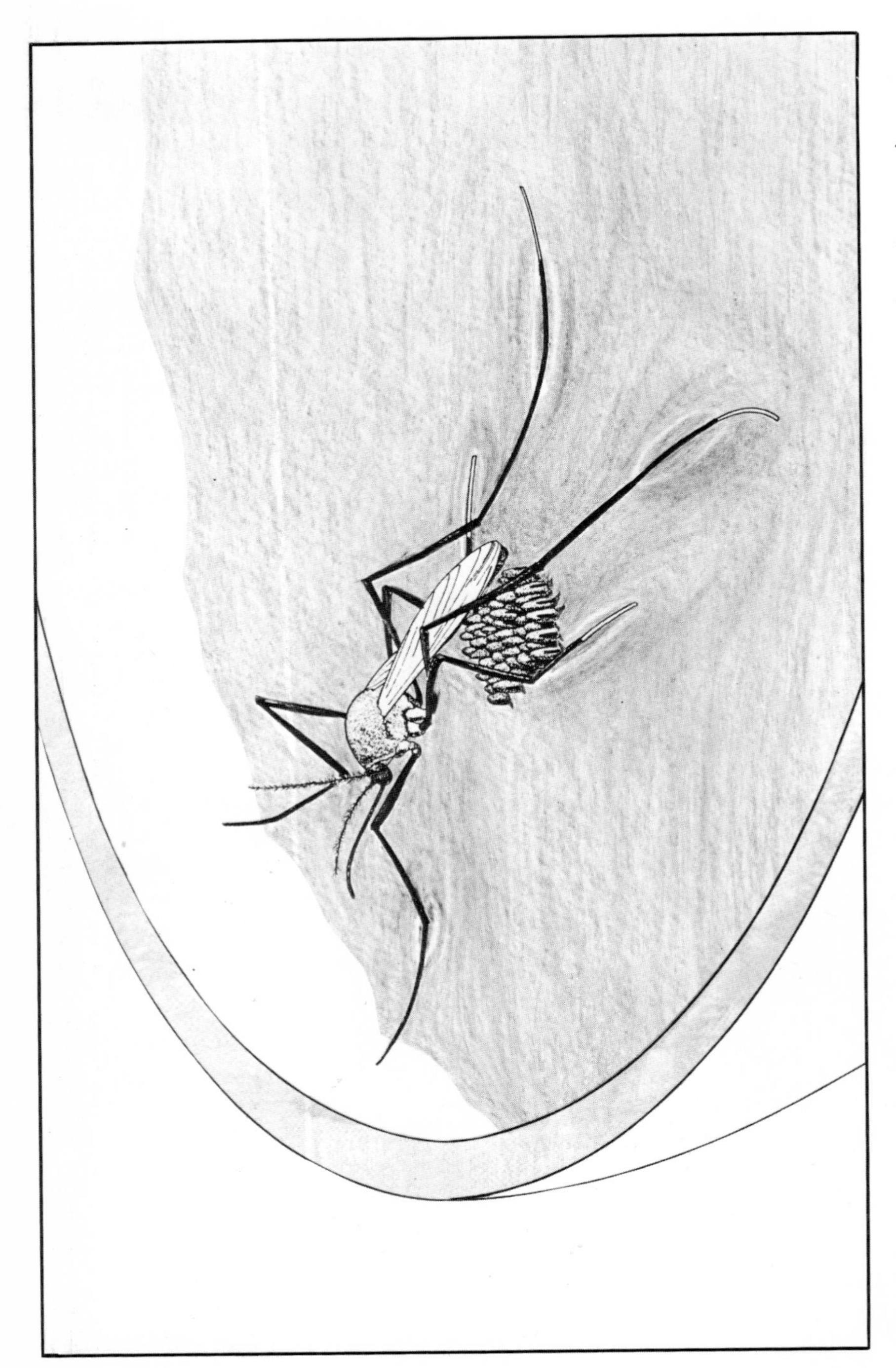

Figure 11.2. A typical posture, drawn from a photograph, of a female *T. digitatum* brooding an egg raft on the water in a cacao husk. The eggs are held between mesothoracic tarsi and connected at vertical "embrasures." Forelegs may be used for holding onto the container wall. Hindlegs may be used to sweep the water surface to recover lost eggs.

being covered with waxy, mushroom-like bodies as well as three "embrasures" (Mattingly 1974) in which arise filamentous hooks that link eggs together in a raft (Lounibos and Linley, unpublished). The loose, polygonal array of eggs in a raft is maintained both by the attachments at embrasures and the tarsi of the mesothoracic legs of brooding females (Figure 11.2).

Several lines of evidence support the conclusion that egg brooding by *T. digitatum* involves no individual recognition of eggs. If a disturbed female abandons a raft, we have observed that her eggs may be incorporated into the raft of, and brooded by, another female in the same cacao husk. Stray eggs are incorporated into preexisting rafts with a sweeping motion of a brooding female's long metathoracic legs (Figure 11.2). Females have also been observed adding to their rafts eggs of *Culex corniger* Theobald, the only other mosquito species that deposits its eggs in cacao husks at Panaquire (Machado-Allison et al. 1985).

The period of egg brooding, slightly longer than 1 day in nature, is apparently governed by a "clock" internal to the female and is not regulated by, although usually synchronous with, the hatching of her larvae. We reached this conclusion from a series of experiments in which brooding females were deceived by replacment of their eggs with another raft at a different stage of embryonic development. Brooding females given eggs approximately 12 hr older than their own remained guarding these rafts ca 12 hr after larvae had hatched. Conversely, females given eggs younger than their own abandoned rafts before foster eggs had hatched, usually flying away about 24 hr after the onset of brooding (Lounibos, unpublished).

Survival Value of Brooding Behavior

Field Experiments

During July (1983), customarily the wettest month of the year at Panaquire (Machado-Allison et al. 1983), the survivorship of brooded and unbrooded rafts in cacao husks was compared. Rafts ca 2 hr old were collected in husks set as oviposition traps. A female on her raft was transferred in a teaspoon to a fresh husk with rainwater. A companion husk of the same size and "cut" was seeded with an unbrooded raft. The husks of each pair were placed within 0.3 m of one another in a discard pile and examined four times in the following 24 hr. Thirty hours after placement, the experimental husks were examined for the presence or absence of newly hatched first instar larvae. The experiment was repeated four times between July 18–29, 9–16 paired comparisons per occasion (Table 11.1).

For every experimental series, two to five comparisons were removed from consideration either because (a) a brooding female had left her raft, or (b) an unbrooded raft had assumed a "nursemaid" during the 24-hr surveillance period, or (c) one of the experimental husks had leaked its fluid contents. Among ac-

Table 11.1 Egg survival in brooded versus unbrooded cacao husks

Experiment date (1983)	Weather	No. accepted pairs	Husks with larvae[a]		p (Fisher's Exact test[b])
			Brooded	Unbrooded	
VII/18–19	Dry	11	11	9	0.4762
VII/22–23	Rain	10	10	3	0.0031
VII/26–27	Dry	5	5	4	1.0000
VII/28–29	Dry	13	13	13	1.0000
	Totals	39	39	29	0.0101

[a]Entries are numbers of successful clutches.
[b]Test of null hypothesis that egg survivorship is independent of brooding.

ceptable comparisons, survivorship of brooded rafts was significantly greater than that of unbrooded rafts during only one of four trial periods (Table 11.1). This single instance of differential survival coincided with a 24-hr interval of intense rainfall. During this trial seven out of ten unbrooded rafts were washed out of husks within the first 10 hr after placement, but corresponding brooded rafts were maintained by females through the duration of the storm. There was no evidence of predation on rafts in any observations of brooded or unbrooded eggs.

We assume that eggs washed out of husks perish because the porous soils at Panaquire rarely support standing water for more than a few days even in the rainy season. In a survey of all known collection records of immatures of *T. digitatum,* Zavortink et al. (1983) reported no incidences of this species from ground pools.

Of the 39 acceptable paired comparisons, 20 were in open "boat" husks and 19 in the less-exposed "crowns". All 10 pairs in which only the brooded husk yielded newly hatched larvae were boats. The probability that survival of unbrooded rafts is independent of husk type is 0.0013 (Fisher's exact test, two-tailed). Evidently, brooding behavior confers a greater survival benefit in more exposed containers.

Laboratory Experiments

To observe the responses of brooding females and rafts to water droplets falling onto their containers, we created mild rain storms in the laboratory. "Rain" passed through holes 0.4–0.8 mm in diameter drilled in a container suspended 18 cm over an artificial habitat. The catchment container was a square, plastic vessel, of 1000 cc capacity, which on one side was cut away 2 cm from the top, providing a spillway for overflow. At the initiation of each observation, water was added to approach but not exceed the level of the spillway, and an unbrooded and brooded raft were placed side by side on the water surface. Females and eggs were derived from a laboratory colony maintained in an outdoor cage with minor modifications of the husbandry methods of Aitken et al.

(1968). Drop rate was measured with a stop watch. Ambient temperatures during the periods were 18 ± 2°C.

Fourteen, 15-min observations were made at the most intense drip rate; five of these were discarded because the brooding female flew away before either raft was affected by rainfall. In the remaining nine comparisons, the unbrooded raft was the first to be swept out of the container twice as often as the brooded raft (Table 11.2). However, in seven of these nine trials, the brooded raft was also swept out within 15 min; in all cases the female maintained her position atop the raft even after it landed on the ground below.

In subsequent trials, drip intensity was reduced to ca 30–40% of the first trial. In these two series the unbrooded raft was the first to be washed out in 94% of the trials (Table 11.2). At the lower drop rates, a brooding female abandoned her raft only once and was swept out with it five times in 20 observations. We observed several instances of brooding females holding on to the sides of containers with their prothoracic legs while rain disturbed the water surface; this behavior secured the female and her raft from being washed out of the container.

These laboratory experiments, although providing some support for the protective value of brooding behavior against rainfall, are poor simulations of conditions beneath the under story of the forest canopy where cacao husks provide habitat to *T. digitatum*. The kinetic energy released on impact of a rain drop is a product of droplet mass and velocity (Laws 1941), both of which are modified in complex ways by throughfall from the canopy and driptips (e.g., Williamson 1981) on understory vegetation.

Rainfall, Egg Mortality, and the Evolution of Brooding

Perhaps because rainfall is usually associated with increased mosquito abundance, little attention has been paid to mosquito mortality caused by rain. Careful studies have shown, however, that rainfall is responsible for severe egg mortality among species inhabiting phytotelmata in the tropics and subtropics. In both Kenya and the Central African Republic, *Aedes simpsoni* (Theobald)

Table 11.2. Egg raft survival in artificial rainfall

Series	Droplet rate (drops/sec; $x \pm$ SD)	First rafts washed out[a]		
		Brooded	Unbrooded	
I	1.02 ± 0.25	3	6	
II	0.39 ± 0.17	0	7	
III	0.29 ± 0.11	1	8	
	Totals	4	21	$p < 0.001$ (sign test)

Protocol is described in text.
[a]Includes only observations in which one or both rafts were swept out.

sensu lato suffers population depressions during rainy periods because intense precipitation washes eggs out of their preferred habitat, the axils of banana leaves (Teesdale 1957; Pajot 1976). Eggs of *Wyeomyia vanduzeei* Dyar and Knab float on the surface of water held in the axils of *Tillandsia* spp. bromeliads and are readily washed out of their epiphytic habitats by subtropical rainstorms common in southern Florida (Frank and Curtis 1977a). The buoyancy of *W. vanduzeei* eggs is in part attributable to a waxy, tuberculated outer coating (Frank et al. 1981) analogous to the hydrofuge layer on *T. digitatum* eggs. Rainfall also washed floatable eggs of *Toxorhynchites amboinensis* (Doleschall) out of man-made containers in an urban release program in coastal Louisiana (Focks and Sackett 1985), and we have documented egg mortality of *Toxorhynchites haemorrhoidalis* (Fabricius) in *Heliconia* bracts during rain storms in Panaquire, Venezuela.

These records indicate that mosquito species that lay floatable eggs on or above the waterline in natural or man-made containers in the tropics or subtropics are at special risk to egg mortality due to rainfall. We suggest that egg brooding by *T. digitatum* may have evolved as one solution to this hazard. The current geographical distribution of *T. digitatum* (Zavortink et al. 1983) conforms strikingly with the distribution of lowland rainforest in the neotropics (Richards 1966). The lesser-known species of this genus are also confined to rain-forested areas in Central and South America.

Trichoprosopon digitatum is the only described species of the tribe Sabethini, composed of more that 350 spp., that lays its eggs in a raft. Two undescribed members of the *T. digitatum* complex, however, also form a small raft (Mattingly 1974; Zavortink, personal communication). Although nothing is known about maternal behavior in these undescribed relatives, raft formation in the *T. digitatum* complex is an obvious precursor to the evolution of brooding behavior.

We suggest that the evolutionary "success" of *T. digitatum*, which includes the exploitation of diverse, abundant, yet potentially perilous phytotelmata, is attributable in part to the incorporation of egg brooding into its life cycle. Selection for brooding would have been greatest in open containers where the risk of egg loss was highest.

Although cacao husks are the primary habitats of *T. digitatum* at our study site, we have also observed brooding by this species in *Heliconia* bracts and rain-filled calabashes of *Crescentia cujete* with as many as 21 females simultaneously guarding rafts in a large shell of the latter. There is no evidence that egg brooding evolved to exploit man-made habitats, even though a large proportion of the breeding sites noted for *T. digitatum* (Zavortink et al. 1983) have been directly or indirectly created by man. Rather, brooding behavior was probably an evolutionary preadaptation, effective in natural phytotelmata, that facilitated successful colonization of man-made containers.

To assess how strong a selective pressure might be rainfall-induced egg mortality, we calculated the expected reduction in net reproductive rate (R_0) for *T. digitatum,* considering the possibilities that rainfalls of two intensities, 2.5 or 5.0 cm per day, washed 70% of unbrooded eggs from their habitats. We examined daily rainfall records from Gatun, Panama (average annual precipitation = 2600 mm) from 1979–1982 and calculated the mean annual and wet-season (May-August) frequencies of rainfall >2.5 cm/day and >5.0 cm/day. R_0 was calculated = $\Sigma l_x m_x$, where x defined the pivotal age of three developmental periods: oviposition to hatching, hatching to pupation, and pupation to first reproduction. Survivorship during the first period was assumed to be 100% in the absence of rain. Larval survivorship was estimated as 88% from the midpoint of the regression in Figure 11.1. As no data were available on survivorship of adult *T. digitatum,* we used an average daily survival estimate of 77% based on data in Frank and Curtis (1977b) on *W. vanduzeei,* the only neotropical sabethine for which survival estimates in nature are available. Finally, we assumed that a female reached reproductive age 6 days after adult eclosion.

If a rainfall of 2.5 cm washed 70% of unbrooded eggs from their habitats, R_0 would be reduced from 8.6% to 15.0% per generation, depending on the use of annual or seasonal rainfall frequency data (Table 11.3). If 5.0 cm are required to effect 70% egg mortality, R_0 would be reduced from 2.7% to 6.2% per generation. In spite of obvious limitations, we regard these estimates as a useful start towards quantifying the selection differential that may have led to the incorporation of egg brooding into the *T. digitatum* life cycle.

Mosquito Maternity in the Old World Genus *Armigeres*

Convergent evolution of behavioral traits in tropical American, African, and Asian culicid inhabitants of phytotelmata has been noted previously (Lounibos 1983). Certain species of the Asian genus *Armigeres* glue their eggs to their hind tarsi, and this behavior and subsequent attendance by the female may be analagous to egg guarding in *T. digitatum.* Strickland (1917), the first observer of eggs glued to the tarsi of *Armigeres,* concluded that females used their legs to deposit eggs in such habitats as water-filled bamboos with tiny access holes, which would preclude ordinary oviposition. MacDonald (1960) disagreed with Strickland's conclusion because some *Armigeres* that oviposit in open microhabitats also glue their eggs to their legs. In observations of a laboratory colony of *Armigeres annulitarsis* (Leicester), Bailey et al. (1975) noted that females after oviposition rested with their legs in contact with the water surface for approximately 2 days, until the eggs hatched.

Although egg attendance by *Armigeres* has not yet been confirmed in nature,

Table 11.3. Hypothetical reductions (%) in R_0 if 70% of eggs are lost due to rainfalls of two intensities

	Rainfall frequency (Gatun, Panama 1979–1982)		Percent reduction in R_0	
	>2.5 cm/day	>5.0 cm/day	>2.5 cm/day	>5.0 cm/day
Annual	0.123	0.039	8.6	2.7
Wet season	0.210	0.089	15.0	6.2

it seems likely that oviposition on the tarsi evolved to protect eggs from some environmental peril. Egg parasites have been recorded from *Armigeres*, and Mattingly (1972a) originally postulated that attending females may protect their embryonic progeny from tiny, hymenopterous parasitoids. Subsequently, Mattingly (1972b) recognized the egg parasite as a fungus. On the other hand, glueing of eggs to a female's legs would also afford security from rainfall in the phytotelm habitats that *Armigeres* frequent.

Summary

Females of *Trichoprosopon digitatum* brood their egg rafts from oviposition until hatch, a period of ca 30 hr in natural habitats. Field and laboratory experiments demonstrate that guarding females can prevent eggs from being washed from container habitats during rainfall. Egg mortality due to rainfall is significant for subtropical and tropical mosquitoes that lay floatable eggs in phytotelmata. The absence of brooding could reduce, according to our estimates, the net reproductive rate of *T. digitatum* by 3–15%. By any account, *T. digitatum* is among the most ecologically "successful" of phytotelmata inhabiting tropical mosquitoes. Egg brooding, we suggest, is partly responsible and has allowed this species to exploit open containers without incurring reduced survivorship of eggs. Egg brooding apparently evolved independently in the tropical Asian genus *Armigeres* whose females glue eggs to their hind legs.

Acknowledgments We are grateful to Jack Petersen for supplying the rainfall data, to Ramón Matos for hospitality at the cacao plantation, to Jim Newman for drawing Figure 11.2, and Howard Frank, John Linley, Tom Zavortink, and contributors to this volume for criticisms of a draft of this paper. Research was supported by NSF INT-8212581 of the USA (LPL) and CONICIT S1-1332 of Venezuela (CM-A). This is U. of Florida IFAS Journal Series No. 6590.

References

Aitken, T.H.G., Hingwan, J.O., Manuel R., Hosein, H.: Laboratory colonization of *Trichoprosopon digitatum* (Rondani) (Diptera: Culicidae). Mosq. News 28, 445–454 (1968).

Bailey, C.W., Vongpradist, S., Mongkolpanya, K.: Observations of an unusual oviposition habit in *Armigeres (Leicesteria) annulitarsis* (Leicester) Mosq. News 35, 564–565 (1975).

Bro Larsen, E.: On subsocial beetles from the salt-marsh, their care of progeny and adaptation to salt and tide. In: Transactions of the 11th International Congress of Entomology, 1952, pp. 502–506.

Focks, D.A., Sackett, S.R.: Some factors affecting interaction of *Toxorhynchites amboiensis* with *Aedes* and *Culex* in an urban environment. In: Ecology of Mosquitoes: Proceedings of a Workshop. Lounibos, L.P., Rey, J.R., Frank, J.H. (eds.). Vero Beach, Florida: Florida Medical Entomology Laboratory, 1985, pp 55–64.

Frank, J.H., Curtis, G.A.: On the bionomics of bromeliad-inhabiting mosquitoes. IV. Egg mortality of *Wyeomyia vanduzeei* caused by rainfall. Mosq. News 37, 239–245 (1977a).

Frank, J.H., Curtis, G.A.: On the bionomics of bromeliad-inhabiting mosquitoes. V. A mark-release-recapture technique for estimation of population size of *Wyeomyia vanduzeei*. Mosq. News 37, 444–452 (1977b).

Frank, J.H., Curtis, G.A., Erdos, G.W., Ellis, E.A.: On the bionomics of bromeliad-inhabiting mosquitoes. VIII. The flotational structure of *Wyeomyia vanduzeei* eggs (Diptera: Culicidae). J. Med. Entomol. 18, 337–340 (1981).

Hinton, H.E.: The Biology of Insect Eggs. Oxford: Pergamon Press, Vol. 1, 1981.

Howard, L.O., Dyar, H.G., Knab, F.: The Mosquitoes of North and Central America and the West Indies. Vol. 3, Systematic Description. Washington: Carnegie Inst., 1915.

Knight, K.L., Stone, A.: A Catalog of the Mosquitoes of the World. Baltimore: Thomas Say Foundation, 1977.

Lamb, R.J.: Parental behavior in the Dermaptera with special references to *Forficula auricularia* (Dermaptera: Forficulidae). Can. Entomol. 108, 609–619 (1976).

Laws, J.O.: Measurement of fall velocity of water drops and rain drops. Trans. Am. Geophys. Union 22, 709–721 (1941).

Lounibos, L.P.: Behavioral convergences among fruit-husk mosquitoes. Fla. Entomol. 66, 32–41 (1983).

Lounibos, L.P., Machado-Allison, C.E.: Oviposition and egg brooding by the mosquito *Trichoprosopon digitatum* in cacao husks. Ecol. Entomol. 8, 475–478 (1983).

Lounibos, L.P., Machado-Allison, C.E.: Female brooding protects mosquito eggs from rainfall. Biotropica 18 (in press).

MacDonald, W.W.: Malaysian parasites, XXXVIII. On the systematics and ecology of *Armigeres* subgenus *Leicesteria* (Diptera, Culicidae). Stud. Inst. Med .Res. Malaya 20, 110–153 (1960).

Machado-Allison, C.E., Rodriguez, D.J., Barrera-R.,R. Gomez-Cova, C.: The insect community associated with inflorescences of *Heliconia caribaea* Lamarck in Venezuela. In: Phytotelmata: Terrestrial Plants as Hosts for Aquatic Insect Communities. Frank, J.H., Lounibos, L.P. (eds.). Medford, New Jersey: Plexus, 1983, pp. 247–270.

Machado-Allison, C.E., Barrera-R.,R. Frank, J.H., Delgado, L., Gomez-Cova, C.: Mosquito communities in Venezuelan phytotelmata. In: Ecology of Mosquitoes: Proceedings of a Workshop. Lounibos, L.P., Rey, J.R., Frank, J.H. (eds.). Vero Beach, Florida Medical Entomology Laboratory, 1985, pp. 79–93.

Maguire, B.: Phytotelmata: biota and community structure determination in plant-held waters. Annu. Rev. Ecol. Syst. 2, 439–464 (1971).

Mattingly, P.F.: Mosquito eggs. XVII. Further notes on egg parasitization in the genus *Armigeres*. Mosq. Syst. 4, 1–8 (1972a).

Mattingly, P.F.: Mosquito eggs. XX. Egg parasitism in *Anopheles* with a further note on *Armigeres*. Mosq. Syst. 4, 84–86 (1972b).

Mattingly, P.F.: Mosquito eggs. XXVI. Further descriptions of sabethine eggs. Mosq. Syst. 6, 231–238 (1974).

Pajot, F.-X.: Contribution à l'étude écologique d'*Aedes* (*Stegomyia*) *simpsoni* (Theobald, 1905) (Diptera, Culicidae). Observations concernant les stades préimaginaux. Cah. ORSTOM, Sér. Entomol. Méd. Parasit. 16, 129–150 (1976).

Pawan, J.L.: The oviposition of *Joblotia digitatus* Rondani (Diptera, Culicidae). Insecutor Inscitiae Mens. 10, 63–65 (1922).

Richards, P.W.: The Tropical Rain Forest. London: Cambridge University Press, 2nd ed., 1966.

Seifert, R.P.: Mosquito fauna of *Heliconia aurea*. J. Anim. Ecol. 49, 687–697 (1980).

Seifert, R.P., Barrera-R.,R.: Cohort studies on mosquito (Diptera: Culicidae) larvae living in the water-filled floral bracts of *Heliconia aurea* (Zingiberales: Musaceae). Ecol. Entomol. 6, 191–197 (1981).

Strickland, C.: A curious adaptation of habit to its environment of a Malayan mosquito. J. Straits Br. Roy. Asiatic Soc. 75, 39 (1917).

Tallamy, D.W.: Insect parental care. Bioscience 34, 20–24 (1984).

Tallamy, D.W., Denno, R.F.: Maternal care in *Gargaphia solani* (Hemiptera: Tingidae). Anim. Behav. 29, 771–778 (1981).

Teesdale, C.: The genus *Musa* Linn. and its role in the breeding of *Aedes* (*Stegomyia) simpsoni* (Theo.) on the Kenya coast. Bull. Entomol. Res. 48, 251–260 (1957).

Williamson, G.B.: Driptips and splash erosion. Biotropica 13, 228–231 (1981).

Wilson, E.O.: The Insect Societies. Cambridge, Mass.: Belknap Press of Harvard University, 1971.

Zavortink, T.J.: Species complexes in the genus *Trichoprosopon*. Mosq. Syst. 13, 82–85 (1981).

Zavortink, T.J., Roberts, D.R., Hoch, A.L.: *Trichoprosopon digitatum*—Morphology, biology, and potential medical importance. Mosq. Syst. 15, 141–149 (1983).

Part III

Mechanisms of Insect Life Cycle Evolution

Chapter 12
The Evolution of Insect Life Cycle Syndromes

HUGH DINGLE

Insect life cycle syndromes are suites of traits that covary and function together. These suites involve four main physiological or behavioral elements: reproduction, growth and development, dormancy, and migration (Tauber and Tauber 1981). Each must be synchronized with appropriate seasonal and habitat conditions, and the behavior and physiology ensuring breeding at the proper time and place are important in conferring the requisite flexibility (Dingle 1984). Flexibility can and does occur in all phases of the life cycle, and I shall explore here some of the ways it does so. But the notion of a syndrome of co-functioning traits suggests that changes at one stage of a life cycle or in one trait may have consequences at other stages and in other traits. If so, important evolutionary questions concern both which characters are coupled or most intimately interwoven and which mechanisms promote the associations. For evolution to act on syndromes, there must obviously be genetic variation for individual traits and genetic covariation among traits. I shall, therefore, be concerned here with both genotypes and phenotypes.

Because of their own diversity and the diversity of the habitats they occupy, insects encounter an extraordinary variety of seasonal and biotic conditions and must modify life cycle syndromes accordingly. A major problem in dealing with these conditions is timing, and I shall begin with a brief discussion of life cycle timing, some consequences for more traditional views of life history evolution, and an example of phenotypic flexibility of response. I then consider some difficulties insects face because environments are usually uncertain, and the consequences of this uncertainty especially for the genetic architecture of life histories. I shall also consider how some species can influence the responses of their progeny through maternal influences and protect their progeny from biotic factors such as predators, and finally examine briefly the possible constraints placed on life history evolution by previous phylogenetic history. It should be apparent that a complex of interacting factors influences life history traits and the variation and covariation among them. The resultant intricacies of pattern in life cycle evolution are thus to be expected. I shall try to indicate throughout profitable approaches evolutionary biologists might use to sort out the elements of these patterns.

Timing and Developmental Flexibility

The problem of timing in life histories has been considered in an interesting way by Taylor (1980a,b and this volume). He notes that at the end of a particular period of population growth, say a temperate zone summer, an insect population may be confined to a relatively narrow age interval which he calls the "critical interval." With respect to diapause, for example, the critical interval includes not only the diapausing stage but also all ages that can make it into this stage before the onset of winter. Similarly, winged migrants would have to reach a stage capable of reaching adulthood before winter or other form of habitat deterioration. Taylor then analyzes the consequences of the critical interval for the life history traits of a model insect whose initial age distribution, survivorship, and maternity functions are specified and are reasonable approximations for those of a "typical" seasonal insect.

Several consequences for the life cycle are apparent from the analysis. Perhaps the most interesting one concerns generation length. In contrast to the general notion in population biology, based on exponential growth in deterministic environments, that a decrease in generation length *increases* fitness when other traits remain constant (Cole 1954; Lewontin 1965; Stearns 1976), in environments of limited duration a decrease in generation length can *decrease* fitness. This is illustrated in Figure 12.1 in which the critical interval is defined as the last 10 units of the immature period. The age at first reproduction (α) determines the generation time, and in Figure 12.1 is shown the influence of α on Taylor's model insect when the population must enter the critical period no later than 90 time units after the start of the growing season (i.e., they must diapause by 100 units to survive the winter). A local maximum in fitness is reached at α of approximately 30 units, and fitness decreases at lower and higher values of α, quite contrary to predictions derived from simple exponential growth. At still lower values of α (<26 units), a surge of population growth enters the critical interval, and fitness increases rapidly. It is quite possible, however, that because of physiological or phylogenetic constraints (see below), the 4-unit interval may prove an insurmountable barrier with the result that selection could act to increase α when it falls below 30 (arrow in Figure 12.1). A somewhat similar model for the evolution of patterns of voltinism has been developed by Roff (1983).

A good example of how a real insect may adjust its life cycle to meet the requirements of a critical interval is provided by the lawn cricket *Allonemobius fasciatus*, studied by Tanaka in Oregon (1983; Tanaka and Brookes 1983). This species spans a wide range of altitudes (and latitudes) and was studied by Tanaka at two sites 76 m and 1100 m above sea level, respectively. The life cycles at these two altitudes are indicated in Figure 12.2; in both cases overwintering occurs in diapause in the egg stage. How then does the cricket arrive at the

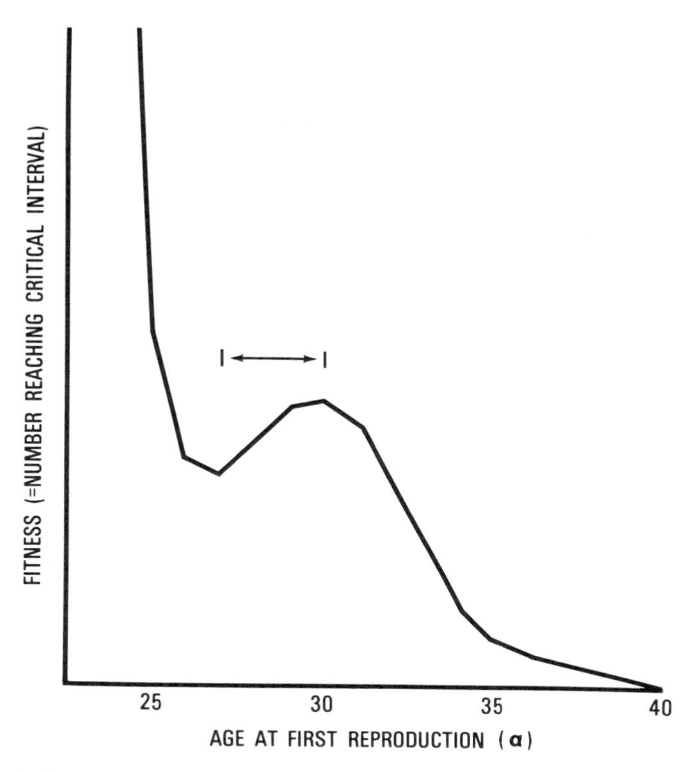

Figure 12.1 Influence of age at first reproduction on fitness in an insect that must reach a "critical interval" at the appropriate developmental stage in order to enter diapause before the onset of winter. Note that in this example a local maximum in fitness occurs at $\alpha = 30$ and that in the interval indicated by the arrow fitness actually declines with earlier ages of first reproduction. (After Taylor 1980b.)

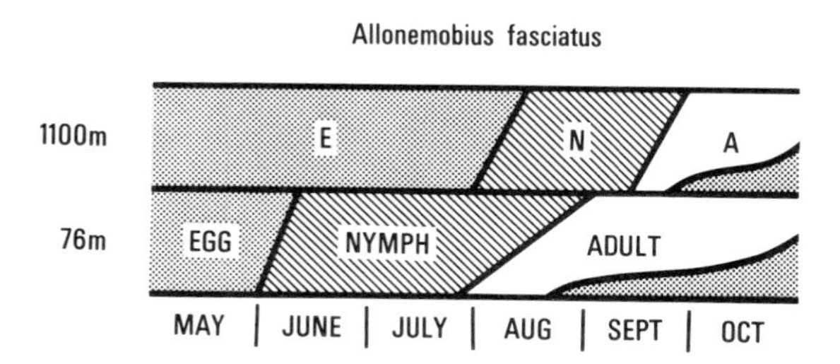

Figure 12.2 Timing in life cycles of high (1100 m)- and low (76 m)-altitude populations of the lawn cricket, *Allonemobius fasciatus*, in Oregon. (After Tanaka 1983.)

appropriate diapausing egg when the lengths of the growing seasons and the life cycle patterns displayed are so different? The answer turns out to be the remarkable developmental flexibility inherent in the *A. fasciatus* life history.

This flexibility occurs in all three stages of the life cycle: egg, nymph, and adult. Females, for example, influence the intensity of diapause through maternal influences (see also below). Those reared at low temperature (20°C), as would occur in late September or October, lay eggs that display direct development of the embryo until the stage when it enters winter diapause. The intensity of this diapause is greater in eggs laid earlier in the season, when they might pass through transient warm periods that could break low intensity diapause, than it is later. These diapause characteristics occur in virtually all eggs laid by 1100 m crickets and in the later eggs of the low-altitude animals. The females of the 76 m crickets are initially subjected to higher temperature, and at 30°C there was a stage of developmental suppression in their eggs resulting in a summer diapause. These eggs later resumed development until the embryos reached the stage where they entered winter diapause which thus represents a second egg diapause for these individuals. Eggs from both high and low altitudes can exhibit both kinds of diapause so the differences are a phenotypic response and are not due, at least in any major way, to gene differences between the two populations. The net result is that eggs reach the critical interval in the right stage no matter when laid.

Similarly the nymphs at the two altitudes differ in their developmental timing. At low altitude they occur early in the season, and the problem is to avoid reaching adulthood and the reproductive stage too soon when high temperatures might disrupt diapause in the eggs produced. The opposite problem is faced by high-altitude nymphs. The solution in both cases is differential response to photoperiod (Figure 12.3). Under an LD 14:10 cycle, as would occur when high-altitude nymphs are present, development is most rapid. On longer days, when low-altitude nymphs are present, development takes longer. The result is that there is much more temporal overlap in the adults (early September onwards) than there is in the nymphs (August only and only late low altitude nymphs), and so there is greater synchrony of reproduction (Figure 12.2). These responses are not without trade-offs, however, as the more slowly developing low-altitude nymphs pass through more instars and are larger. The influences these latter differences may have on reproduction or fecundity have yet to be explored.

Phenotypic and genotypic variation in diapause is known for a number of insects, and several examples are given in Dingle (1978a) and Brown and Hodek (1983). There are also other cases of variation in developmental rate, usually as a function of density, perhaps best known in various cockroaches (Willis et al. 1958; Wharton et al. 1967; Woodhead and Paulson 1983), locusts and grasshoppers (Kennedy 1961; Uvarov 1966; Dingle and Haskell 1967; Nolte 1974), and some lepidopterans (Iwao 1962, 1968). In the cockroach, *Blatella germanica*, selection experiments for high and low instar number suggest that gene influences may also contribute (Kunkel 1981). Suffice it to say, it is apparent that insects have an array of mechanisms available to them to arrive at the critical interval at the proper time and to meet other problems of timing imposed by

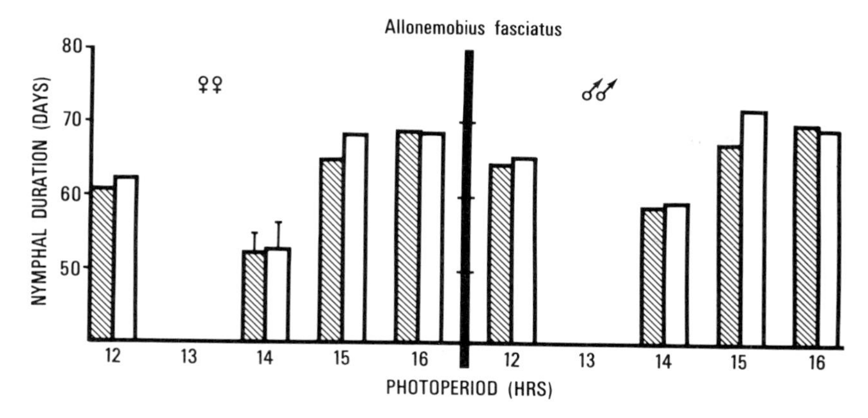

Figure 12.3 Nymphal duration as a function of photoperiod in high-altitude (shaded bars) and low altitude (open bars) populations of the lawn cricket, *Allonemobius fasciatus*. Standard deviations are indicated for females at 14:10 and were similar across all conditions. No significant differences between populations occur. High-altitude nymphs occur in the field under 14:10 photoperiods and so develop rapidly; low-altitude nymphs occur when days are longer and so develop more slowly (After Tanaka 1983.)

the environment. A set of problems largely for future inquiry concerns which mechanisms are most likely to be matched with which circumstances.

Environmental Uncertainty

Solutions to timing in insect life histories are often complicated by the fact that environments are uncertain, and individuals cannot respond in a way guaranteed to maximize fitness (Istock 1978, 1982, 1984; Dingle 1984). Even photoperiod, the most reliable of the available cues, may not be predictably associated with the variables such as temperature, which control physiology and development (Hoffmann 1978). Two examples should serve to illustrate the nature of the resulting conundrum faced by an insect unable to forecast future conditions with accuracy.

The first case is that of the milkweed leaf beetle, *Labidomera clivicollis*, studied by Palmer (1982) in central Texas. This insect oversummers in adult summer diapause and initiates reproduction with the onset of autumnal rains; egg laying occurs from late August through most of October. The early eggs mature to adults that reproduce and whose offspring reach the critical interval in time to enter winter diapause cued by photoperiod. The late eggs oviposited in October mature into adults that proceed directly to winter diapause without reproducing. The uncertainty comes for the adults maturing from the eggs produced during September. In about half the years autumn conditions are such that reproduction will lead to a third generation of adults, which subsequently enter winter diapause (as do the offspring of reproducing adults from early eggs). In the remaining years, however, conditions are not favorable for reproduction, and the best course is to enter diapause without producing a third generation. The result of these unpredictable selection pressures is a high variance in the tendency of the September egg adults to respond to short days and enter diapause.

A somewhat similar situation is faced by the milkweed bug, *Oncopeltus fasciatus,* in California. In years of mild winters, bugs continuing reproduction into October, or perhaps even November, may succeed in producing a complete additional generation, for nymphs can complete development almost all winter long (personal observation). In other winters, however, there may be one or more frosts of sufficient severity to kill the nymphs (although the adults can usually survive), and diapause in response to short days is the best option. This uncertainty results in extreme variability in the diapause response of California *O. fasciatus* (Evans 1982; and Figure 12.4). Even in the American Midwest, where *O. fasciatus* populations cannot overwinter but must diapause and migrate to survive (Dingle 1978b), unpredictability in the onset of severe frost can contribute to a variable response. Both flight and diapause are genetically variable with apparently high additive (polygenic) components (Dingle 1968; Dingle et al. 1977). The unpredictably fluctuating action of selection and population mixing from migration are the most likely contributors to this genetic flexibility (Dingle et al. 1977; see also Istock 1978, 1981, 1984).

Genetic Architecture

The demonstration of genetic variability for life cycle traits such as diapause and migration raises an important issue in the study of life history evolution; namely, to what extent is the commonly observed phenotypic variability genetically based? The question concerns both genetic variation for individual characters and the patterns of genetic covariation among traits contributing to particular life cycle syndromes.

We (Hegmann and Dingle 1982; Dingle 1984) have tried to address this problem by analyzing the migratory midwestern population of *Oncopeltus fasciatus.* We used the quantitative genetic technique of sibling analysis (Falconer 1981) in which each member of one sex (males in this case) are mated with several members of the other sex. This produces both full-sib and half-sib families and allows the partitioning of genetic variance into its additive and other components and the determination of additive genetic correlations. We found, among others, strong positive genetic correlations among size measurements such as wing length and body length and life table characters such as clutch size, hatchability of the eggs, and development rate. What was particularly interesting was that we found no correlation between any of these traits and age at first reproduction (α), measured under nondiapausing conditions as the interval from adult eclosion to first oviposition. Age at first reproduction is, of course, important in conferring flexibility in timing, and this lack of genetic correlation with other traits means it is free to vary without altering other aspects of the life cycle. It is as

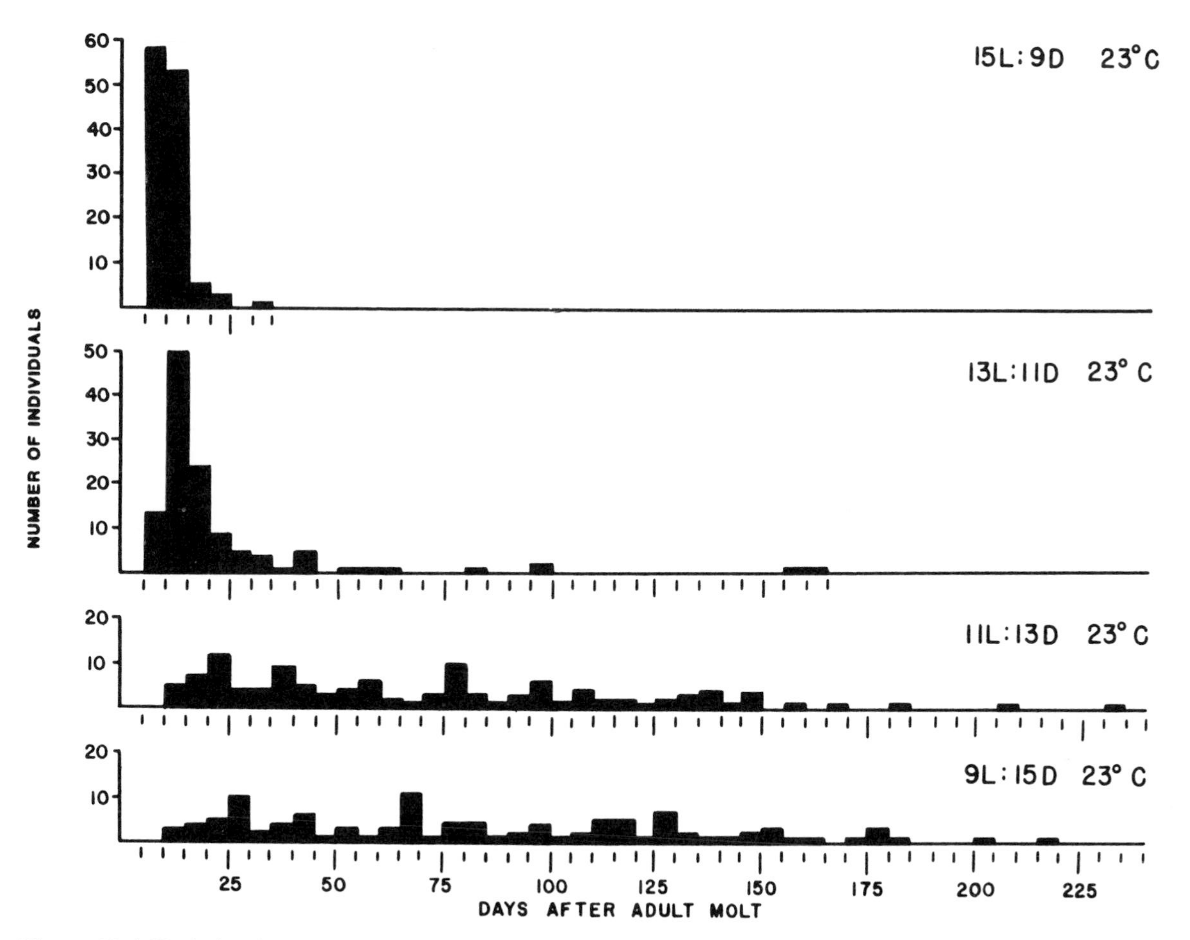

Figure 12.4 Variation in age at first reproduction in a California population of *Oncopeltus fasciatus* as a function of photoperiod. Ordinate indicates number of females reproducing at given ages following adult eclosion. (After Evans 1982.)

if α is free of a "cost of correlation" (Dingle 1984). The absence of genetic correlation of α with wing length and the presence of positive correlations of other traits, including flight duration, with wing length was confirmed by Palmer (1984; Dingle et al. 1984) in an experiment involving artificial selection. Derr (1980) similarly found no correlation of α (her τ) with other life history characters in the cotton stainer bug, *Dysdercus bimaculatus*. Flexible timing is thus maintained in life cycles since the set of genetically correlated traits does not constrain an important aspect of timing in the form of age at first reproduction (see Dingle 1984 for detailed discussion).

Genetic correlations in life history traits involving timing and reproductive output have also been noted in other insects. Istock (1978; Istock et al. 1976) selected for fast and slow development and for diapause in the pitcher plant mosquito, *Wyeomyia smithii*. He found a very high correlation of fast development with nondiapause and of slow development with diapause, probably because slow development increases the period of sensitivity to photoperiod (Saunders and Bradley 1984). In the latter case individuals avoid "risking" reproduction and survival in the current year, when slow development is likely to take them into a winter thermal regimen. They instead postpone adulthood and reproduction to the following spring. In the fleshfly, *Sarcophaga bullata,* selection for late pupariation resulted in a line that not only pupariated later but also developed more slowly. Further, those in diapause inducing conditions were both more likely to diapause and to remain in diapause longer than the unselected line (Henrich and Denlinger 1982). In the migratory flour beetle, *Tribolium castaneum,* selection for "dispersal" reveals a syndrome involving positive correlations of body size, development rate, egg production and fertility, and egg size with tendency to migrate (Ritte and Lavie 1977; Lavie and Ritte 1978; Wu 1981). This "colonizing syndrome" contrasted with the relatively nonmigratory *T. confusum,* which did not display this pattern of correlations. Different life cycles thus reveal their own genetic correlation structures, presumably functioning as complex adaptations (Frazzetta 1975) for given sets of environmental contingencies.

In many ways the most interesting of these presumptive life cycle adaptations is the alteration of sexual and asexual generations as seen, for example, in aphids. In most aphid life cycles a parthenogenetic summer generation or series of generations functions primarily to populate the summer habitat and spread individuals and their descendants over the available resources. The sexual generation serves to produce zygotic offspring (usually eggs) that survive the vagaries of winter. But Lynch and Gabriel (1983) point out another consequence of alternating phases. During the unisexual phase there may be storage of a great deal of hidden genetic variation in the form of negative correlations between characters determined by linked alleles. When these linkages are broken by sex and recombination, tremendous amounts of genetic variation can be released allowing great flexibility of response to selection and higher rates of per-generation phenotypic evolution than can be attained by obligate bisexuality (Figure 12.5). In the long run this short-term advantage is balanced because recombinants are not produced during the asexual phase, but it could lead to

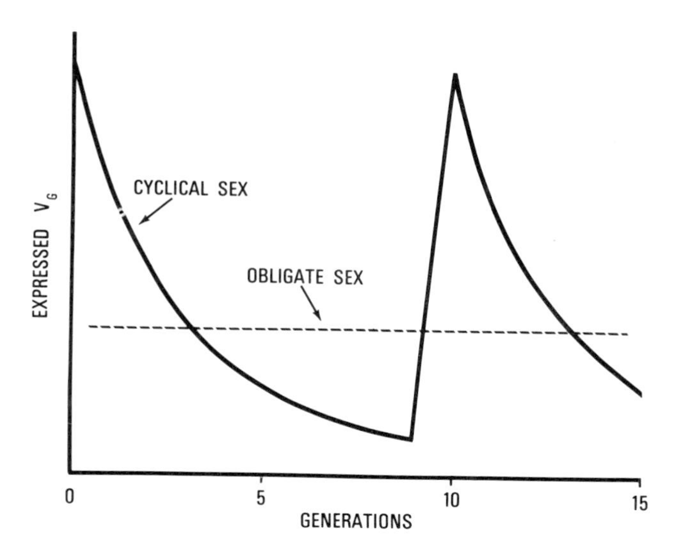

Figure 12.5. Theoretical patterns of expressed genetic variance (V_G) in a species or population with cyclical sex alternating with parthenogenesis as contrasted to a species with obligate sex. Very high levels of genetic variance are exposed to natural selection at periodic intervals. (After Lynch and Gabriel 1983.)

dramatic response to selection in the short run. Studies of *Daphnia* populations qualitatively support these theoretical conclusions (Lynch 1984). It has also been noted by Istock (1978) that sex in general has a direct role in fitness because continuing genetic reassortment and recombination are necessary to produce a "best mixed strategy" for dealing with uncertain environments. This suggests an added role for sex in the life cycles of aphids, for example, that should repay more intensive study.

Maternal Influences

A further aspect of life cycle flexibility in insects is the ability of parents to influence the responses of their offspring environmentally. I have noted with Tanaka's crickets (above) the influence of temperature in the maternal environment on diapause characteristics of the eggs. In many other insects the photoperiod experienced by the mother influences diapause in the offspring (reviewed in Saunders 1982). Similarly crowding, in combination with maternal age, determines the proportion of winged offspring in aphids (MacKay and Wellington 1977). Developmental rates of offspring can be influenced by maternal age in the bug (*Lygaeus kalmii*), with older mothers producing more rapidly growing young (Peter C. Frumhoff, personal communication). In the latter case the adaptive value may be that individuals can thus reach adulthood and diapause in time to increase the probability of winter survival. In *Oncopeltus fasciatus*, offspring of long-day mothers reared under long- and short-day conditions show significantly correlated ages at first reproduction between parents and offspring, whereas offspring of short-day mothers do not (Dingle 1984), possibly because diapause (short-day) mothers tend to produce nondiapausing offspring.

One of the most interesting cases of maternal influences on life histories comes from a study of the collembolan, *Orchesella cincta*, in Holland by Janssen (1984). In this species there are two generations per year, and the winter and summer generations differ with respect to several life history traits. These differences are produced by strong negative maternal effects (Figure 12.6), so that individuals display traits opposite to those of their parents but resembling those of their grandparents. Janssen advances the reasonable hypothesis that these maternal effects are adaptations to a relation between generation time and an alternating seasonal habitat such that grandparents and grandchildren encounter similar environments, whereas parents and offspring encounter dissimilar environments (winter and summer). A given individual thus encounters the environment of its grandparents, not its parents, and the negative maternal effects allow appropriate adjustment to it. Maternal effects would thus play a major role in ecological fine tuning. The genetics of such maternal effects would be a most interesting line of future investigation in insect life cycles. One would

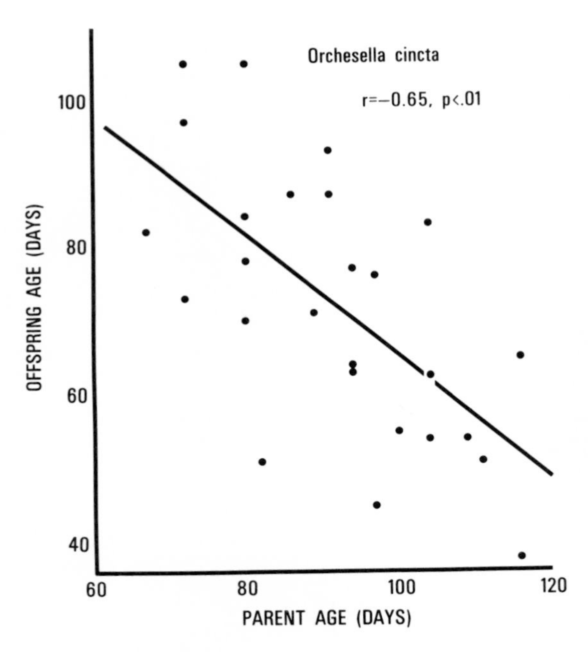

Figure 12.6. Regression of offspring age at maturity (reproduction) on parental age at maturity in the collembolan *Orchesella cincta*. The strong negative correlation results from maternal effects producing different seasonal life history attributes in this bivoltine insect. (After Janssen 1984.)

like to know if natural selection has produced gene differences among populations or species in, say, the intensity of maternal influences.

Oviposition Behavior and Parental Care

The behavior of parent insects with respect to protection of the young may also be an important factor in life history adaptations. One illustrative example occurs in the clutch sizes and oviposition behaviors of females of *Oncopeltus fasciatus* from Iowa and Puerto Rico (Dingle 1981). In Iowa these bugs produce about 25–30 eggs per clutch whereas those from Puerto Rico produce 45–50 eggs. These differences may be due to different causes of egg mortality. In the field Iowa females most often lay their eggs in leaf axils and similar sites, whereas in Puerto Rico the eggs are usually deposited in spider webs or in the chambers formed by the webs of leaf rolling caterpillars. In Iowa the primary source of mortality, as revealed by observation of natural clutches and of clutches placed out in the field, is predation by chrysopid larvae. These larvae search until they find an egg mass and then remain beside it until they have consumed the entire batch. A "bet-hedging" tactic of scattering small egg clutches thus seems most appropriate so that some eggs remain undiscovered. In Puerto Rico predation is much less severe, but jostling by rain drops and wind can cause eggs too loosely packed to fall out of their protective chambers. Experiments with clutches placed out in the field into the appropriate sites demonstrated higher survivorship for clutches of 50 eggs in Puerto Rico and clutches of 25 eggs in Iowa (Dingle 1981).

In the laboratory this difference in oviposition behavior is manifested as a tendency on the part of Puerto Rico females to deposit eggs in crevices rather than in the balls of cotton provided as oviposition sites (Leslie and Dingle 1983). The difference is due to polygenic influences although there has been the evolution of a "dominance deviation" toward the Puerto Rico population. This same tendency is also seen in a number of other life history traits differing between the two populations (Dingle et al. 1982; Leslie, J.F. manuscript in-preparation). *Oncopeltus fasciatus* may have been present in Puerto Rico for less than 500 years, arriving with the introduction of one of its major host plants, *Asclepias curassavica,* by European colonists (Blakley and Dingle 1978). If so, this has been long enough for natural selection to produce the gene differences observed in both clutch size and oviposition behavior.

Perhaps the most effective way to protect offspring from environmental adversity is through direct parental care. (See Lounibos and Machado-Allison in this volume.) This has arisen in several insect groups and is viewed as an important step on the way to eusociality (Wilson 1971; Tallamy 1984). But parental care often has costs in the form of decreased fecundity or survival for the parent so that its evolution requires special preadaptations and strong selection pressures (Tallamy 1984). Nevertheless, where it has arisen, it can be extremely effective in protecting the young. Tallamy (1984; Tallamy and Denno 1981), for example, has demonstrated that in the presence of predators, females of the tingid lace bug, *Gargaphia solani,* more than double the survival of their offspring by guarding them; in the absence of predators there is little difference in the survival of guarded and unguarded young. Again, environmental variation appears to maintain genetic variance for the trait capable of permitting further response to natural selection (Tallamy and Dingle 1984). Some females avoid the costs of parenting by depositing their eggs among those of others who then guard them (Tallamy 1984). An extreme and very interesting form of induced step-parenting occurs in the treehopper, *Publilia reticulata,* studied by Bristow (1983). Here females actually increase both the survivorship of their offspring and their own fecundity (by producing more broods of young) by turning over the care of offspring to attendant ants when the opportunity arises, providing a rather extraordinary example of behavioral and life history flexibility. In the absence of females only 3.1% of the nymphs survive. This is increased to 10.7% with parental care and increases to 27.1 and 42.5%, respectively, in first and second offspring broods when ants are in attendance.

Phylogenetic Constraints

The above discussion has stressed life history flexibility, but is not meant to suggest that in an evolutionary sense insects can in effect do anything they want. Recent authors have emphasized the fact that natural selection does not produce some ideal phenotype but rather does the best with what is available and what is possible given design constraints (Stearns 1976; Gould and Lewontin 1979). The best known design constraints in insects are the exoskeleton and the independent respiratory and circulatory systems, but the point can also be illustrated by examining a life history trait, clutch size, in two groups with which I am familiar, cotton stainer bugs (*Dysdercus*) and milkweed bugs (*Oncopeltus*), as illustrated in Figure 12.7. The cotton stainers generally produce large clutches with the exception of two relatively small species, *D. mimus* and *D. mimulus*. In this group clutch size is a function of body size both within and among species (Derr et al. 1981), and clutches are oviposited every 3–5 days or so. This pattern of producing large clutches every few days seems to be part of an overall life history *Bauplan* of *Dysdercus,* and there are probably limits to how much it could be modified by natural selection independently of body size.

A similar plan seems to be present in the *Oncopeltus* subgenus of the genus *Oncopeltus* represented by *O. varicolor* (large), *O. sexmaculatus* (medium), and *O. cayensis* (small). Again large clutches are produced every few days, and clutch size is a function of body size. By contrast, in the subgenus *Erithrischius* relatively small clutches are generally produced daily and display no relationship to body size even though sizes overlap completely those of the other subgenus. Rather the apparent phylogenetic constraint of producing clutches at frequent intervals limits the number of eggs that can develop in time for each oviposition. A circadian rhythm of oviposition with a narrow daily window may prevent the production of still smaller clutches (Rankin et al. 1972).

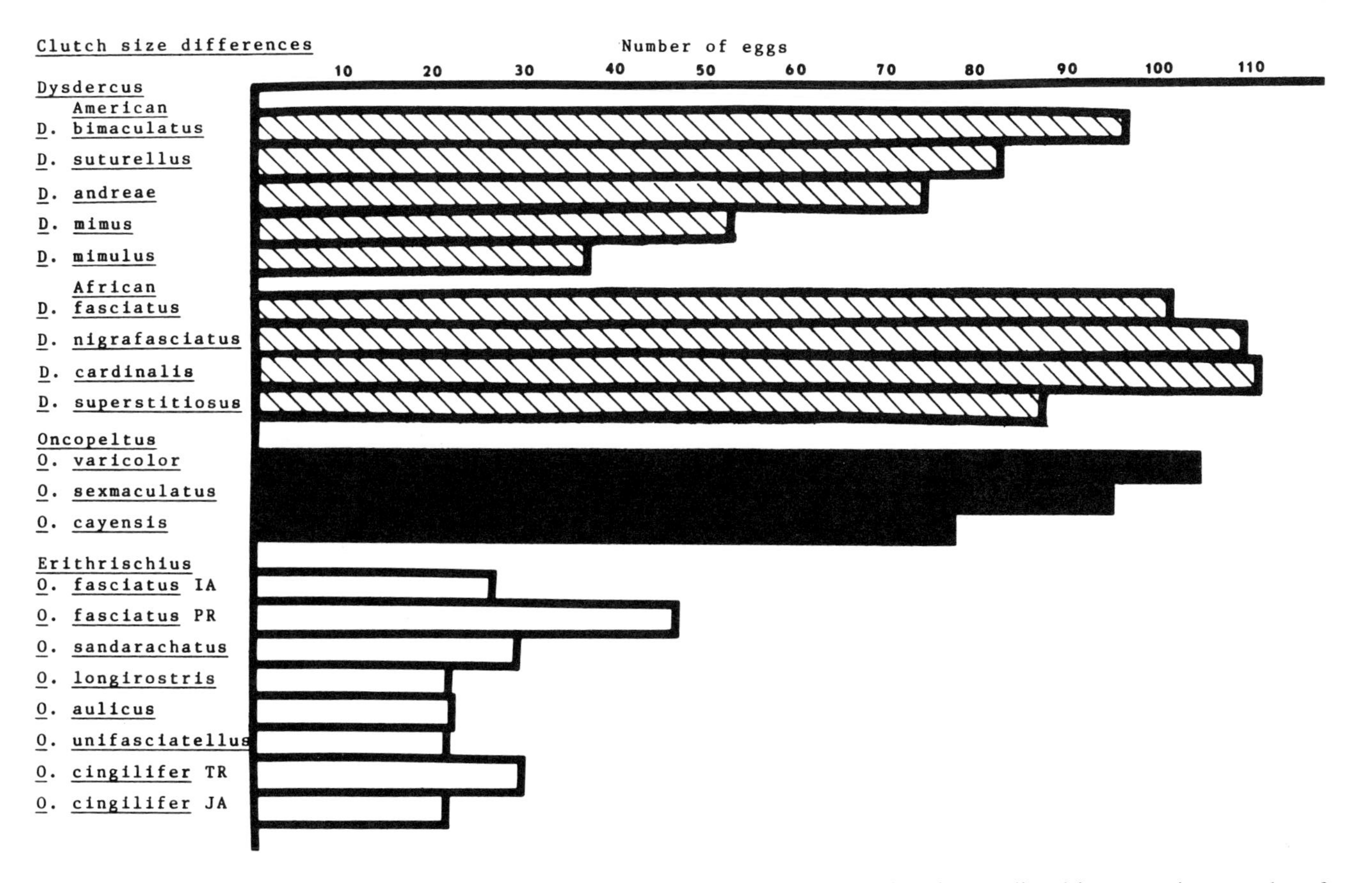

Figure 12.7. Clutch sizes in the heteropteran genera *Dysdercus* and *Oncopeltus*. Species are listed in approximate order of decreasing size. (There is little size difference among the African *Dysdercus* except for the smaller *D. superstitiosus*.) Clutch sizes are a function of body size both within and between species except in *Oncopeltus* (Erithrischius) where there is no relation between clutch and body size among species or populations.

In some locations in the American tropics members of both subgenera occur together, but local selection pressures seem not to produce convergence in clutch size (or divergence, for that matter). The conclusion would seem to be that in these taxa clutch size and oviposition behavior are phylogenetic constraints with limited potential for modification by selection (see also discussion in Denno and Dingle 1981).

Some Final Comments

Metamorphosis is a characteristic of insect life cycles, and separates the different stages into distinct units of developmental time. Each of these units must face environmental exigencies often quite different from those of the others, and frequently there must be a match between life stage and a particular environmental event, such as winter. Insects have solved the problems of synchronizing life cycle stage to seasonal periodicities and of responding to other biotic and abiotic factors with an impressive array of tactical alternatives. These include flexibility in diapause, development rates, maternal influences, parental care, and even sex. Life cycle adjustments to the contingencies presented by the environment result from natural selection acting on genetic variability. Maintenance of the latter by environmental uncertainties has also been a subject of my discussion. The ensuing remarkable diversity in the genetic and phenotypic architecture of insect life histories provides rich material for fundamental studies of the nature of the evolutionary process. The various theoretical and experimental chapters in this book attest to the progress we are making in understanding both the possibilities and the limits involved in life history evolution. It should be evident that insect life history studies are now firmly established in the mainstream of current evolutionary biology (see also Dingle 1984; Istock 1984).

As entomologists we are also interested in how life history studies may help us design appropriate management strategies for the many insects with which we either compete or cooperate. The flexibility I have outlined, both genetic and phenotypic, is bound to have consequences for control or management systems. The capacity of insects to respond to environmental change is all too apparent. We have barely begun to consider those consequences let alone analyze them in detail, yet it is arguable, certainly, that here is where the most understanding is required (see e.g., Barfield and O'Neil 1984). The current and often desperate need to reconcile our requirements for basic sustenance with our proclivity for fouling our world should provide sufficient impetus to pursue such a reconciliation. There is much room for cooperation between entomologists and evolutionary biologists to the benefit of both.

Acknowledgments I am particularly grateful to Bill Bradshaw, who first suggested this symposium and who contributed so much to our enjoyable stay in Hamburg. I owe special debts also to Seiji Tanaka for a copy of his thesis from which Figures 12.2 and 12.3 are taken, to Gerard Janssen for the data from his thesis in Figure 12.6, and to Ken Evans for the data in Figure 12.4. Ken Evans, Peter Frumhoff, Francis Groeters,

Beth Jakob, Randy Snyder, Susan Scott, and Fritz Taylor all made useful comments on the manuscript.

References

Barfield, C.S., O'Neil, R.J.: Is an ecological understanding a prerequisite for pest management? Fla Entomol. 67, 42–49 (1984).

Blakley, N.R., Dingle, H.: Competition: butterflies eliminate milkweed bugs from a Caribbean island. Oecologia 37, 133–136 (1978).

Bristow, C.M.: Treehoppers transfer parental care to ants: a new benefit of mutualism. Science 220, 532–533 (1983).

Brown, V.K., Hodek, I., (eds.): Diapause and Life Cycle Strategies in Insects. The Hague: Junk, 1983.

Cole, L.C.: The population consequences of life history phenomena. Q. Rev. Biol. 29, 103–137 (1954).

Denno, R.F., Dingle, H.: Considerations for the development of a more general life history theory. In: Insect Life History Patterns: Habitat and Geographic Variation. Denno, R.F. and Dingle, H. (eds.). New York: Springer-Verlag, 1981, pp. 1–6.

Derr, J.A.: The nature of variation in life history characters of *Dysdercus bimaculatus* (Heteroptera: Pyrrhocoridae), a colonizing species. Evolution 34, 548–557 (1980).

Derr, J.A., Alden, B.A., Dingle, H.: Insect life histories in relation to migration, body size, and host plant array: a comparative study of *Dysdercus*. J. Anim. Ecol. 50, 181–193 (1981).

Dingle, H.: The influence of environment and heredity on flight activity in the milkweed bug *Oncopeltus*. J. Exp. Biol. 48, 175–184 (1968).

Dingle, H. (ed.): Evolution of Insect Migration and Diapause. New York: Springer-Verlag (1978a).

Dingle, H.: Migration and diapause in tropical, temperate, and island milkweed bugs. In: Evolution of Insect Migration and Diapause. Dingle, H. (ed.). New York: Springer-Verlag, 1978b, pp. 254–276.

Dingle, H.: Geographic variation and behavioral flexibility in milkweed bug life histories. In: Insect Life History Patterns: Habitat and Geographic Variation. Denno, R.F., and Dingle, H. (eds.). New York: Springer-Verlag, 1981, pp. 55–73.

Dingle, H.: Behavior, genes, and life histories: Complex adaptations in uncertain environments. In: A New Ecology: Novel Approaches to Interactive Systems. Price, P.W., Slobodchikoff, C.N., and Gaud, W.S .(eds.). New York: Wiley, 1984, pp. 169–194.

Dingle, H., Haskell, J. B.: Phase polymorphism in the grasshopper, *Melanoplus differentialis*. Science 155, 590–592 (1967).

Dingle, H., Brown, C.K., Hegmann, J.P.: The nature of genetic variance influencing photoperiodic diapause in a migrant insect, *Oncopeltus fasciatus*. Am. Natur. 111, 1047–1059 (1977).

Dingle, H., Blau, W.S., Brown, C.K., Hegmann, J.P.: Population crosses and the genetic structure of milkweed bug life histories. In: Evolution and Genetics of Life Histories. Dingle, H., and Hegmann, J.P. (eds.). New York: Springer-Verlag, 1982, pp. 209–229.

Dingle, H., Leslie, J.F., Palmer, J.O.: Behavior genetics of flexible life histories in milkweed bugs (*Oncopeltus fasciatus*.) In: Evolutionary Genetics of Invertebrate Behavior. Huettel, M. (ed.). New York: Plenum Publishers, 1986, (in press).

Evans, K.E.: Migration and reproduction in Northern California populations of the milkweed bug, *Oncopeltus fasciatus*. Ph.D. Thesis, University of California, Berkeley (1982).

Falconer, D.S.: Introduction to Quantitative Genetics. 2nd ed. London and New York: Longman, 1981.

Frazzetta, T.H.: Complex Adaptations in Evolving Populations. Sunderland, Massachusetts: Sinauer, 1975.

Gould, S.J., Lewontin, R.C.: The spandrels of San Marco and the Panglossian paradigm: a critique of the adaptationist programme. Proc. R. Soc. Lond. B 205, 581–598 (1979).

Hegmann, J.P., Dingle, H.: Phenotypic and genetic covariance structure in milkweed bug life history traits. In: Evolution and Genetics of Life Histories. Dingle, H., and Hegmann, J.P. (eds.). New York: Springer-Verlag, 1982, pp. 177–185.

Henrich, V.C., Denlinger, D.L.: Selection for late pupariation affects diapause incidence and duration in the flesh fly, *Sarcophaga bullata*. Physiol. Ent. 7, 407–411 (1982).

Hoffmann, R.J.: Environmental uncertainty and evolution of physiological adaptation in *Colias* butterflies. Am. Natur. 112, 999–1015 (1978).

Istock, C.A.: Fitness variation in a natural population. In: Evolution of Insect Migration and Diapause. Dingle, H. (ed.). New York: Springer-Verlag, 1978, pp. 171–190.

Istock, C.A.: Natural selection and life history variation: Theory plus lessons from a mosquito. In: Insect Life History Patterns: Habitat and Geographic Variation. Denno, R.F., and Dingle, H. (eds.). New York: Springer-Verlag, 1981, pp. 113–128.

Istock, C.A.: The extent and consequences of heritable variation for fitness characters. In: Population Biology: Retrospect and Prospect. King, C.R., Dawson, P.S. (eds.). Oregon State Univ. Colloq. Pop. Biol. New York: Columbia University Press, 1982.

Istock, C.A.: Boundaries to life history variation and evolution. In: A New Ecology: Novel Approaches to Interactive Systems. Price, P.W., Slobodchikoff, C.N., Gaud, W.S. (eds.). New York: Wiley-Interscience, 1984, pp. 143–168.

Istock, C.A., Zisfein, J., Vavra, K.J.: Ecology and evolution of the pitcher plant mosquito. 2. The substructure of fitness. Evolution 30, 548–557 (1976).

Iwao, S.: Studies on the phase variation and related phenomena in some lepidopterous insects. Mem. Coll. Agric., Kyoto Univ. No. 84, pp. 1–80 (1962).

Iwao, S.: Some effects of grouping in lepidopterous insects. Colloq. Intern. C.N.R.S. No. 173, pp. 185–212 (1968).

Janssen, G.M.: Genetical ecology in springtails—maternal effects, heritabilities and genetic correlations in life history traits. Ph.D. Thesis, Free University of Amsterdam and State University of Utrecht (1984).

Kennedy, J.S.: Continuous polymorphism in locusts. Roy. Entomol. Soc. Symp. 1, 80–90 (1961).

Kunkel, J.G.: A minimal model of metamorphosis—fat body competence to respond to juvenile hormone. In: Current Topics in Insect Endocrinology and Nutrition. Bhaskaran, G., Friedman, S., Rodriguez, J. (eds.). New York: Plenum Press, 1981, pp. 107–129.

Lavie, B., Ritte, U.: The relation between dispersal behavior and reproductive fitness in the flour beetle *Tribolium castaneum*. Can. J. Genet. Cytol. 20, 589-595 (1978).

Leslie, J.F., Dingle, H.: A genetic basis for oviposition preference in the large milkweed bug, *Oncopeltus fasciatus*. Entomol. Exp. Appl. 34, 215–220 (1983).

Lewontin, R.C.: Selection for colonizing ability. In: The Genetics of Colonizing Species. Baker, H.G., Stebbins, G.L. (eds.). New York: Academic Press, 1965, pp. 77–94.

Lynch, M.: The limits to life history evolution in *Daphnia*. Evolution 38, 465–482 (1984).

Lynch, M., Gabriel, W.: Phenotypic evolution and parthenogenesis. Am. Natur. 122, 745–764 (1983).
MacKay, P.A., Wellington, W.G.: Maternal age as a source of variation in the ability of an aphid to produce dispersing forms. Res. Pop. Ecol. 18, 195–209 (1977).
Nolte, D.J.: The gregarization of locusts. Biol. Rev. 49, 1–14 (1974).
Palmer, J.O.: Photoperiodic control of reproduction in the milkweed leaf beetle, *Labidomera clivicollis*. Physiol. Entomol. 8, 187–194 (1982).
Palmer, J.O.: Ecological genetics of wing length, flight propensity, and early fecundity in a migratory insect. In: Migration: Mechanisms and Adaptive Significance. Rankin, M.A. (ed.). Contrib. Marine Science, Vol. 27 (Suppl.) 653–663 (1985).
Rankin, M.A., Caldwell, R.L., Dingle, H.: An analysis of a circadian rhythm of oviposition in *Oncopeltus fasciatus*. J. Exp. Biol. 56, 353–359 (1972).
Ritte, U., Lavie, B.: The genetic basis of dispersal behavior in the flour beetle *Tribolium castaneum*. Can. J. Genet. Cytol. 19, 717–722 (1977).
Roff, D.: Phenological adaptation in a seasonal environment: a theoretical perspective. In: Diapause and Life Cycle Strategies in Insects. Brown, V.K., Hodek, I. (eds.). The Hague: Junk, 1983, pp. 251–270.
Saunders, D.S.: Insect Clocks. 2nd ed. Oxford: Pergamon, 1982.
Saunders, D.S., Bradley, H.: Long-night summation and programming of pupal diapause in the flesh-fly, *Sarcophaga argyrostoma*. Ciba Foundation Symposium 104. London: Pitman, 1984, pp. 65–89.
Stearns, S.C.: Life history tactics: a review of the ideas. Q. Rev. Biol. 51, 3–47 (1976).
Tallamy, D.W.: Insect parental care. Bioscience 34, 20–24 (1984).
Tallamy, D.W., Denno, R.F.: Maternal care in *Gargaphia solani* (Hemiptera: Tingidae). Anim. Behav. 29, 771–778 (1981).
Tallamy, D.W., Dingle, H.: Genetic variation in the maternal defensive behavior of *Gargaphia solani* (Hemiptera: Tingidae). In: Evolutionary Genetics of Invertebrate Behavior. Huettel, M.D. (ed.). New York: Plenum Press (in press) (1986).
Tanaka, S.: Seasonal and altitudinal regulation of embryonic and nymphal development in *Allonemobius fasciatus* (Orthoptera: Gryllidae). Ph. D. Thesis, Oregon State University, Corvallis (1983).
Tanaka, S., Brookes, V.J.: Altitudinal adaptation of the life cycle in *Allonemobius fasciatus* DeGeer (Orthoptera: Gryllidae). Can. J. Zool. 61, 986–990 (1983).
Tauber, C.A., Tauber, M.J.: Insect seasonal cycles: genetics and evolution. Annu. Rev. Ecol. Syst. 12, 281–308 (1981).
Taylor, F.: Timing in the life histories of insects. Theor. Pop. Biol. 18, 112–124 (1980a).
Taylor, F.: Optimal switching to diapause in relation to the onset of winter.Theor. Pop. Biol. 18, 125–133 (1980b).
Uvarov, B.: Grasshoppers and Locusts, Vol. 1. London: Cambridge University Press, 1966.
Wharton, D.R.A., Lola, J.E., Wharton, M.L.: Population density, survival, and growth of the American cockroach. J. Insect Physiol. 13, 699–716 (1967).
Willis, E.R., Riser, G.R., Roth, L.M.: Observations on reproduction and development in cockroaches. Ann. Entomol. Soc. Am. 51, 53–59 (1958).
Wilson, E.O.: The Insect Societies. Cambridge, Massachusetts: Harvard University Press, 1971.
Woodhead, A.P., Paulson, C.R.: Larval development of *Diploptera punctata* reared alone and in groups. J. Insect Physiol. 29, 665–668 (1983).
Wu, A.-C.: Life history traits correlated with emigration in flour beetle populations. Ph.D. Thesis, University of Illinois at Chicago Circle (1981).

Chapter 13

Evolution of Wing Polymorphism and Its Impact on Life Cycle Adaptation in Insects

DEREK ROFF

Wing length dimorphism is very common among the Insecta, particularly in the large orders Coleoptera, Orthoptera, Hemiptera, and Homoptera. There is no single genetic mechanism underlying the trait, both single locus systems and polygenic systems being found (Table 13.1). The evolution and maintenance of wing dimorphisms is undoubtedly due in many, if not most, cases to spatial and temporal heterogeneity in the environment. The importance of such factors has been well demonstrated for pond skaters, *Gerris* spp. (Vepsäläinen 1973, 1974a,b, 1978) and leafhoppers (Denno 1976, 1978, 1979, Denno et al., 1980). In both these groups (Hemiptera and Homoptera) wing form appears to have a polygenic basis with considerable genotype–environment interaction (Honek 1976a,b, Harrison 1980). There is reasonable evidence that wing dimorphism in some coleopteran species is maintained by environmental heterogeneity (den Boer 1970, 1971, 1981; Meijer 1974; Stein 1977). In this order the inheritance of wings may be either controlled by a single locus, two-allele system or by several loci showing environmental interactions (Table 13.1).

In this chapter I shall present an hypothesis concerning the evolution of the genetic basis of wing dimorphism and examine how life cycle events may be modified as a result of selection for changes in the proportion of macropterous individuals in a population.

The Genetic and Physiological Basis of Wing Determination

It is reasonable to suppose that microptery, or brachyptery, could originate with a single mutation, giving rise to a single-locus, two-allele mechanism. Such a mutation must not, however, disrupt development to such an extent as to reduce viability significantly, as happens in the vestigial mutants of *Drosophila melanogaster, Pieris napi* (Bowden 1963) or *Bombyx mori* (Tazima 1964). In fact, brachypterous individuals usually start breeding prior to macropterous individuals and have a higher fecundity (Dingle 1979, 1982; Roff 1986a). Whether these attributes are the result of the original mutation or result from subsequent mutations is unknown. Brachypterous individuals, being nondispersers, can begin breeding immediately and utilize the energy otherwise devoted to wing muscle production and maintenance for reproduction. As a result, mutations that cause early gonad development should be rapidly selected once a mutation for brachyptery has arisen. Mutations for brachyptery and early gonad maturation are likely to become linked and inherited as a unit since individuals with only one of the features will be at a selective disadvantage.

It has been suggested that the suppression of wing production occurs when some factor, such as juvenile hormone, exceeds a critical threshold value during some particular stage of development (Southwood 1961; Wigglesworth 1961). Consider a single-locus, two-allele system with microptery dominant (Table 13.1). We may postulate that each allele acts additively to determine the level of the wing suppressor substance. If the sum of the alleles in the heterozygote exceeds the threshold level, a micropterous individual results and the allele appears dominant (Figure 13.1a). Extension of this model to a polygenic system is obvious; instead of two alleles summing we have a large number of genes whose effects are additive (Figure 13.1b).

Although both single-locus and polygenic systems occur, I shall argue that polygenic modes of inheritance will be found to be generally more common because single-locus systems, giving only limited offspring combinations, do not permit the production of the optimal frequency of macropterous types. (The percent macropterous from any cross involving *BB*, where *B* is dominant brachyptery, is 0%, from *Bb* × *Bb* it is 25%, from *Bb* × *bb*, 50%, and 100% from *bb* × *bb*.) "Optimal" is used in this chapter in the sense of able to resist invasion by some alternate strategy (i.e., evolutionary stable strategy). Quantitative modes of inheritance, in contrast, are capable of producing any frequency of dispersal type. Thus, the likelihood that the limited frequency distribution produced by single-locus systems will be able to resist invasion by a polygenic system seems low. Consider a habitat colonized by a single, impregnated macropterous female. There are three possible crosses, *bb* × *bb*, *bb* × *BB*, and *bb* × *Bb* (where *b* is the recessive allele and *B* the dominant). In the first case, all the offspring will be macropterous and the habitat will be evacuated in the next generation, hardly an efficient means of colonization. The offspring of the second cross will all be micropterous. Assuming random mating, with the macropterous individuals dispersing before oviposition, the trajectory of percent macropterous offspring (prior to dispersal) will be as shown in Figure 13.2b. The frequency of micropterous individuals rises rapidly, after an initial drop due to random mating of the heterozygotic first generation. The trajectory for the cross *bb* × *Bb* is similar except that it starts from 50% micropterous. If the habitat persists for a reasonably long period these changes in disperser production may be adequate, but they are rarely likely to be optimal. Furthermore, if the persistence time of a habitat is on the order of three to five generations a much higher rate of dispersal in later generations might be optimal.

The trajectories resulting from a quantitative mode of wing inheritance are

Table 13.1. The probable genetic basis of pterygomorphism in a variety of insect species. Adapted from Roff (1986a)

Order	Family	Species	Mode of Inheritance	Method of Analysis	Reference
Coleoptera	Curculionidae	*Sitona hispidula*	Single locus, 2 alleles, Brachyptery dominant	1	Jackson 1928
Coleoptera	Curculionidae	*Apion virens*	Single locus, 2 alleles, Brachyptery dominant	2	Stein 1973; Roff 1986a
Coleoptera	Carabidae	*Pterostichus anthracinus*	Single locus, 2 alleles, Brachyptery dominant	1	Lindroth 1946; Roff 1986a
Coleoptera	Carabidae	*Bembidion lampros*	Single locus, 2 alleles, Macroptery dominant(?)	1	Langor and Larson 1983, Roff 1986a
Coleoptera	Carabidae	*Calathus erythroderus*	Single locus, 2 alleles, Brachyptery dominant	1	Aukema 1984
Coleoptera	Carabidae	*Calathus melanocephalus*	Polygenic	1	Aukema 1984
Coleoptera	Ptiliidae	*Ptinella apterae*	Polygenic	2	Taylor 1981
Coleoptera	Bruchidae	*Callosobruchus maculatus*	Polygenic	3	Utida 1972

Diptera	Sphaeroceridae	*Apterina pedestris*	Single locus, 2 alleles, Brachyptery dominant	1	Guibé 1939
Hymenoptera	Bethylidae	*Cephalonomia gallicola*	diploid, apterous, haploid, 2 alleles	1	Kearns 1934
Hymenoptera	Formicidae	*Harpagoxenus sublaevis*	diploid, 2 alleles, Brachyptery dominant	1	Buschinger 1978
Hemiptera	Pyrrhocoridae	*Pyrrhocoris apterous*	Polygenic	2,3,4	Honek 1976a, b, 1979
Hemiptera	Gerridae	*Gerris lacustris*	Polygenic	2	Vepsäläinen 1974b Pers. Comm.
Hemiptera	Gerridae	*Limnoporus caniculatus*	Polygenic	1	Zera et al. 1983
Orthoptera	Gryllidae	*Gryllus pennsylvanicus*	Polygenic	1	Harrison 1979
Orthoptera	Gryllidae	*Gryllus firmus*	Polygenic	1	Roff 1984, 1986b
Orthoptera	Gryllidae	*Gryllodes sigillatus*	Polygenic	3	Ghouri and McFarlane 1958
Orthoptera	Acrididae	*Melanoplus lakinus*	Insufficient data	2	Bland and Nutting 1969
Homoptera	Delphacidae	*Laodelphax striatellus*	Polygenic	2	Mahmud 1980
Homoptera	Delphacidae	*Nilaparvata lugens*	Polygenic	2	Mochida 1975
Homoptera	Delphacidae	*Javesella pellucida*	Polygenic	2	Ammar 1973
Homoptera	Aphididae	*Schizaphis gramimum*	Polygenic	4	Kvenberg and Jones 1974
Homoptera	Aphididae	*Acyrthosiphon pisum*	Polygenic	5	Lamb and McKay 1979

1. Data from individual crosses.
2. Grouped data from known parent morphs.
3. Selection for wing morphs.
4. Comparison of different geographic strains.
5. Comparison of clones.

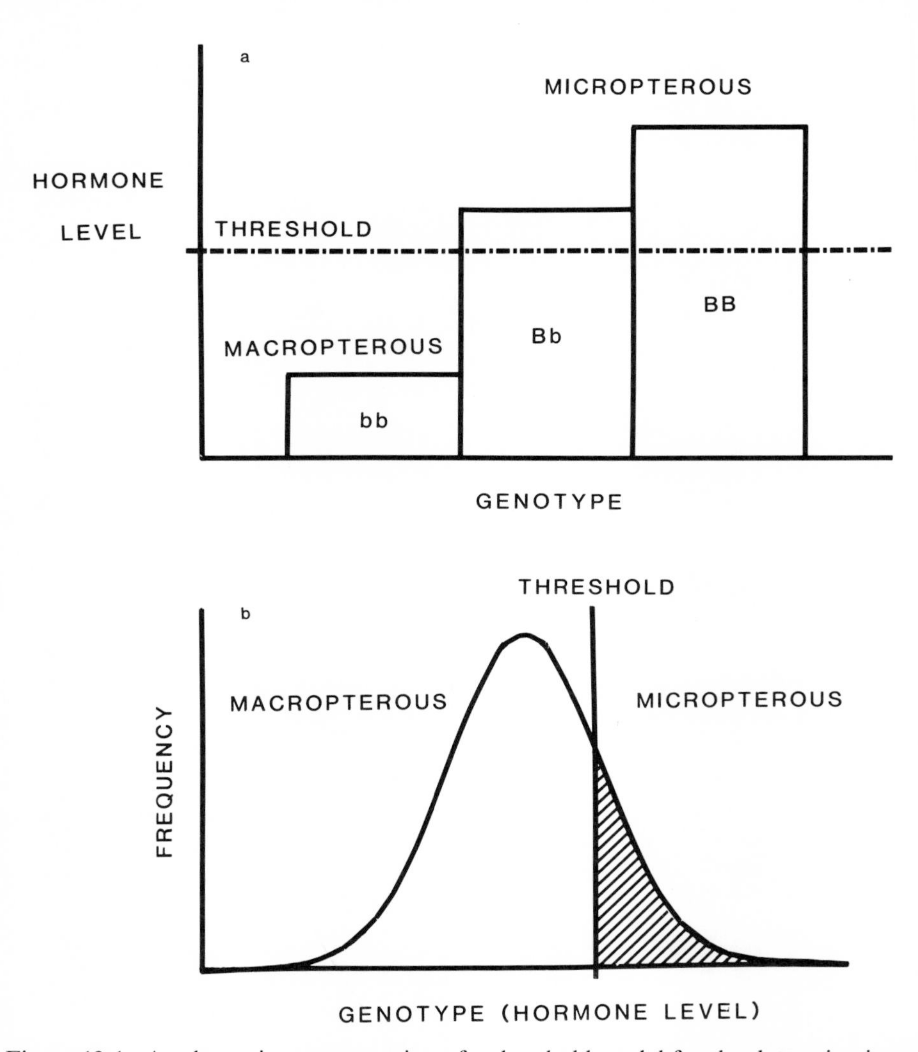

Figure 13.1. A schematic representation of a threshold model for the determination of wing morph (a) single locus with two alleles and (b) polygenic system. (From Roff 1986a.)

both more numerous and variable than from a single locus system (Figure 13.2a). Two factors determine the trajectory; the genetic constitution of the immigrant female and the heritability of the trait (h^2). In the illustrated example (see Appendix for the method of estimating the trajectory) the female is assumed to come from a population comprising 50% macropterous types. The genotypic frequency in the source population will itself depend on the heritability of the character. By changing the heritability, a wide range of trajectories is possible.

The stability characteristics of a habitat can be divided into two categories. First, there are those characteristics associated with the temporal sequence of

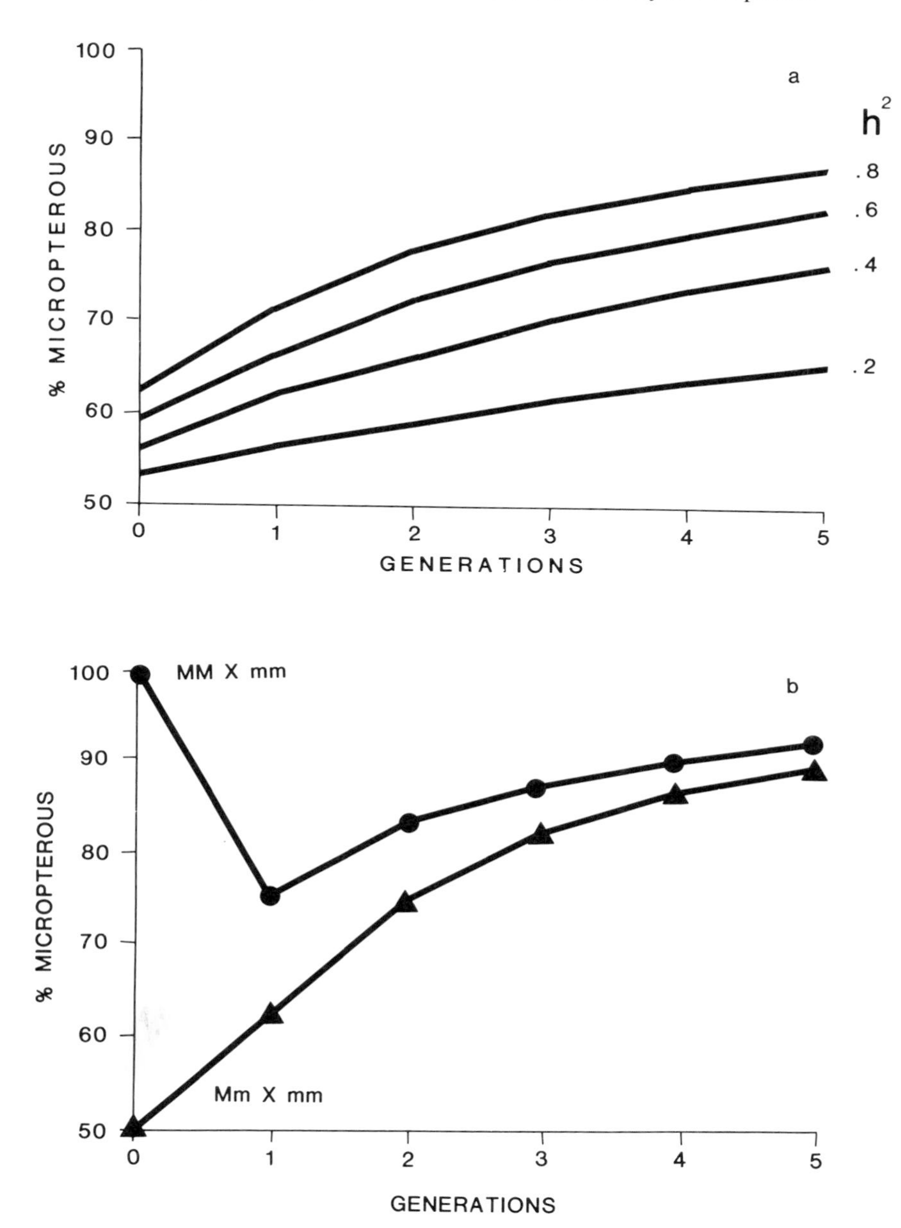

Figure 13.2. Percentage of micropterous individuals expected in a population colonized by a single impregnated female. In (a) a single-locus model is assumed and in (b) a polygenic model. For mathematical details see Appendix.

events during succession and, second, there are those associated with stochastic fluctuations in habitat suitability. In this latter group can also be placed, for convenience, environmental fluctuations that directly affect the reproductive success of an individual. As succession proceeds, the suitability of a habitat will eventually decline and increased emigration will be favored. On the other hand, stochastic fluctuations will favor a constant rate of emigration. The proportion of nondispersers will actually decrease over time, however, as the frequency of the macropterous gene(s) decreases due to emigration. If all habitats have the same stability characteristics, selection will favor a zero heritability since this will produce a constant rate of macropterous individuals which, under the circumstances outlined above, is the best that can be achieved. If, however, the habitats vary in their characteristics, a different rate of emigration will be required at each site: I conjecture that under such conditions selection will favor an intermediate heritability.

Thus, selection should favor the evolution of a polygenic system with the heritability of the trait varying with environmental uncertainty. There are basically two ways in which heritability is altered: first, by "gene–gene interactions" such as dominance, epistasis, etc., and, second, by genotype–environment interactions. The first restricts the rate at which the genotype can respond to selective pressures since nonadditive effects generally require relatively long and sustained pressure to change. On the other hand, genotype–environment interactions offer the possibility of preserving a high heritability in some circumstances and a low heritability in others. Furthermore, a low heritability does not imply the loss of genetic variation since this is being masked by the genotype–environment interactions and not lost owing to fixation.

It has been demonstrated in a wide variety of insect species that the presence or absence of wings is strongly influenced by environmental factors, particularly temperature and photoperiod (Honek 1976a,b). Environmental factors are also known to greatly influence diapause induction. Although these two phenomena (production of wings and induction of diapause) seem rather disparate, they may share a common physiological basis. In the moth *Orgyia thyellina,* for example, brachyptery and production of diapausing eggs are controlled by the same photoperiodic regimen and appear to be coupled (Kimura and Masaki 1977). The following model, based in part on one proposed for diapause induction in *Sarcophaga argyrostoma* (Gibbs 1975), demonstrates how genotype–environment interactions might occur. The basic proposition is that wing production (or diapause) is induced or suppressed if a hormone exceeds a threshold value during a critical period in development. Because physiological processes proceed at different rates with environmental factors, hormone production will not be sufficient to exceed the threshold level under all environmental conditions. This may result from direct effects on the rate of hormone production or from changes in the rate of development being different from changes in the rate of hormone production. Thus, a change in environmental conditions might not affect hormone production but could greatly accelerate or retard development. To illustrate this, consider the following simple model (Figure 13.3) in which hormone

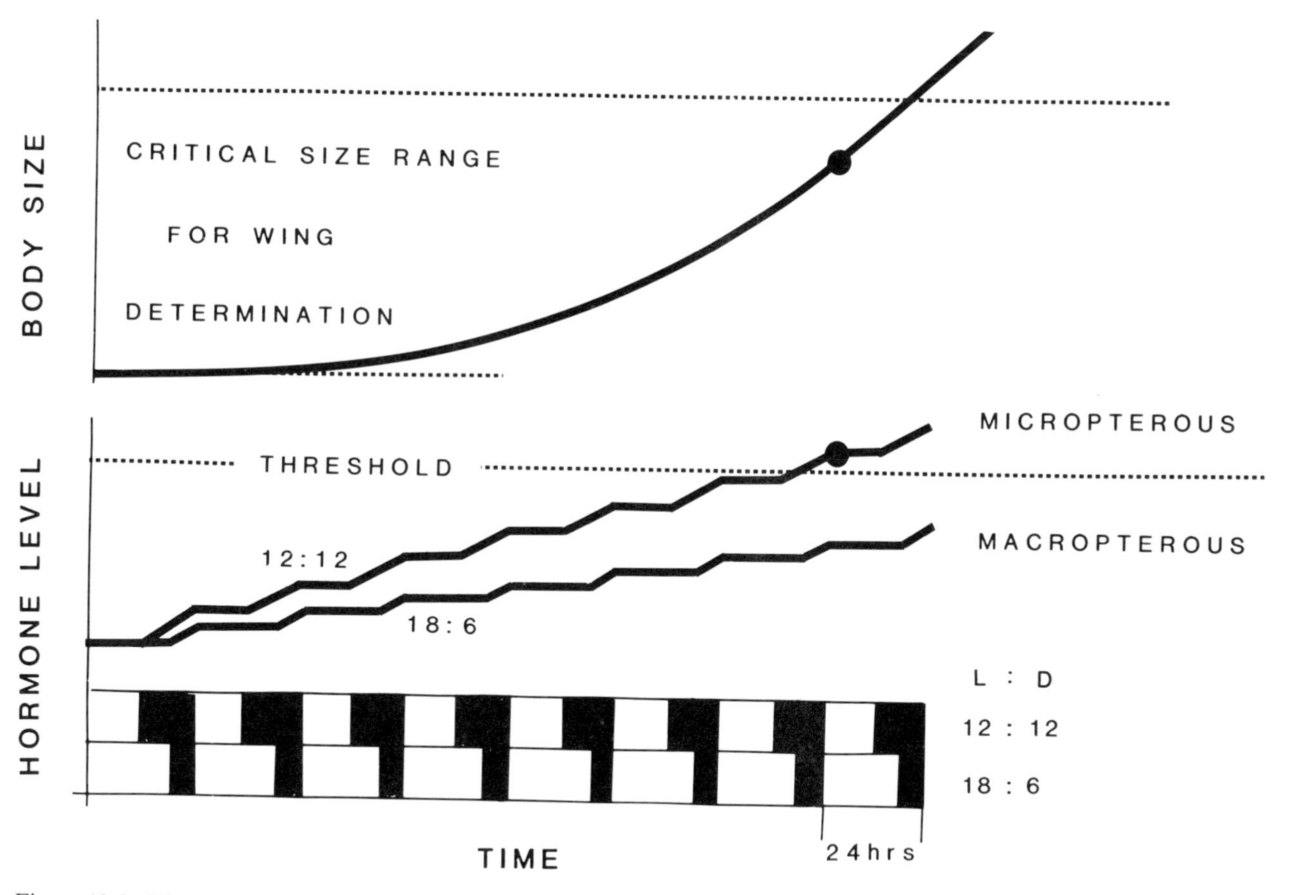

Figure 13.3 A hypothetical model for the physiological determination of wing morph when the appearance of each morph is environmentally and genetically cued.

production, but not growth, is affected by a particular change in environmental conditions, in this instance, photoperiod.

The production of the wing suppressor compound, say juvenile hormone, is switched on in the dark and off in the light. In long photoperiods (light regime), insufficient hormone is produced by the time the animal has passed through the critical period (here shown as a size range) for wing determination and the insect develops wings. In a short photoperiod, the hormone level increases sufficiently fast that the threshold level is surpassed before the end of the critical period and a micropterous individual is produced (Figure 13.3). Genetic differences between individuals are produced because of genetic variation in the rate of production of the suppressor substance. Thus "hormone level" and "rate of production" are synonymous under a given set of environmental conditions. Under any given environmental regimen, a proportion of the genotypes will produce macropterous phenotypes, the proportion changing with the environment. Because the rate of hormone production and development rate do not change in synchrony with changes in environmental factors, the relationship between percent macroptery and, say, photoperiod may be either monotonic (Figure 13.4a) or modal (Figure 13.4b). In the particular response shown, heritability will be zero at the extreme photoperiodic conditions, the population being either 100% winged or 100% micropterous. At intermediate conditions, the heritability is greatest, subject, of course, to nonadditive effects that will always make it < 1.

Wing Dimorphism and Life Cycle Adaptation

Let us suppose that the phenology of a particular population is such that its critical period for wing determination occurs at the point indicated in Figure 13.4, and that this results in a stable gene frequency. Now suppose the habitat changes such that a lower frequency of dispersers is favored. There are four types that may be at an advantage under these conditions. First, genotypes that have a higher production rate of suppressor hormone will be favored, shifting the response curve to the right (Figure 13.4).

The remaining three responses involve shifts in the critical period for wing determination so that it occurs at a shorter photoperiod. The first manner in which this may occur is by the selection of faster developing individuals, which may, in turn, occur in the following two ways. The mean rate of development might be increased by the actual increase in the rate of growth (Figure 13.5a) or by a shortening of the duration of development (Figure 13.5b) and reduction of the critical size threshold (Figure 13.3). In the first instance no reduction in body size will occur whereas in the second, body size is reduced (I assume that the critical size range for wing determination is some constant fraction of the adult body size). The factors that set the limit on development rate, in the sense of the rate of change in weight, are unclear. One possible factor may be the trade-off between decreased development time or increased size and the problems of obtaining sufficient nutrient to maintain this rate. It is sufficient,

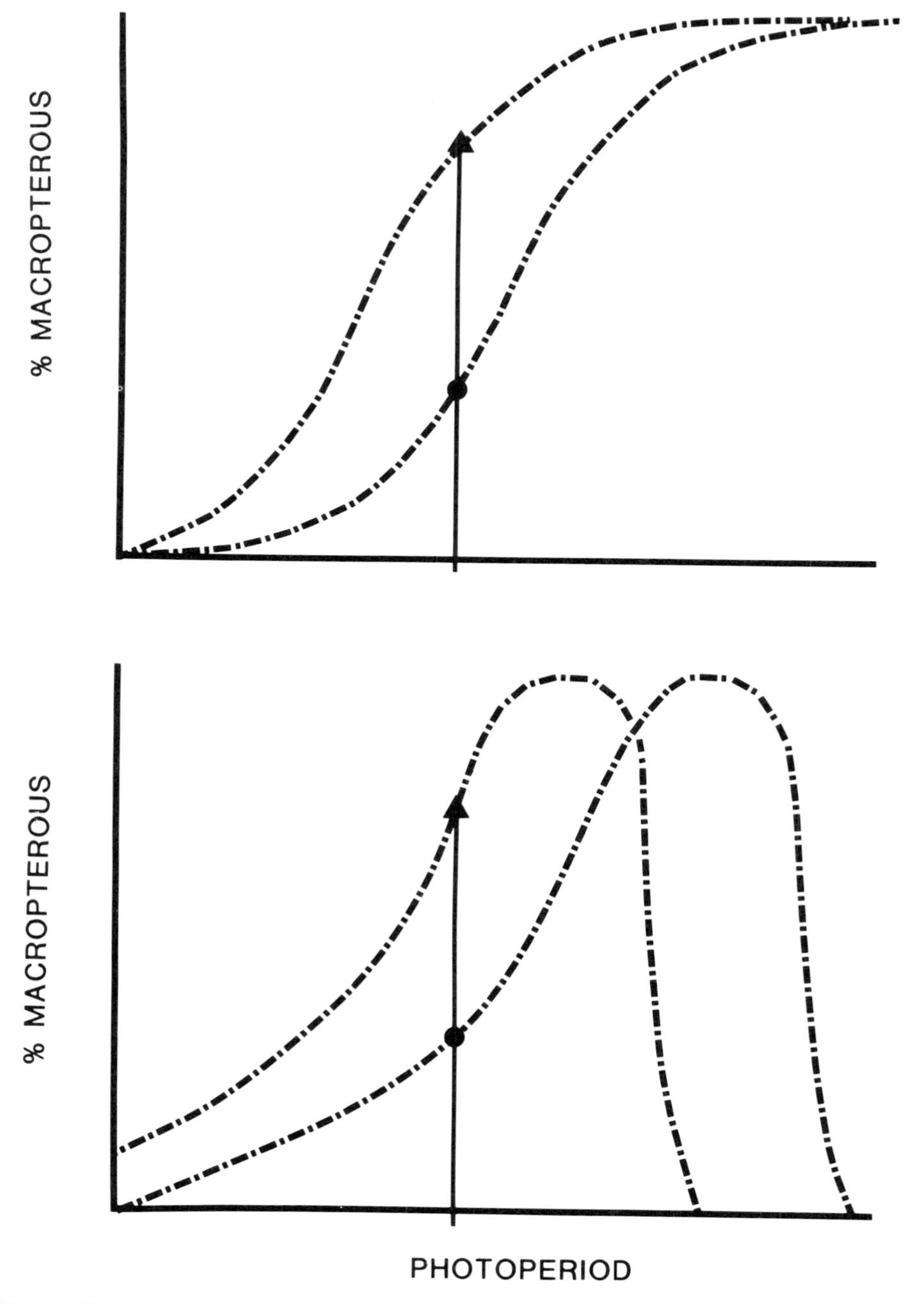

Figure 13.4. Possible relationships between percent macroptery and photoperiod. Selection for a reduced proportion of dispersers from ▲ to ● will shift the curve to the right, from ▲ to ●.

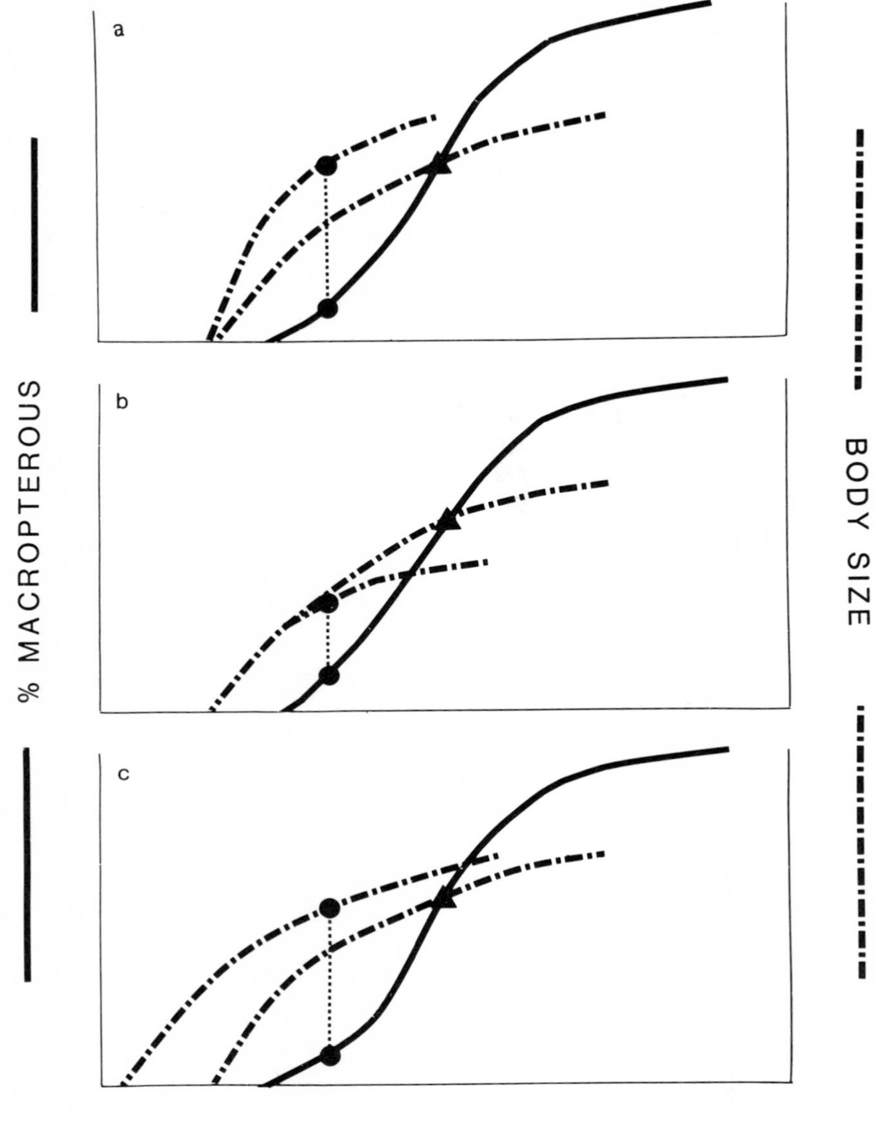

Figure 13.5. Possible life cycle responses to selection for a reduced dispersal rate. Selection favors a change from ▲ to •. The solid line designates the relationship between percent macroptery and photoperiod (see Figure 13.4a). The dashed lines designate the change in body size over time, the critical size for wing determination being denoted by ▲ or •.(a) Proportion of dispersers reduced by selection for faster growing individuals. (b) Proportion reduced by selection for earlier maturing individuals. (c) Proportion reduced by selection for earlier emerging individuals.

however, that there be genetic variation for development rate. Provided this exists, selection will increase the rate, the extent of the change depending on the factors that oppose this change.

The fourth and final adaptation is selection for earlier emergence (Figure 13.5c), thereby shifting the critical period back, though possibly increasing the probability of early mortality due to unfavorable environmental conditions. These four responses are not mutually exclusive and a combination is most likely to occur since this would be minimally opposed by those factors determining the original equilibrium set of life history parameters.

The Temporal Pattern of Morph Frequencies

There is rarely, if ever, truly synchronous development of a family cohort. Completion of development will typically be distributed over time due to differences in the time of first laying, genetic differences between individuals, differences in microhabitat, or chance. This variation can be quite considerable (Figure 13.6) and may have a significant impact on the proportion of macropterous individuals produced. The selected responses outlined in the previous section will clearly depend on the extent and nature of this variation. Apart from affecting the response of the population, a consequence of life cycle asynchrony will be a temporal pattern in the production of macropterous individuals. Two examples are shown in Figure 13.7. There are clear differences in the pattern of production of macropterous individuals both temporally and with

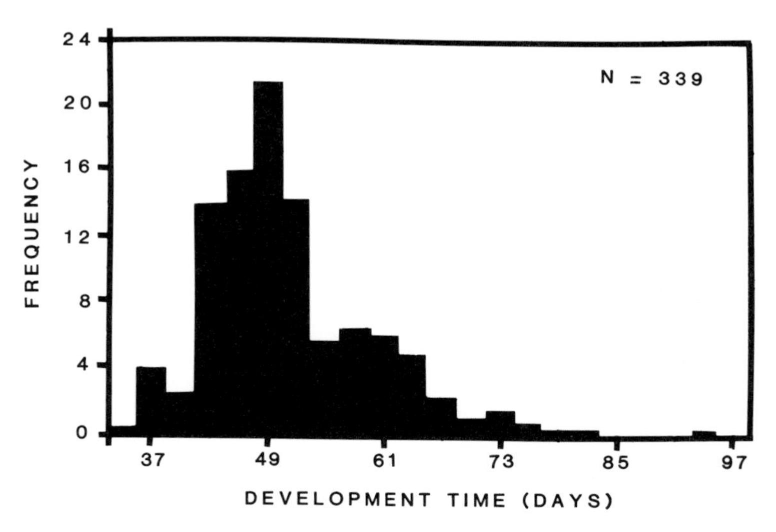

Figure 13.6. Frequency distribution of time from hatching to final moult in *Gryllus firmus* raised at 30°C in 17 hr L: 7 hr D. Details of rearing technique given in Roff (1986b).

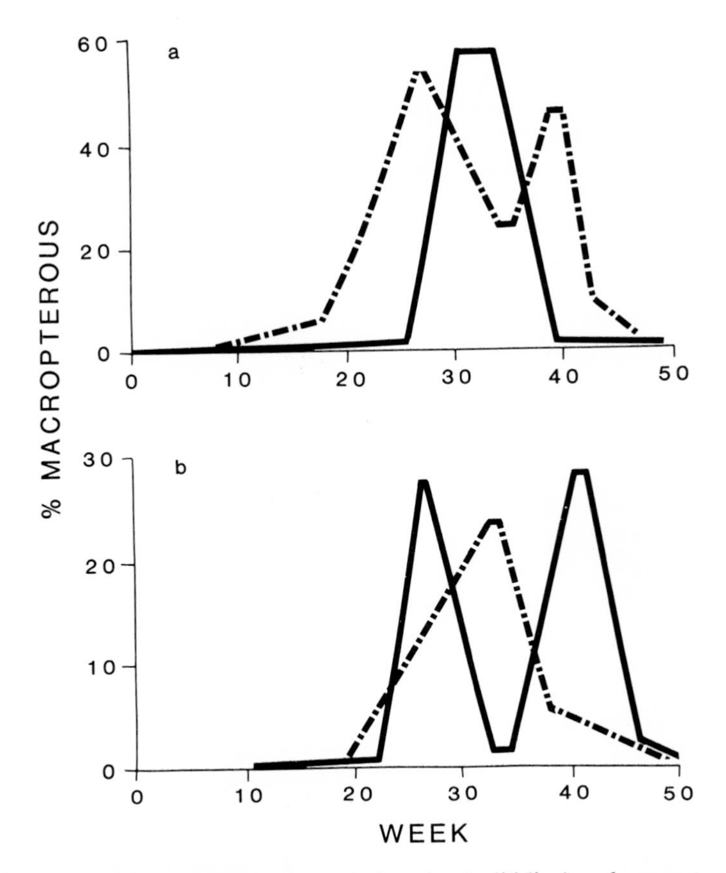

Figure 13.7. Temporal distribution of population size (solid line) and percent macroptery (dashed line) in two species of *Gryllus* (a: *G. rubens,* b: *G. firmus.*) The curves have been smoothed to illustrate the main trends. (Data from Veazey et al. 1976.)

respect to population density. At first glance one might be tempted to search for some adaptive significance in the pattern itself. In some instances there are obvious advantages in a particular pattern. Certain leafhoppers, for example, migrate to trees where they overwinter; production of macropterous individuals at the end of the "season" is of obvious selective advantage. It is not, however, necessary that there be a selective advantage for a pattern to exist. A distribution of production is an inevitable consequence of asynchrony in life cycle events and selection may operate simply to produce the optimum percentage of macropterous individuals per family. The particular temporal pattern produced may have little or no selective advantage in itself. Thus the differences in pattern between *G. rubens* and *G. firmus* may be accidental in the sense that it is a result only of the particular set of genotypes initially present.

Conclusions

Many insect species are characteristically polymorphic with respect to the possession of flight capability, generally recognizable by the presence of long- and short-winged forms. Experiments suggest that in some cases brachyptery is dominant, the trait being determined by a single locus with two alleles. In other cases, perhaps most, the polymorphism is controlled by a polygenic system. I have argued in this chapter that, although the origin of brachyptery may in some cases be due to a single gene mutation, selection will, in general, strongly favor the evolution of a polygenically controlled trait. This is because there are only a limited number of temporal patterns of dispersal possible with a single-locus system whereas with a polygenic system many patterns can be generated. A simple model is presented that can account for such effects as photoperiod and temperature on the expression of the trait. Selection for a change in dispersal rate may be accomplished by a shift in the genotype–environmental interaction or by shifts in life cycle events. The generation of different ratios of long- and short-winged morphs under the influence of changes in, say, photoperiod and temperature may be timed to occur at specific times during the year. This timing may not in itself be of adaptive advantage. In some cases at least it may represent a simple mechanism for the generation of short- and long-winged morphs.

References

Ammar, E.D.: Factors related to the two wing forms in *Javesella pellucida* (FAB) (Homoptera: Delphacidae). Z. Ang. Entomol. 74, 211–216 (1973).

Aukema, B.: Winglength determination in relation to dispersal by flight in two wing dimorphic species of *Calathus bonelli* (Coleoptera, Carabidae) Abstract, XVII Int. Congr. Entomol. (1984).

Bland, G., Nutting, W.L.: A histological and biometrical study of wing development in the grasshopper *Melanoplus lakinus*. Ann. Entomol. Soc. Am. 62, 419–436 (1969).

Bowden, S.R.: A recessive lethal, "wingless" in *Pieris napi* L. (Lep., Pieridae). Entomologist 96, 52 (1963).

Boer, P.J. den: On the significance of dispersal power for populations of carabid-beetles (Coleoptera, Carabidae). Oecologia 4, 1–28 (1970).

Boer, P.J. den: (ed.). Dispersal and dispersal power of Carabid beetles. Misc. Papers 8, Wageningen: Veenman and Zonen N.V. (1971).

Boer, P.J. den: On the survival of populations in a heterogeneous and variable environment. Oecologia 50, 39–53 (1981).

Buschinger, A.: Genetish Bendigte entstehung geflügelter weibchen bei der sklavenhaltenden ameise *Harpagoxenus sublaevis* (Nyl.) (Genetically induced origin of alate females in the slavemaker ant *Harpagoxenus sublaevis* Hymenoptera formicidae). Insectes Soc. 25(2), 163–172 (1978).

Crow, J.F., Kimura, M.: An Introduction to Population Genetics Theory. New York: Harper & Row, 1970.

Denno, R.F.: Ecological significance of wing polymorphism in Fulgoroidea which inhabit tidal salt marshes. Ecol. Entomol. 1, 257–266 (1976).

Denno, R.F.: The optimum population strategy for planthoppers (Homoptera: Delphacidae) in stable marsh habitats. Can. Entomol. 10, 135–142 (1978).

Denno, R.F.: The relation between habitat stability and the migratory tactics of planthoppers (Homoptera: Delphacidae). Misc. Publ. Entomol. Soc. Am. 11, 41–49 (1979).

Denno, R.F., Raupp, M.J., Tallamy, D.W., Reichelderfer, C.F.: Migration in heterogeneous environments: differences in habitat selection between the wing forms of the dimorphic planthopper, *Prokelisia marginata* (Homoptera: Delphacidae). Ecology 61, 859–867 (1980).

Dingle, H.: Adaptive variation in the evolution of insect migration. In: Movement of Highly Mobile Insects: Concepts and Methodologies in Research. Rabb, R.L., Kennedy, G.G., (eds.). North Carolina State University: University Graphics, 1979, pp. 64–87.

Dingle, H.: Function of migration in the seasonal synchronization of insects. Entomol. Exp. Appl. 31, 36–48 (1982).

Falconer, D.S.: Introduction to quantitative genetics. 2nd Ed. New York: Longman, 1981.

Ghouri, A.S.K. and McFarlane, J.E.: Occurrence of a macropterous form of *Gryllodes sigillatus* (Walker) (Orthoptera: Gryllidae) in laboratory culture. Can. J. Zool. 36, 837–838 (1958).

Gibbs, D.: Reversal of pupal diapause in *Sarcophaga argyrostoma* by temperature shifts after puparium formation. J. Inst. Phys. 21, 1179–1186 (1975).

Guíbe, J.: Contribution a l'étude d'une espece: *Apterina pedestris* Meigen (Diptera). Bull. Biol. France Belg. Suppl.26, 1–112 (1939).

Harrison, R.G.: Flight polymorphism in the field cricket *Gryllus pennsylvanicus*. Oecologia 40, 125–132 (1979).

Harrison, R.G.: Dispersal polymorphism in insects. Annu. Rev. Ecol. Syst. 11, 95–118 (1980).

Honek, A.: Factors influencing the wing polymorphism in *Pyrrhocoris apterus* (Heteroptera, Pyrrhocoridae). Zool. Jb. Syst. Bd. 103, 1–22 (1976a).

Honek, A.: The regulation of wing polymorphism in natural populations of *Pyrrhocoris apterus* (Heteroptera, Pyrrhocoridae). Zool. Jb. Syst. Bd. 103, 547–570 (1976b).

Honek, A.: Independent response of 2 characters to selection for insensitivity to photoperiod in *Pyrrhocoris apterus*. Experientia, 35, 762–763 (1979).

Jackson, D.T.: The inheritance of long and short wings in the weevil, *Sitona hispidula,* with a discussion of wing reduction among beetles. Trans. R. Soc. Edinburgh 55, 665–735 (1928).

Kearns, C.W.: Method of wing inheritance in *Dephalonomia gallicola* Ashmead (Bethylidae: Hymenoptera). Ann. Entomol. Soc. Am. XXVII, 533–541 (1934).

Kimura, T., Masaki, S.: Brachypterism and seasonal adaptation in *Orgyia thyellina* Butler (Lepidoptera, Lymantriidae). Kontŷu 45, 97–106 (1977).

Kvenberg, J.E. and Jones, P.A.: Comparison of alate offspring produced by two biotypes of the greenbug. Environ. Entomol. 3, 407–408 (1974).

Lamb, R.J., MacKay, P.A.: Variability in migratory tendency within and among natural populations of the pea aphid, *Acyrthosiphon pisum*. Oecologia 39, 289–299 (1979).

Langor, D.W., Larson, D.J.: Alary polymorphism and life history of a colonizing ground beetle, *Bembidion lampros*. Herbst (Coleoptera: Carabidae) Coleop. Bull. 37, 365–377 (1983).

Lindroth, C.H.: Inheritance of wing dimorphism in *Pterostichus anthracinus* Ill. Hereditas 32, 37–40 (1946).

Mahmud, F.S.: Alary polymorphism in the small brown planthopper *Laodelphax striatellus* (Homoptera: Delphacidae). Entomol. Exp. 28, 47–53 (1980).
Meijer, J.: A comparative study of the immigration of carabids (Coleoptera, Carabidae) into a New Polder. Oecologia 16, 185–208 (1974).
Mochida, O.: A strain producing abundance brachypterous adults in *Nilaparvata lugens* (Homoptera: Delphacidae). Entomol. Exp. Appl. 18, 465–471 (1975).
Roff, D.A.: The cost of being able to fly: a study of wing polymorphism in two species of crickets. Oecologia 63, 30–37 (1984).
Roff, D.A.: The evolution of wing dimorphism in insects. Evolution (in press) (1986a).
Roff, D.A.: The genetic basis of wing dimorphism in the sand cricket, *Gryllus firmus* and its relevance to the evolution of wing dimorphism in insects. Heredity (in press) (1986b).
Southwood, T.R.E.: A hormonal theory of the mechanism of wing polymorphism in heteroptera. Proc. R. Entomol. Soc. Lond. (A) 36, 63–66 (1961).
Stein, W.: Zur Vererbung des Flugeldimorphismus bei *Apion virens* Herbst (Col., Curculinoidae). Z. Ang. Entomol. 74, 62–63 (1973).
Stein, W.: Die Beziehung zwischen Biotop-Alter und Auftreten der Kurzflugeligkeit bei populationen dimorpher Rüsselkäfer-Arten (Col., Curculionidae). Zeitschrift Angew. Entomol. 83 (1), 37–39 (1977).
Taylor, V.A.: The adaptive and evolutionary significance of wing polymorphism and parthenogenesis in *Ptinella* Motschulsky (Coleoptera: Ptiliidae) 6, 89–98 (1981).
Tazima, Y.: The genetics of the silk worm. London: Logos Press, 1964.
Utida, S.: Density dependent polymorphism in the adult of *Callosobrachus maculatus* (Coleoptera, Brachidae). J. Stored Prod. Res. 8, 111–126 (1972).
Veazey, J.N., Kay, C.A.R., Walker, T.J. and Whitcomb, W.H.: Seasonal abundance, sex ratio and macroptery of field crickets in northern Florida. Ann. Entomol. Soc. Am. 69, 374–380 (1976).
Vepsäläinen, K.: The distribution and habitats of *Gerris* Fabr. species (Heteroptera, Gerridae) in Finland. Ann. Zool. Fennici 10, 419–444 (1973).
Vepsäläinen, K.: The life cycles and wing lengths of Finnish *Gerris* Fabr. species (Heteroptera, Gerridae). Acta Zool. Fennica 141, 1–73 (1974a).
Vepsäläinen, K.: Determination of wing length and diapause in water striders (*Gerris* Fabr., Heteroptera). Hereditas 77, 163–175 (1974b).
Vepsäläinen, K.: Wing dimorphism and diapause in *Gerris:* determination and adaptive significance. In: Evolution of Insect Migration and Diapause. Dingle, H. (ed.). New York: Springer-Verlag, 1978.
Wigglesworth, V.B.: Insect polymorphism—a tentative synthesis. Symp. R. Entomol. Soc. Lond. 1, 103–113 (1961).
Zera, A.J., Innes, D.T., Saks, M.E.: Genetic and environmental determinants of wing polymorphism in the water strider, *Limnoporus canaliculatus*. Evolution 37, 513–522 (1983).

Appendix

Single Locus Model

Let the two alleles be represented by *B* and *b,* where *B* is dominant, brachyptery. Further let the frequencies of *BB, Bb,* and *bb* prior to dispersal be p_1, p_2, and p_3, respectively. Assuming that mating occurs prior to dispersal and that all macropterous individuals leave, the frequency of genotypes prior to dispersal in the next generation (p_i') is derived as follows:

The probability of the male parent "donating" an allele *B* is $p_1 + \frac{1}{2} p_2$ and allele *b*, $p_3 + \frac{1}{2} p_2$. Since all females of genotype *bb* leave, the proportion of BB females in the female population is $p_1/(1 - p_3)$ and the proportion of *Bb* females is $p_2/(1 - p_3)$. From these considerations we can construct the trellis diagram shown in Table A.1. Summing down each column gives the offspring frequencies,

$$p_1' = (p_1^2 + p_1p_2 + 0.25\, p_2^2)/(1 - p_3)$$

$$p_2' = (p_1p_2 + p_1p_3 + 0.5\, p_2^2 + 0.5\, p_2p_3)/1 - p_3)$$

$$p_3' = (0.25\, p_2^2 + 0.5\, p_2p_3)/(1 - p_3)$$

Obviously the genotype of the original immigrant female must be *bb*. However, because of prior mating her offspring may be any of the three possible types.

Table A.1. Trellis diagram showing frequency of offspring genotypes from non-dispersing females

		Offspring genotypes		
Genotype	Alleles	BB	*Bb*	*bb*
BB	*B*	$\frac{p_1(p_1+0.5p_2)}{1-p_3}$	$\frac{p_1(p_3+0.5p_2)}{1-p_3}$	0
Bb	*B*	$\frac{0.5p_2(p_1+0.5p_2)}{1-p_3}$	$\frac{0.5p_2(p_3+0.5p_2)}{1-p_3}$	0
	b	0	$\frac{0.5p_2(p_1+0.5p_2)}{1-p_3}$	$\frac{0.5p_2(p_3+0.5p_2)}{1-p_3}$

Polygenic Model

For a discussion of threshold characters see Falconer (1981, p. 270) or Crow and Kimura (1970, p. 225).

The distribution of genotypes is assumed to be normally distributed with mean *m* and standard deviation $\sigma = 100$. Genotypes whose value exceeds $T = 1000$ are micropterous. The proportion of the population exceeding this value is given by:

$$p = \int_T^\infty f(m)dx$$

where

$$f(m) = \frac{1}{\sqrt{2\pi}} e^{-(x - m)^2/2\sigma}$$

The mean genotypic value of macropterous individuals is:

$$G_1 = m - \frac{100\, f(m)}{(1 - p)}$$

and of micropterous individuals it is:

$$G_2 = m + \frac{100\, f(m)}{p}$$

To compute the mean genotypic value of offspring from those remaining in the habitat, we first compute the separate crosses possible.

Cross	Genotypic value of offspring	
Mic × Mic	$m(1 - h^2) + h^2\, G_2$	$= g_1$
Mic × Mac	$m(1 - h^2) + h^2\, (G_1 + G_2)/2$	$= g_2$

where h^2 is the heritability of the trait. The mean genotypic value of the next generation, m′, is thus:

$$m' = pg_1 + (1 - p)g_2$$

Chapter 14
Prolonged Development in Cicadas

RICHARD KARBAN

Compared to other insect groups, the cicadas require an extremely long time to complete nymphal development. Although there are over 1500 species of cicadas (Myers 1929), development times are known for only 16 of these species and range from 2 years to 17 years. The so-called "annual cicadas" are misnamed since none of the species, for which we know the development requirement, can mature in a single year, under natural conditions.

Seventeen years is the longest known development time for any insect. This trait is shared by three distinct species that are periodic and emerge as adults in any locality in eastern North America only once every 17 years. The three species are sympatric and their emergences co-occur with tight synchrony. Despite the often repeated observation that periodical cicadas emerged only every 17 years in a particular locale, it required C. L. Marlatt's 17-year experimental rearing to convince skeptics of this remarkable life cycle (Marlatt 1907). Each of the 17-year periodical species has a 13-year equivalent (Alexander and Moore 1962) that is able to accelerate early growth by 4 years (White and Lloyd 1975). Other cicada species are not periodical; some individuals emerge as adults in every year (or at least most years).

Cicadas are record holders not only for development time but also for body size, as well, being much larger than other Homopterans and most other insects. The giant Malaysian cicada, *Formotosena seebohmi,* measures over 7 cm from head to abdomen and 21 cm across the wings.

This chapter considers two hypotheses for prolonged development in cicadas and the evidence in support of each. The first posits that development may be constrained by nutritional factors. The second is a demographic argument that prolonged development may be favored by an extremely high mortality among young nymphs coupled with very low mortality among older, established nymphs.

Nutrition and Prolonged Development

> An ass envied greatly the cicadas' musical talent, and with rather vulgar estimate of the origin of genius, enquired on what food they lived to give them such beautiful voices. They replied, "Dew". In following this alimentary regimen the ass died.
>
> Ascribed to Aesop (620–560 B.C.)

Cicadas feed exclusively on xylem fluid as nymphs and as adults (Cheung and Marshall 1973; White and Strehl 1978). Xylem sap contains very dilute concentrations of amino acids and only trace quantities of carbohydrates (Anderssen 1929; Wiegert 1964; Tromp and Ovaa 1967; Cheung and Marshall 1973; Horsfield 1977). Although there is disagreement about whether xylem feeding insects are limited by energy (Raven 1983) or limited by nitrogen (Mattson 1980), all workers agree that concentrations of essential nutrients are much lower in xylem fluid than in other plant tissues.

During much of the day, xylem fluid is under tension or negative pressure as it moves up the plant. As a result, cicadas must expend energy to extract xylem fluid and they are equipped with relatively enormous cibarial muscles, which fill most of their head.

Cicada growth rates may be limited because they are feeding on an extremely dilute food that is energetically expensive to procure (Lloyd and Dybas 1966; White and Strehl 1978; Lloyd 1984). Cicadas may feed for only a small portion of the time when nutritional quality and/or plant water relations make this activity profitable. The size of the cibarial pump may limit the rate of ingestion even if feeding is always profitable. Feeding may be associated with a risk of the plant shedding the particular rootlet that is being parsitized by the cicada nymph. Rootlet shedding is a common event and has been associated with parasitism by insects, nematodes, and fungi (Head 1973). I was unable to demonstrate, however, that periodical cicadas reduced the standing crop of living rootlets of oak trees (Karban 1985). Cicada nymphs form feeding cells of unknown permanence around themselves and their host rootlet, and establishing a new feeding site is probably extremely expensive. Therefore, nymphs may refrain from feeding at a rate that kills their host rootlet.

Two "natural experiments" support the hypothesis that cicada growth rates are limited by food. *Mogannia minuta* required 1–5 years to complete development in *Miscanthus* grassland on Okinawa (Table 14.1) (Ito and Nagamine 1981). Most individuals matured in 3 years. Recently this species has colonized sugarcane fields that are heavily fertilized where most individuals matured in only 2 years. Similarly *Diceroprocta apache* develops in 3 years on desert shrubs in Arizona and northern Mexico (L. D. Anderson personal communication and E. L. Nigh ms.). On fertilized storage roots of asparagus, in contrast, nymphs feed throughout the winter and 60% mature in 1 year and another 20–35% complete development in the second year (E. L. Nigh ms. and personal communication).

Periodical cicadas generally emerge synchronously, but occasionally indi-

Table 14.1. Development time for *Mogannia minuta*[a]

	Percentage of nymphs eclosing to adults	
Year	Fertilized sugarcane	Miscanthus grassland
1	6.7	7.3
2	67.1	28.5
3	19.6	42.7
4	6.6	14.2
5	0	7.2

[a]Data from Ito and Nagamine (1981).

viduals shorten their development time. Individuals that do emerge either a few days or a few years before most of the cohort, however, stand little chance of escaping avian predation (Karban 1982). If growth of periodical cicadas is limited by food, we might expect individuals from better feeding sites to be larger and/or more fecund. Nymphs in a fertilized apple orchard were significantly larger than those in adjacent forests (Maier 1980). White and Lloyd (1975) observed that "when digging up these nymphs, one gets the distinct impression that the larger nymphs come from the better feeding sites."

Unfortunately, "natural experiments" vary in many factors besides presumed differences in host nutrition. While these observations are interesting and suggestive, they do not allow conclusions about the factors that caused the differences in developmental period or size.

Interspecific comparisons provide further evidence supporting the hypothesis that growth rates of cicadas are limited by nutrition or time available to feed. Using life history information for all cicada species for which reliable data on development time was available (Table 14.2), I found a significant negative correlation between latitude and growth rate (Kendall coefficient of rank correlation, $N = 64$, $t = 2.20$, $n = 12$, $p < 0.05$) (Figure 14.1). Cicadas that live in more equatorial latitudes add more length per year than species from more polar latitudes. Latitude presumably reflects the length of the growing season, as well as other factors. Other workers have noted that closely related species exhibit more rapid growth in lower latitudes compared to temperate and polar species (Schoener and Janzen 1968; Masaki 1978).

The hypothesis that prolonged development is associated with poor nutrition posits that there is selection for large body size and that, given the nutritional constraints imposed by feeding on xylem fluid, it takes a long time to attain a large size. Less time is required, however, in tropical habitats with longer seasons than in temperate ones. According to this scheme, selection for body size drives evolution of the other correlated life history traits. To argue that small size is associated with shorter time to maturation, I am assuming that xylem fluid and rates of assimilation do not vary greatly for the cicada species. If this hypothesis is correct that selection for large body size is very strong, then

Table 14.2. Life histories of cicadas

Species[a]	Body length[b] (mm)	Dev time[c] (yr)	Growth rate[d] (mm/yr)	Latitude	Range
Cyclochila australasiae	40–50	3[a]	13.3–16.7	18–36	S.E. Australia, Fiji
Diceroprocta apache	20–26	3[b]	6.7–8.7	29–37	S.W. U.S.A., Mexico
Mogannia minuta	12–17	2[c]	6.0–8.5	24–27	Ryukyu Arch.
Platypleura kaempferi	21–23	4[d]	5.3–5.8	5–45	S.E. Asia–N. China
Graptopsaltria nigrofuscata	32–39	7[e]	4.6–5.6	33–42	Japan–Manchuria
Oncotympana coreana	30–38	7[f]	4.3–5.4	25–43	Japan, Korea, China
Munza (=Platypleura) kuroiwae	16–21	4[g]	4.0–5.3	27–29	Ryukyu Arch.
Cicadetta cruentata	14–23	3.5[h]	4.0–4.7	35–44	Australia, New Zealand
Cicadetta calliope	12–15	4[i]	3.0–3.8	30–41	U.S.A.
Okanagana rimosa	20–24	9[j]	2.2–2.7	35–50	North America
Migicicada spp. (13 yr)	24–28	13[k]	1.8–2.2	30–40	U.S.A.
Magicicada spp. (17 yr)	24–28	17[k]	1.4–1.6	34–44	U.S.A.

[a]Nomenclature follows Metcalf (1963).

[b]Body length from head to abdomen in mm. From published reports and measurements of museum specimens.

[c]References are as follows: *a*–Froggatt (1903); *b*–L.D. Anderson, pers. comm. and E.L. Nigh ms.; *c*–Nagamine et al. (1975); *d*–Hirose (1977); *e*–Kato (1961); *f*–Kato (1961); *g*–Nagamine et al. (1982) and Y. Ito, pers. comm.; *h*–Cumber (1952); *i*–Beamer (1928); *j*–Soper et al. (1976); *k*–Marlatt (1907).

[d]Growth rate = body length/development time.

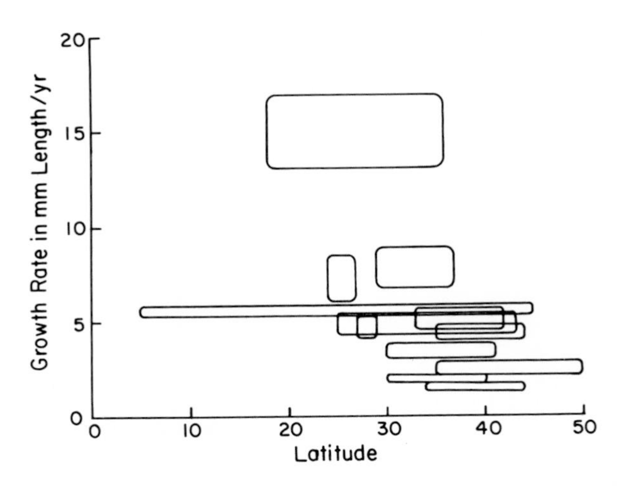

Figure 14.1. Latitude and growth rate. (Data from Table 14.2).

species that complete development in a shorter time should grow more rapidly but should not be smaller than species that require more time to complete development. In other words, development time should be more variable for cicadas than body size and development time should be correlated with growth rate but not with size. These predictions were supported by the interspecific comparisons. Growth rate and development time were significantly negatively correlated (Kendall coefficient of rank correlation test: $N = 82$, $t = 2.86$, $n = 12$, $p < 0.05$) (Figure 14.2). There was no relationship, however, between

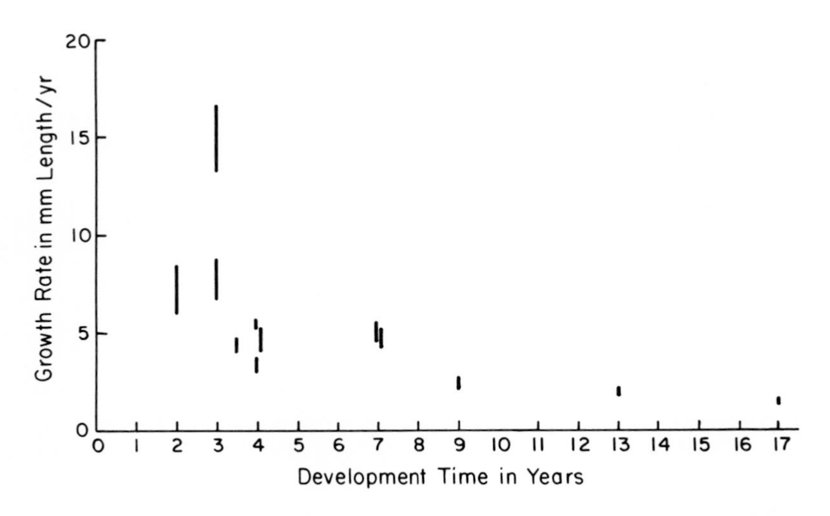

Figure 14.2. Development time and growth rate. (Data from Table 14.2).

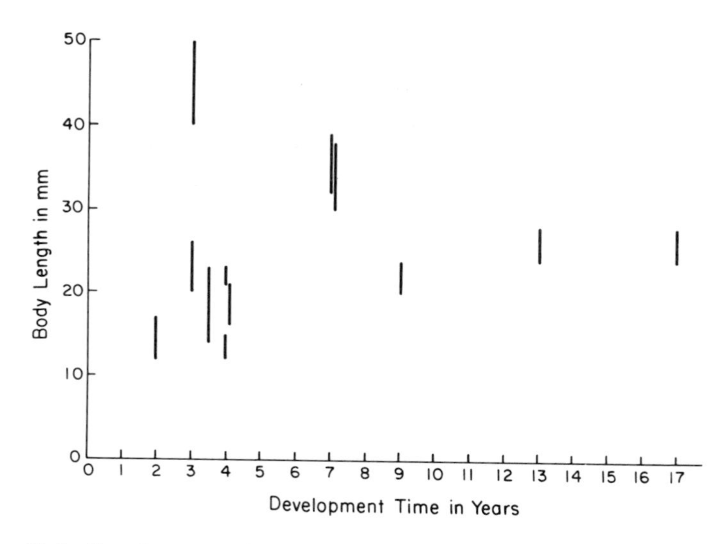

Figure 14.3. Development time and adult body length. (Data from Table 14.2).

size and development time (Kendall coefficient of rank correlation test: $N = 28$, $t = 0.99$, $n = 12$, NS.) (Figure 14.3). Species that developed in few years were not smaller than those that took longer to mature.

This line of reasoning assumes that large body size is strongly favored by selection and the following observations support this idea. All cicada species drum or sing and a large resonating abdomen is essential for loud sound production (Pringle 1957; Lloyd and Dybas 1966). Singing is an important part of mate selection in cicadas (Darwin 1871; Dunning et al. 1979). Larger males produce a lower-pitched song, which carries farther than song produced by smaller cicadas (MacNally and Young 1981; D. Young, personal communication). Larger males of *Magicicada cassini* were more successful in achieving copulations with females than smaller individuals (Karban 1983). There is some evidence that larger females may have greater fecundity as well. During nymphal development, female periodical cicadas accumulate resources used for eggs (Brown and Chippendale 1973) and females from "better" sites contained more vitellogenic follicles than those from sites of presumed lesser quality (Karban 1981).

There may be many other advantages of large adult body size (see reviews by Mattson (1980) for insects with low nitrogen diets and Blau (1981) for a general discussion). Large body size is often associated with increased physiological control over the environment. Larger size may make adult cicadas less vulnerable to predation, either because they are too difficult for predators to handle or because they can escape more readily. Large individuals of *Magicicada cassini,* however, were no more successful at avoiding sparrow predation in an aviary than were smaller individuals (Karban 1983).

Responding to this hypothesis that nutritional constraints result in long development for cicadas, Slansky (1980) argued that a spittlebug, *Philaenus spumarius,* feeds on xylem fluid and grows more rapidly than cicadas. *P. spumarius* attains an adult length of 5.5–6.0 mm, is univoltine throughout its worldwide range (Halkka et al. 1967), and has a relative growth rate of about 0.1 mg dry wt/mg dry weight/day (Wiegert 1964). Other xylem-feeding spittlebugs grow even more rapidly. *Aphrophora permutata* feeds on xylem fluid from pines in California and achieves an adult length of 9–12 mm. This spittlebug completes two generations per year. Calculating growth rate, as I have done for cicadas in Table 14.2, *A. permutata* grows 18–24 mm/year, which is a growth rate of 50% faster than that of any of the known cicada species. Although this is far from a definitive test, the relatively fast growth rates of spittlebugs, which like cicadas feed on xylem fluid, sheds doubt on the hypothesis that the growth of cicadas is limited solely by poor nutrition.

Selection for Prolonged Development

Previous discussions of prolonged development of cicadas have assumed that selection favors early reproduction rather than delayed reproduction (Lloyd and Dybas 1966; Lloyd 1984). Although selection favors a young age of first reproduction for individuals in populations that are increasing (Lewontin 1965), there is no advantage to reproducing at an early age for individuals in a stationary or declining population (Mertz 1971). Caswell and Hastings (1980) have calculated the trade-off in fecundity and development times for individuals in populations that are experiencing different rates of growth. The arguments that follow can be thought of as a special case of their model.

I will make the following two assumptions, which will be discussed later:

1. The population is not growing, i.e., $\lambda = 1$.
2. There is only one bout of reproduction and the adult dies after reproducing. This semelparous behavior is true for all cicada species.

Net reproductive rate, R_o, is the number of females surviving to age x (l_x) multiplied by fecundity at age x (m_x), summed for the entire life span.

$$R_0 = \sum_{x=0}^{\infty} l_x m_x$$

Since cicadas only reproduce once at maturity, age a, net reproductive rate is simply the probability of reaching adulthood (l_a) times fecundity at that age (m_a):

$$R_0 = l_a m_a$$

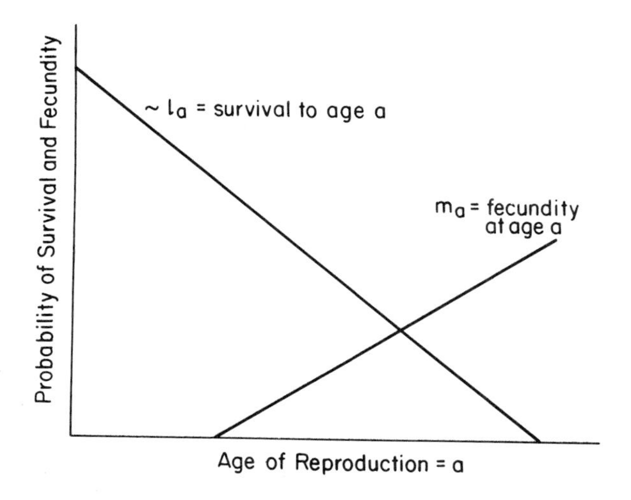

Figure 14.4. Survival and fecundity as a function of development time.

Any development time that maximizes the product of $l_a \times m_a$ should be-favored.

As the development time to sexual maturity is increased, the probability of prereproductive mortality will increase, although fecundity will also increase (Figure 14.4). If, however, the mortality schedule is shaped as in Figure 14.5, then selection will favor increasing the age of reproduction (development time). The situation depicted in Figure 14.5 may be extreme with the probability of

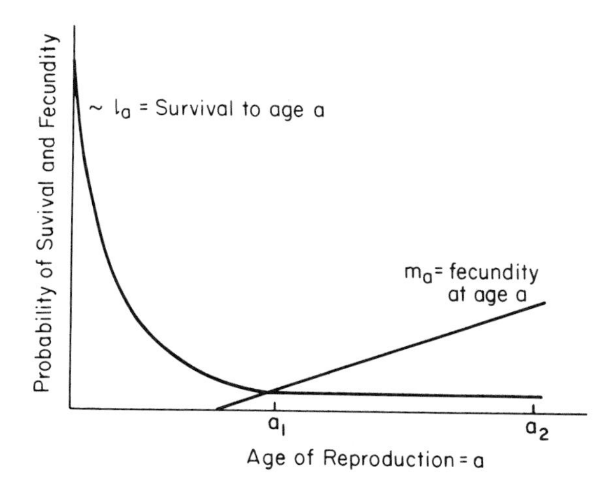

Figure 14.5. Survival and fecundity curves hypothesized for cicadas.

mortality being very great early in development and thereafter approaching zero. If l_a changes little over some range of possible ages (a_1 to a_2 in Figure 14.5) during which m_a increases, then there is an advantage in prolonging development from a_1 to a_2. In general, selection will favor prolonged development as long as m_a increases proportionately more rapidly than l_a decreases so that their product increases.

As long as the m_a increases proportionately faster than the l_a decreases, maturation should be delayed. Cicadas don't take even longer to mature because presumably either m_a begins to flatten out or l_a decreases rapidly. Cicadas maybe close to the maximum size that is physiologically possible.

The two species of cicadas for which juvenile survival has been studied show survivorship curves approximating the l_a curve in Figure 14.5. Mortality for *Mogannia minuta* is very great during early nymphal establishment, but becomes relatively flat following the second instar (Figure 14.6). The survivorship curve is similar for the 17-year cicada *Magicicada septendecim* (Figure 14.7). Even though the actual number of adults that will survive their 17-year development

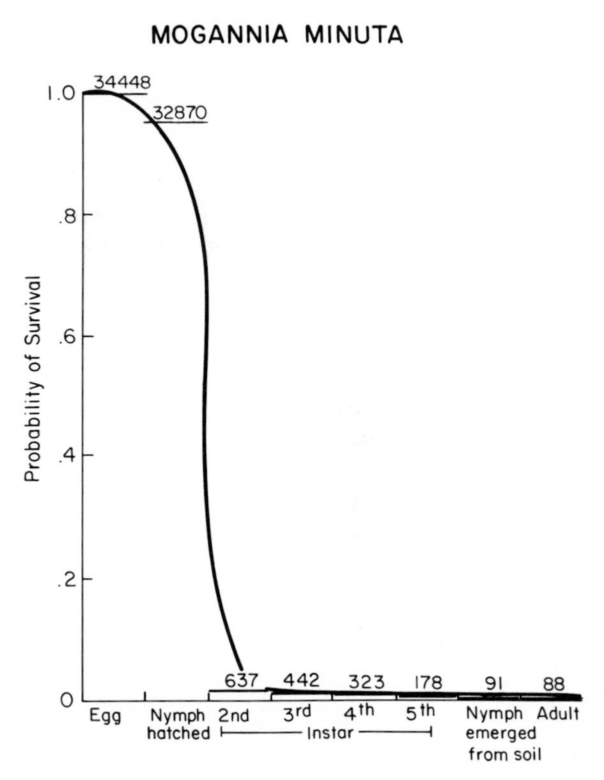

Figure 14.6. Stage-specific survival for *Mogannia minuta* on *Miscanthus*. Numbers shown represent actual population size of a cohort. (Data from Ito and Nagamine 1981.)

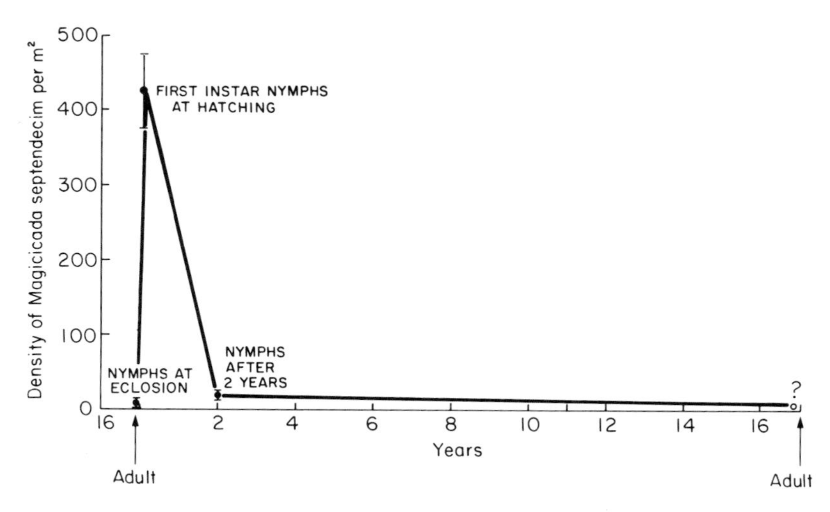

Figure 14.7. Survival of *Magicicada septendecim*. Means and standard errors are shown for three sampling times. The actual number of cicadas at eclosion in year 17 is unknown. (Data from Karban 1984.)

is unknown, the majority of mortality has already occurred by the second year of life (Karban 1984). Unfortunately, age-specific fecundity data are not available for cicadas, but it is plausible that fecundity will increase as development time increases. Egg number and adult body size are correlated for many other insects (Southwood 1978). It is hoped that age specific survival and fecundity schedules will be collected for several species of cicadas in the near future.

The two available long-term studies of cicada population dynamics support the assumption that populations of cicadas are not growing. Since 1965–1967 when *Mogannia minuta* was first recorded on Okinawa, population densities have remained nearly constant for seven or eight generations (Ito and Nagamine 1981). Densities of *Platypleura kaempferi* remained relatively stable from 1953 to 1969 (four generations) in a study site monitored by Hirose (1977). Populations of cicadas are thought to fluctuate much less than those of other insects (Ito 1978).

Even if cicada population sizes remain nearly constant on a wide spatial scale over many generations, they will experience local fluctuations in the short term. The theoretical result described above is still valid, however, under some conditions when the strict assumption of zero population growth is relaxed and the growth rate of the population fluctuates about zero (Hastings 1984; Prout and Christiansen, ms.).

Spittlebug life histories can also be compared with predictions of the demographic model developed for cicadas and presented above. Spittlebugs feed on xylem fluid and exhibit faster growth rates and shorter development periods than cicadas. If prolonged development is favored for cicadas because the risk of mortality associated with long life is relatively small, then the theory predicts that rapidly developing spittlebugs will risk increased mortality as immature development time is increased. The probability of mortality does remain relatively constant for all stages of *Neophilaenus lineatus* at two sites in England (Whittaker 1971), and for *Philaenus spumarius* at two sites in North America (Wiegert 1964; Karban unpublished). In other words, prolonged development for these two spittlebugs is associated with substantially increased risk of mortality. Thus consideration of spittlebug life histories supports the demographic hypothesis, but not the nutritional hypothesis, for prolonged development.

Conclusions

Cicadas require a relatively long development time to reach sexual maturity and they are semelparous, undergoing a single, short bout of reproduction, followed by death. Two hypotheses are considered to explain the very long-development times.

Cicada development may be limited by nutrition. Cicada nymphs feed exclusively on xylem fluid, an extremely nutrient-poor food that is expensive to procure. Two anecdotal examples of accelerated development of two cicada species associated with fertilization of the host support this hypothesis. Interspecific comparisons of all cicada species for which development time is known reveals that growth rate (body length added per year) is inversely correlated with latitude; i.e., tropical species grow more rapidly than temperate species although they are no different in size. This suggests that a long development is required to attain a large body size. The nutritional hypothesis is not supported by the observation that xylem feeding spittlebugs grow at a faster rate than any known species of cicada.

Prolonged development may be selectively favored if prolonged development is associated with increased fecundity and little risk of mortality. The two cicada species for which the shape of the survivorship schedule is available, exhibit high mortality during nymphal establishment but minimal mortality thereafter. Survival curves having this form lend credibility to the hypothesis, although knowledge of the effects of the length of development on fecundity is still lacking.

Acknowledgments I benefited from correspondences with and unpublished data of Lauren Anderson, Henry Dybas, Yosiaki Ito, Thomas Moore, Edward Nigh, Timothy Prout, and Allen Young. Measurements of cicada specimens were graciously provided by Michel Boulard, Masami Hayashi, Yosiaki Ito, Thomas Moore, Philip Ward, and Mick Webb. Earlier drafts of the manuscript were improved by Hugh Dingle, Alan Hastings, Daniel Janzen, Monte Lloyd, Leo Luckinbill, Timothy Prout, David Reznick, Chris Simon, Frank Slansky, Fritz Taylor, JoAnn White, and Truman Young. This research was supported by grants from the University of California and NSF DEB 79-14039.

References

Alexander, R.D., Moore, T.E.: The evolutionary relationships of 17-year and 13-year cicadas, and three new species (Homoptera: Cicadidae: *Magicicada*). Misc. Publ. Mus. Zool. Univ. Mich. 121, 1–59 (1962).

Anderssen, F.G.: Some seasonal changes in the tracheal sap of pear and apricot trees. Plant Physiol. 4, 459–476 (1929).

Beamer, R.H.: Studies on the biology of Kansas Cicadidae. Univ. Kansas Sci. Bull. 18, 155–230 (1928).

Blau, W.S.: Latitudinal variation in the life histories of insects occupying disturbed habitats: a case study. In: Insect Life History Patterns: Habitat and Geographic Variation. Denno, R.F. and Dingle, H. (eds.). NewYork: Springer-Verlag, 1981, pp. 75–95.

Brown, J.J., Chippendale, G.M.: Nature and fate of the nutrient reserves of the periodical (17-year) cicada. J. Insect Physiol. 19, 607–614 (1973).

Caswell, H., Hastings, A.: Fecundity, development time, and population growth rate: an analytical solution. Theor. Pop. Biol. 17, 71–79 (1980).

Cheung, W.W.K., Marshall, A.T.: Water and ion regulation in cicadas in relation to xylem feeding. J. Insect Physiol. 19, 1801–1816 (1973).

Cumber, R.A.: Notes on the biology of *Melampsalta cruentata* Fabr. (Hemiptera-Homoptera: Cicadidae), with special reference to the nymphal stages. Trans. Royal Entomol. Soc. Lond. 103, 219–237 (1952).

Darwin, C.: The Descent of Man, and Selection in Relation to Sex. London: Murray, 1871.

Dunning, D.C., Byers, J.A., Zanger, C.D.: Courtship in two species of periodical cicadas, *Magicicada septendecim* and *Magicicada cassini*. Anim. Behav. 27, 1073–1090 (1979).

Froggatt, W.W.: Cicadas and their habits. Agr. Gaz. N.S. Wales 14, 334–341, 418–425 (1903).

Halkka, O., Raatikainen, M., Vasarainen, A., Heinonen, L.: Ecology and ecological genetics of *Philaenus spumarius* (L.) (Homoptera) Ann. Zool. Fenn. 4, 1–18 (1967).

Hastings, A.: Evolution in a seasonal environment: simplicity lost? Evolution 38, 350–358 (1984).

Head, G.C.: Shedding of roots. In: Shedding of Plant Parts. Kozlowski, T. T. (ed.). New York: Academic Press, 1973, pp. 237–293.

Hirose, Y.: Annual population fluctuation of *Platypleura kaempferi* (Fabricius) (Homoptera, Cicadidae). Kontyu 45, 314–319 (1977).

Horsfield, D.: Relationship between feeding of *Philaenus spumarius* L. and the amino acid concentration in the xylem sap. Ecol. Entomol. 2, 259–266 (1977).

Ito, Y.: Comparative Ecology. Cambridge: Cambridge University Press, 1978.

Ito, Y., Nagamine, M.: Why a cicada, *Mogannia minuta* Matsumura, became a pest of sugarcane: an hypothesis based on the theory of escape. Ecol. Entomol. 6, 273–283 (1981).

Karban, R.: Effects of local density on fecundity and mating speed for periodical cicadas. Oecologia 51, 260–264 (1981).

Karban, R.: Increased reproductive success at high densitites and predator satiation for periodical cicadas. Ecology 63, 321–328 (1982).

Karban, R.: Sexual selection, body size and sex-related mortality in the cicada *Magicicada cassini*. Am. Midl. Natur. 109, 324–330 (1983).

Karban, R.: Opposite density effects of nymphal and adult mortality for periodical cicadas. Ecology 65, 1656–1661 (1984).

Karban, R.: Addition of periodical cicada nymphs to an oak forest: effects on cicada density, acorn production, and rootlet density. J. Kans. Entomol. Soc. 58, 269–276 (1985).

Lewontin, R.C.: Selection for colonizing ability. In: The Genetics of Colonizing Species. Baker, H.G., Stebbins, G.L. (eds.). New York: Academic Press, 1965, pp. 77–91.

Lloyd, M.: Periodical cicadas. Antenna 8, 79–91 (1984).

Lloyd, M., Dybas, H.S.: The periodical cicada problem. II. Evolution. Evolution 20, 466–505 (1966).

MacNally, R., Young, D.: Song energetics of the bladder cicada, *Cystomsoma saundersii*. J. Exp. Biol. 90, 185–196 (1981).

Maier, C.T.: A mole's eye view of seventeen-year periodical cicada nymphs, *Magicicada septendecim* (Hemiptera-Homoptera: Cicadidae). Ann. Entomol. Soc. Am. 73, 147–152 (1980).

Marlatt, C.L.: The periodical cicada. U.S.D.A. Bur. Entomol. Bull. 71, 1–181 (1907).

Masaki, S.: Seasonal and latitudinal adaptations in the life cycles of crickets. In: Evolution of Insect Migration and Diapause. Dingle, H. (ed.). New York: Springer-Verlag, 1978, pp. 72–100.

Mattson, W.J.: Herbivory in relation to plant nitrogen content. Annu. Rev. Ecol. Syst. 11, 119–161 (1980).

Mertz, D.B.: Life history phenomena in increasing and decreasing populations. In: Statistical Ecology. Vol. II. Patil, G.P., Pielou, E.C., Waters, W.E. (eds.). University Park: Penn State University Press, 1971, pp. 361–399.

Metcalf, Z.P.: A general catalogue of the Homoptera. Facicle VIII. Cicadidae. Baltimore: Waverly Press, 1963.)

Myers, J.G.: Insects Singers. A Natural History of Cicadas. London: Routledge, 1929.

Nagamine, M., Ito, Y.: *Platypleura kuroiwae* Matsumura (Hemiptera: Cicadidae), a cicada which became a pest of sugarcane in the Ryukyus. JpnJ. Appl. Entomol. Zool. 26, 80–84 (1982).

Nigh, E.L.: Summary of 1975–1981 Asparagus Research. Asparagus Workshop, Pasco, Washington (1982).

Pringle, J.W.S.: The structure and evolution of the organs of sound production in cicadas. Proc. Linn. Soc. Lond. 167, 144–159 (1957).

Prout, T., Christiansen, F.B.: Some theoretical aspects of selection on age specific fertility in a density regulated population (manuscript).

Raven, J.A.: Phytophages of xylem and phloem: a comparison of animal and plant sap-feeders. Adv. Ecol. Res. 13, 135–234 (1983).

Schoener, T.W., Janzen, D.H.: Notes on environmental determinants of tropical versus temperate insect size patterns. Am. Natur. 102, 207–224 (1968).

Slansky, F.: Eating xylem not enough for longevity. Bioscience 30, 220 (1980).

Soper, R.S., Delyzer, A.J., Smith, L.F.R.: The genus *Massospora* entomopathogenic for cicadas. Ann. Entomol. Soc. Am. 69, 89–95 (1976).

Southwood, T.R.E.: Ecological Methods With Particular Reference to the Study of Insect Populations. Second Edition. London: Chapman and Hall, 1978.

Tromp, J., Ovaa, J.C.: Seasonal variations in the amino acid composition of xylem sap of apple. Z. Pflanzenphysiol. 57, 11–21 (1967).

White, J., Lloyd, M.: Growth rates of 17- and 13-year periodical cicadas. Am. Midl. Natur. 94, 127–143 (1975).

White, J., Strehl, C.E.: Xylem feeding by periodical cicada nymphs on tree roots. Ecol. Entomol. 3, 323–327 (1978).

Whittaker, J.B.: Population changes in *Neophilaenus lineatus* (L.) (Homoptera: Cercopidae) in different parts of its range. J. Anim. Ecol. 40, 425–443 (1971).

Wiegert, R.G.: Population energetics of meadow spittlebugs (*Philaenus spumarius* L.) as affected by migration and habitat. Ecol. Monogr. 34, 217–241 (1964).

Chapter 15
Toward a Theory for the Evolution of the Timing of Hibernal Diapause

FRITZ TAYLOR

Few decisions that temperate-zone insects make during their lifetimes affect fitness more than that of entering hibernal diapause, for an insect diapausing too early in the growing season fails to reproduce when more than enough time remains to replace itself in the overwintering population and one committing itself to reproduction when it should have entered diapause reproduces without enough time for replacement before winter. Furthermore, the time when an individual enters diapause may influence survivorship through the winter and even alter the time when it will commence activity in the following year (Denlinger and Bradfield 1981; Henrich and Denlinger 1982). Given the significance of the timing of diapause induction to the overall success of seasonal life histories, it is of fundamental importance that we understand its evolution.

Because of its ubiquity and obvious importance, diapause induction has been extensively investigated by physiological ecologists. Only recently has it captured the interest of evolutionary ecologists. Our attempts to understand the evolution of this interesting trait will certainly be aided by the extensive background knowledge concerning its physiological and genetic basis. Three additional characteristics of diapause induction make it especially well suited for evolutionary studies. (1) In cases where diapause is induced to avoid high mortality brought on by a hard frost, the history of the selection regimen, i.e., the occurrence of hard frosts, can be reconstructed from local weather records. It is highly unusual for the historical record of selection on a trait to be so readily available. (2) The fitness associated with a particular time of entrance into diapause can be explicitly related to other life history traits such as survivorship, fecundity, and time of emergence from diapause (Taylor 1980, 1986a,b). Thus, one can determine the fitnesses for the various times when individuals are observed to enter diapause in a particular environment and one can also investigate the effects of variation in other life history traits on the pattern of fitness. (3) Also attractive is the likelihood that the timing of diapause induction is often not genetically correlated with other traits except, perhaps, development time (see below). Thus, in modeling the evolution of this trait it may not be necessary to consider simultaneously the effect of selection on other traits (Lande 1982) and this greatly simplifies the generation and testing of evolutionary predictions. In contrast, one of the central difficulties in testing predictions about the evolution of age-specific fecundity or survivorship stems from the likely genetic correlations between early fecundity and subsequent fecundity and survival.

In this chapter, I outline the structure for an evolutionary theory of the timing of diapause induction and review some of the findings from this theory. Then I discuss some further directions for the theory and related experimental studies.

An Overview of the Theory

The problem we shall deal with is that of a population of arthropods adapting to a periodically recurring catastrophe, as might occur with a hard frost in the autumn that kills an insect or its host plant or both, or with the sudden increase in predation on small aquatic invertebrates by fish in ponds in the spring (Hairston and Munns 1984). In such cases, the survivorship of age classes other than the diapause stage is sharply reduced by the catastrophe. An important ingredient in the problem is that the catastrophe be predicted by some event so that it can be prepared for in advance. Typically an arthropod deals with this problem in the following simple manner (Figure 15.1). Each individual passes through a sensitive period during which it assesses those variables in its environment that are in some way predictors of the catastrophe. For hibernal diapause, these predictors are usually photoperiod and temperature. Depending on the values of these variables, the arthropod either completes development and reproduces in the current season or it enters diapause and waits to complete development until conditions become suitable at some later time. The evolutionary problem we shall concentrate on is understanding why arthropods enter diapause when they do.

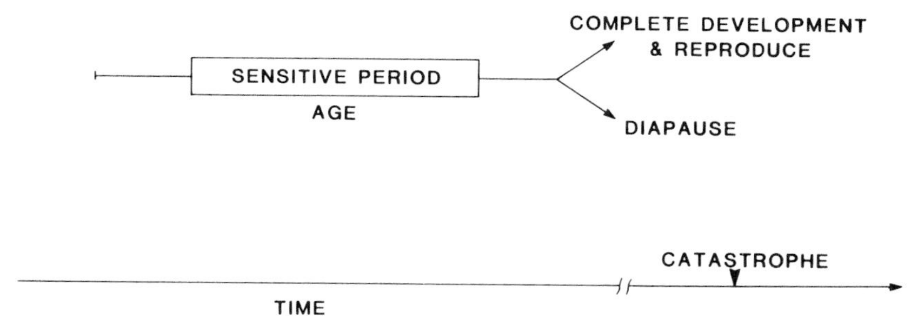

Figure 15.1. The diapause problem. The typical insect passes through its sensitive period during which it assesses its environment, making the decision whether to complete development in the current season or to diapause and wait until the next season to complete development. The decision is made relative to some catastrophe that greatly reduces the survivorship of individuals not in diapause.

The first step toward solving this problem is to define the term *switching time* as follows. Suppose two insects have the same switching time (Figure 15.2). One born early in the season reaches the end of its sensitive period before its switching time, t_s, and, consequently, does not enter diapause. The second, born later, reaches the end of its sensitive period after its switching time and will diapause. Thus, an individual reaching the end of its sensitive period on or after its switching time will diapause. Switching time is not synonymous with the critical photoperiod; for a population is characterized by a distribution of switching times, but only one critical photoperiod, which is the median of its photoperiodic response curve.

The goal of the theory outlined in this chapter is to predict the evolutionary changes in a population's distribution of switching times resulting from selection. The problem of understanding evolutionary changes in the distributions of phenotypic traits in general is fundamental to the study of adaptation. The major building block of the theory of the evolution of switching times consists of the processes occurring in a single growing season (Figure 15.3). At the start of the season, the population is characterized by a set of genotypes coding for a distribution of switching times, which is the culmination of its previous evolutionary history. Three elements of the theory interact in translating this distribution into that at the beginning of the next season. (1) *Physiological mechanism–environment interaction*. The actual distribution exhibited in the current season depends on the interaction between the physiological mechanism for diapause induction and the environment, i.e., events during the sensitive periods of all individuals in the population. (2)*Fitness function*. A fitness can be assigned to each switching time, depending on the particular environmental conditions during the growing season and on other aspects of the life history as explained below. This defines a fitness function. (3) *Mode of inheritance*. Finally, it is necessary to know the mode of inheritance for the trait, i.e., whether it is a simple Mendelian or polygenic trait and whether it is genetically coupled (by pleiotropy or linkage disequilibrium) to other important traits.

In what follows, I shall focus on one component of the theory, the analysis of the fitness function, touching on the physiological mechanism and the mode of inheritance toward the end of the chapter. After outlining the model used

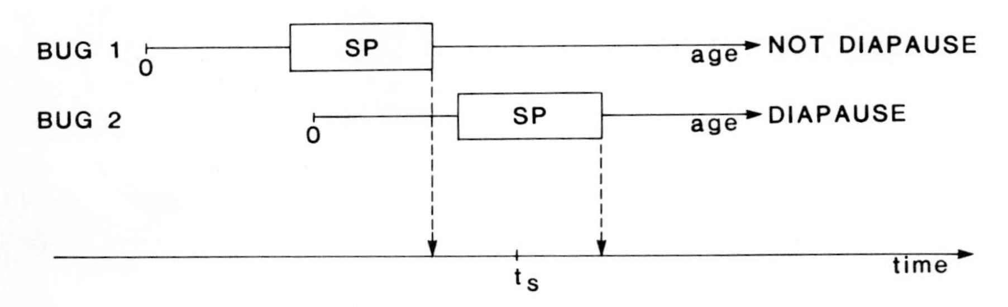

Figure 15.2. Switching time. An insect reaching the end of its sensitive period (SP) before its switching time, t_s, will not diapause, but one reaching this age on or after its switching time will diapause.

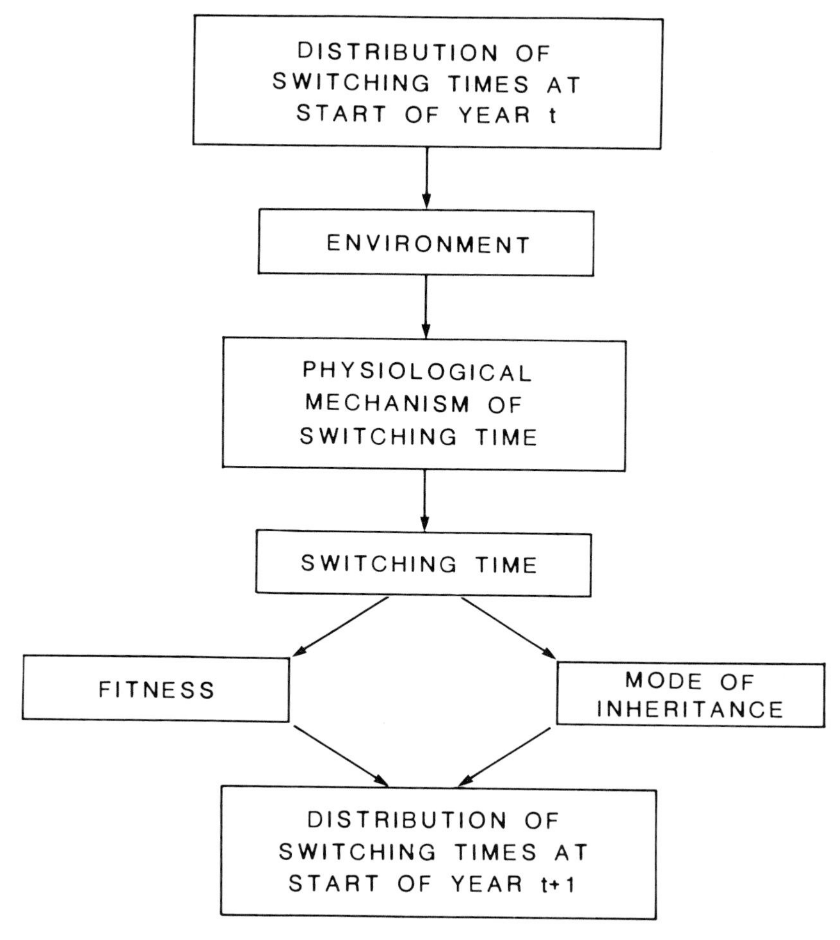

Figure 15.3. Diagram of the theory to explain the evolution of the distribution of switching times. It illustrates the three main requirements of the theory: (1) the physiological mechanism of diapause induction, (2) the fitness associated with each switching time, and (3) the mode of inheritance.

to derive fitness functions, I shall review the results concerning the two important characteristics of the fitness function: the position of its optimum and the determination of its shape.

Fitness Function

The Model

For concreteness, I shall consider insects preparing for hibernal diapause, even though the results from this model are relevant to arthropods entering diapause to avoid any catastrophic event. The form of this simulation model is identical to the renewal equation presented by Taylor (1980) except that for computational purposes it has been converted to the difference equation form (Charlesworth 1980, p. 11) as outlined below.

Both the age and time axes in the model (Figure 15.4) represent a scale of physiological time or developmental accumulation in which 100 developmental units (du) equals the time required to develop from birth to adult emergence, which in most arthropods precedes the age of first reproduction. This scale takes into account the fact that ectotherms develop at different rates depending on their body temperatures (Taylor 1981, 1982). I am using it here for generality, instead of days, because the fitness of a switching time relative to the end of the season must be computed on this scale in an environment in which temperature changes over time.

The model used to derive the fitness function can be divided into three parts (Figure 15.4). The first part includes the demographic components, which are age-specific fecundity and survivorship, and an initial age distribution. The usual convention will be followed of considering explicitly only females. Age-specific

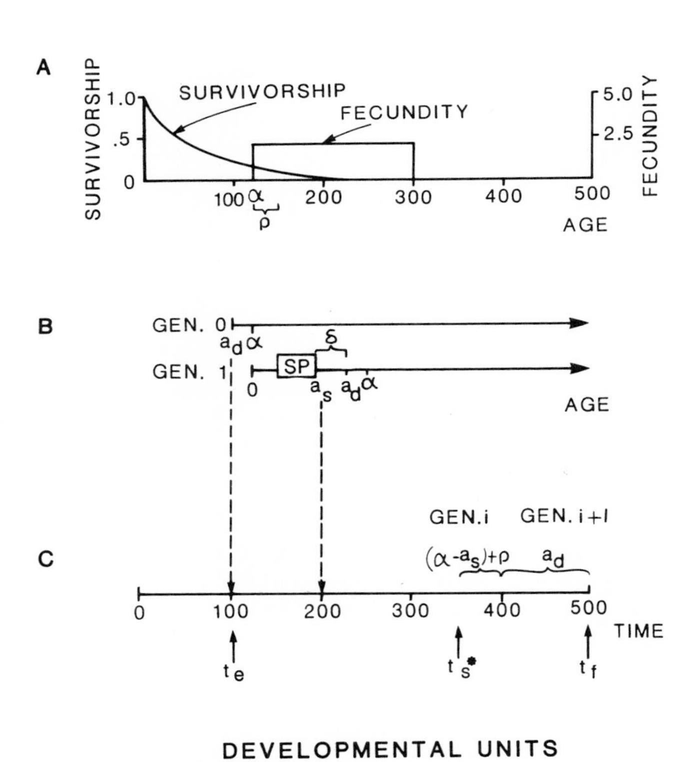

Figure 15.4. The three components of the theory to derive a fitness function. (A) Demographic parameters. (B) Physiological mechanism. (C) Time axis on which the diapause decision is made and the catastrophe occurs. See text for details. (After Taylor 1986a)

fecundity is represented by a step function, $m(a)$, starting at the age of first reproduction, α (Figure 15.4A). The time a female takes to produce one net female offspring, i.e., the number of female offspring such that on the average one will survive to the age of first reproduction, is indicated by the symbol ρ (rho) and will be called the *replacement time*. Age-specific survivorship, $l(a)$, is taken to be negatively exponential, meaning that there is a fixed probability of surviving through each age interval of constant length.

The distribution of emergence times, when individuals break diapause and resume normal development starting from the diapause age, a_d, determines the distribution of ages at the beginning of the season. The process of emergence can be treated by using either a single emergence time, t_e (as indicated on the time axis in Figure 15.4C), or a distribution of emergence times.

The second part of the model provides the physiological mechanism for diapause induction (Figure 15.4B). It is accepted by physiologists studying diapause that the decision whether to enter diapause at the diapause age, a_d, is made during a sensitive period for diapause induction, which ends at some age represented here by a_s (Beck 1980; Saunders 1982; Taylor 1985). There is a delay of length φ from age a_s, at which an individual may decide to enter diapause, until she actually does so at age a_d.

The third part consists of the time axis on which diapause decisions are made (Figure 15.4C). Females in the overwintering generation (generation 0) break diapause at time t_e, their development proceeds from age a_d, and they begin reproducing at age α. Females of generation 1 and subsequent generations develop through their sensitive period to age a_s when they decide whether to enter diapause. They do so by referring to their switching time. This model does not deal explicitly with processes taking place during the sensitive period.

I define switching time relative to age a_s because this is what is measured by the commonly run experiment to characterize a population's photoperiodic response in which samples are reared at a series of constant photoperiods and the percent diapause recorded for each sample. The photoperiodic response curve, thus derived, measures the distribution of day lengths that, when experienced during the sensitive period, will induce diapause, not the day lengths when individuals actually enter diapause. If one can monitor directly whether an individual enters diapause at age a_d, it is convenient to define an individual's *diapausing time*, t_d, which is similar to her switching time, but in this case measured relative to the time she reaches the diapause age, a_d. So $t_d = t_s + \delta$. By analogy, if she reaches age a_d on or after her diapausing time, she enters diapause.

Fitness is defined as the number of individuals in diapause with a particular switching time at the onset of winter relative to the number that had that switching time at the start of the growing season. So fitness measures the annual rate of increase for each switching time, assuming that overwintering survival is independent of switching time. A fitness function is determined when the fitness associated with each feasible switching time has been computed. The fitness functions in Figure 15.5 were computed by starting at time 0 with one female representing each switching time and fitness was measured as the number

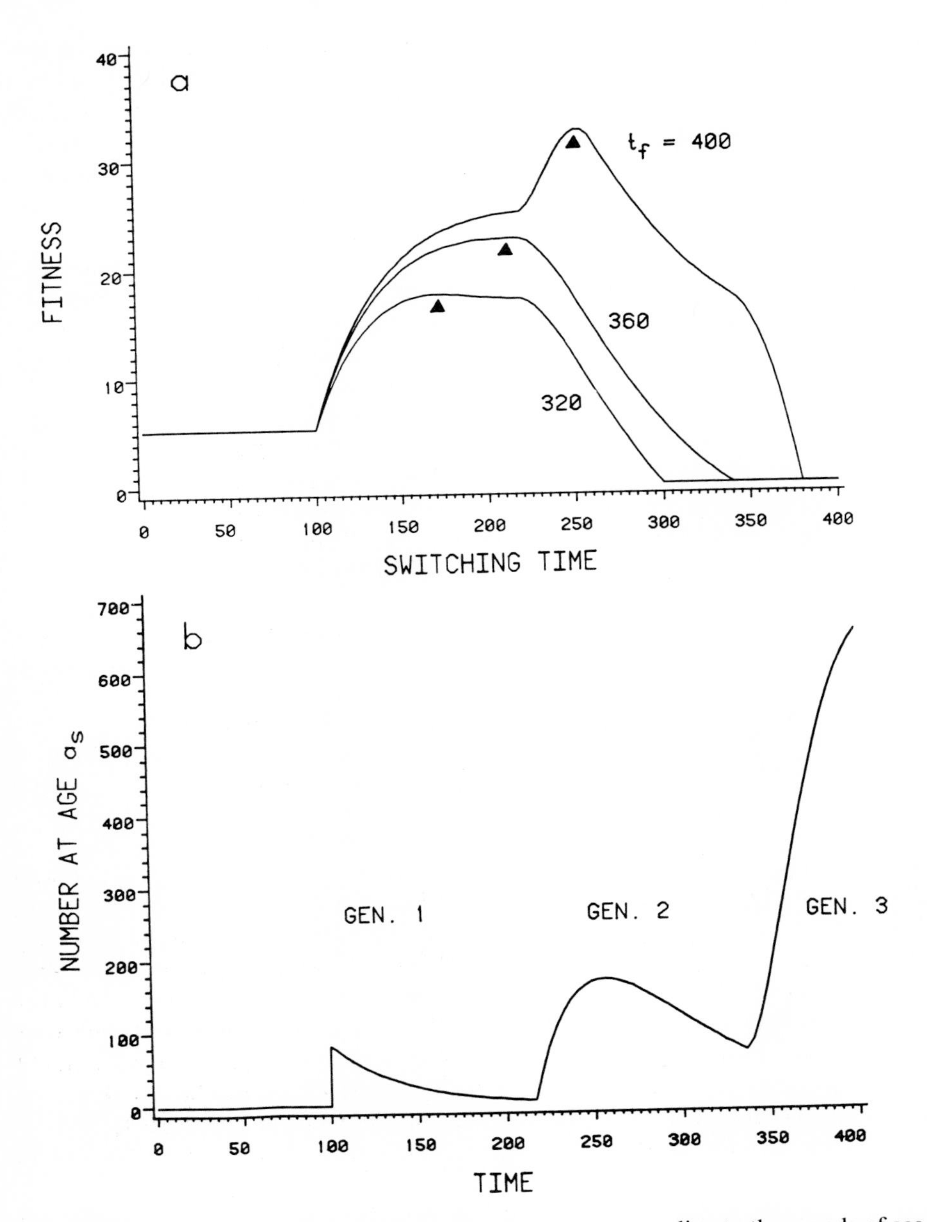

Figure 15.5. Fitness functions. (a) Fitness functions corresponding to three ends of season: t_f = 320, 360, and 400 du. Triangles indicate the optima. (b) Number of individuals at age a_s = 80 du as a function of time. (After Taylor 1986a)

of females with a particular switching time in diapause at the end of the season, time t_f.

We can now write the equations for the processes just outlined. Using the renewal equation model for age-structured population growth (see Taylor, 1980) and assuming population growth begins with all individuals at age a_d at the emergence time, t_e, the number of females of age a_d at each time is given by

$$n(a_d, t) = n_0(t - t_e, t - a_d)\, l_0(t - t_e)\, m_0(t - t_e) \quad (1)$$

$$+ l(a_d) \sum_{a=\alpha}^{\beta} B(t - a_d - a)\, S(t - a_d - a)\, l(a)\, m(a),$$

$$S(t) = \begin{cases} 1 \text{ if } t < t_s - a_s \\ 0 \text{ if } t \geq t_s - a_s \end{cases}$$

for $t - t_e \geq a_d$

where n_0 is the number of individuals in the overwintering cohort (generation 0) with age-specific survivorship and fecundity, l_0 and m_0, respectively. $S(t)$ is a step function that removes diapausing females from the population of reproductives. The first set of terms to the right of the equal sign in the top expression accounts for births due to overwintering females, all of which reproduce, and the second set of terms accounts for births (B) due to females of subsequent generations, some of which will diapause and, therefore, not reproduce. Age-specific survivorship and fecundity are given by $l(a)$ and $m(a)$, respectively. The values for age and time in n_0 arise because we are interested in individuals of generation 1 born at time $t - a_d$, at which time overwintering females were age $a_d + (t - t_e - a_d) = t - t_e$. Fitness for a switching time, t_s, is the number of individuals, N, in diapause at age a_d accumulated at the end of the season, assuming there was one individual with each switching time at the beginning:

$$N(t_s) = \sum_{t = t_s + \delta}^{t_f} n(a_d, t). \quad (2)$$

The model just outlined is appropriate for an asexual population in which offspring inherit exactly the mother's switching time. We have, nonetheless, found the results from this relatively simple model useful in understanding more complicated models incorporating sexual reproduction and variability in the end of the season (Taylor and Spalding, in preparation).

Position of the Optimum

Taylor (1980) has deduced that the optimal switching time, t_s^*, is given by the following formula:

$$t_s^* = t_f - (\alpha + \rho + \delta) \quad (3)$$

As illustrated in Figure 15.4c, the quantity $(\alpha + \rho + \delta)$ is the time, starting at age a_s, for an individual to replace herself with one diapausing offspring in the overwintering population, given that at age a_s the parent decides not to enter diapause. To see this, note that a female from generation i could produce one net offspring by reproducing until age $\alpha + \rho$, which occurs at a time $\alpha + \rho - a_s$ after she reaches age a_s. The last born of these offspring (in generation $i + 1$) would enter diapause at age a_d after a time interval of $\alpha + \rho - a_s + a_d = \alpha + \rho + \delta$ (because $a_d - a_s = \delta$) from when the female in generation i reached age a_s. Thus, $t_s{}^*$ is the first time that a female reaching age a_s *cannot* replace herself with one net offspring before t_f because her offspring reaching age a_d *on* or after t_f are killed. An individual that switches to diapause before $t_s{}^*$ foregoes reproduction when more than enough time remains to replace herself in the overwintering population. An individual committing herself to reproduction on or after $t_s{}^*$ does so when there is not enough time to replace herself in the overwintering population.

The Shape of the Fitness Function

The key to explaining the shape of the fitness function is understanding how it depends on the time course of individuals reaching age a_s at which they decide whether to enter diapause. To see this, we shall examine in detail the fitness function associated with an end of season at time $t_f = 400$ du in Figure 15.5a. The optimum is indicated by the triangle and corresponds to $t_s{}^* = t_f - (\alpha + \rho + \delta) = 400 - 148 = 252$ du as predicted by the corresponding analytical model (Taylor 1980).

Figure 15.5b shows the number of individuals at age $a_s = 80$ du at each time. In using this figure to understand the fitness function in Figure 15.5a, note that each switching time divides the individuals in Figure 15.5b into two groups, those to the left reaching age a_s before their switching time that reproduce and, thus, can be represented in the overwintering population only by their descendants, and those to the right reaching a_s on or after their switching time that diapause and may occur in the overwintering population themselves. The fitness associated with a switching time depends on the combined contribution to the overwintering population from these two groups. Thus, to understand each point in the fitness function in Figure 15.5a, the curve in Figure 15.5b must be considered in its entirety. The particular curve shown in Figure 15.5b was generated by using a very late switching time for which reproduction, or the production of individuals aged a_s, was not curtailed because no individuals had entered diapause. In general, for early switching times the number of individuals at age a_s will be less at later times because some individuals have entered diapause.

The fitness function is flat until switching time 100 for the following reason. All overwintering individuals (generation 0) were age $a_d = 100$ at the emergence time $t_e = 0$. They began to reproduce 20 time units later at age $\alpha = 120$ and their first-born offspring (generation 1) reached age 80 at time 100 (Figure 15.5b).

Thus, all switching times less than or equal to 100 would lead to 100% diapause (because no individuals reach age a_s before time 100, meaning that they all reach a_s on or after their switching time and will diapause) and would, therefore, confer the same fitness (because all of generation 1 would diapause and the model assumes that once in diapause an individual survives until the end of the season).

As switching times become greater than 100, some individuals reach age a_s before their switching times; they will not diapause but will reproduce with more than enough time to replace themselves in the next overwintering population. The fitness associated with later switching times increases because more individuals reproduce with more than enough time to replace themselves in the overwintering population (i.e., the proportion of individuals to the left of the switching time increases in Figure 15.5b as the switching time increases). Note that when there are more individuals passing through age a_s, fitness increases more rapidly per unit increase in switching time. Thus, there is a second increase in the slope of the fitness function beginning at time 220 when individuals from the second generation reach age a_s and still have more than enough time to replace themselves in the overwintering population. For switching times after the optimum, however, some individuals commit themselves to reproduction without enough time for replacement. Consequently, the fitness associated with later switching times declines. The slope of the fitness function becomes more negative at switching time 340 when individuals in generation 3 would begin to commit themselves to unsuccessful reproduction instead of entering diapause. Fitness reaches zero at time $t_f - \delta$ when individuals would die because they can no longer decide at age $a_s = 80$ to enter diapause and reach age $a_d = 100$ before the end of the season. The shapes of the fitness functions in Figure 15.5a for ends of the season at times 360 and 320 can be explained using similar logic. A comparison of these curves shows that the best-defined optima occur soon after rapid increases in the number of individuals passing through the end of the sensitive period.

The Relevance of the Optimum

For testing this theory, it is of interest to know the conditions under which the median switching time would be expected to occur near the optimum. The basic consideration is that if the optimum varies a great deal between years, the median will not correspond to the optimum in any particular year because the population response will reflect past positions of the optimum, which differed greatly from that in the current year. My twofold approach to this problem will be (1) to assess the conditions under which the optimum would vary little between years, so that it is easily tracked by a population, and (2) to try to understand the response of the median when the optimum shows large yearly variation and is not easily tracked. We shall consider in turn the parameters that may be expected a priori to influence the optimum in any year.

Developmental Parameters

The age of first reproduction, α, and the time, δ, between the end of the sensitive period and the diapause stage both represent intervals of development and are, therefore, constants on a physiological time scale. In a given year, the optimal switching time precedes the end of the season at time t_f by an interval of time, $\alpha + \delta + \rho$ (Eq. 3). Between years, α and δ will be constants on a physiological time scale. (Variability in ρ will be considered next.) Thus, as discussed below, variability in t_f can be translated into variability in the optimum by using local temperature records to translate $\alpha + \delta$ into an interval of days. Instances where this simple view may be incorrect include those in which development of the insect is particularly sensitive to variables in its environment other than temperature, e.g., host plant quality, or in which development times are very variable among individuals. In the latter case, the optimum would become highly variable within a year for different individuals in relation to even a well-defined catastrophe. This might significantly increase the variance of the distribution of switching times.

Demographic Parameters

Survivorship and fecundity influence the optimum via the replacement time, ρ, the interval of reproduction necessary for a female to produce enough offspring such that on the average one female survives to the age of first reproduction. Numerical analyses show that ρ, measured on a physiological time scale, and consequently the optimum, is least sensitive to variability in survivorship or fecundity in populations characterized by high fecundity and slow development (Taylor 1986b). At a finer level of analysis, it is of interest to know for particular populations, given their observed levels of fecundity, what levels of juvenile survivorship will lead to variability in ρ. The initial observation is that ρ is not sensitive to variation in juvenile survivorship when this quantity is high but does become sensitive as it becomes low. Or more precisely, ρ is more affected by a given absolute reduction in $l(\alpha)$, the proportion of the population surviving to age α, when $l(\alpha)$ is small than when it is large. For example, when $l(\alpha) = 1$ (i.e., all individuals survive to age α) a female need produce only one female offspring to replace herself, i.e., to have on the average one female offspring that survivesto age α. Reducing $l(\alpha)$ by 1% to 0.99 means that she must produce (1/0.99 =) 1.01 offspring. If, in constrast, $l(\alpha)$ is 0.1, a female must produce 10 offspring for replacement and, if $l(\alpha)$ is again reduced by 1% to 0.09, she must produce 11.1 offspring for replacement (ignoring adult mortality during reproduction). This effect on the sensitivity of ρ to changes in $l(\alpha)$ becomes much greater for yet smaller values of $l(\alpha)$. I analyzed demographic data from 22 arthropod species to determine a *critical level of juvenile survivorship* below which ρ becomes very sensitive to further reductions in $l(\alpha)$. (This was operationally defined as that value of $l(\alpha)$ for which a further reduction in $l(\alpha)$ of 1% gave a 10 du increase in ρ.) Eight out of 11 arthropod species, other than Lepidoptera, showed critical levels of l (α) less than or equal to 6%. Lepidoptera typically had lower critical levels, 10 out of 11 being $\leq$ 3%. A review of levels of juvenile survivorship observed in natural insect populations (Taylor 1986b) demonstrates that many insect populations in nature show temporal variation in juvenile survivorship above these critical levels. Thus, in many cases, variation in fecundity and survivorship should not keep the mean of the distribution of switching times from approaching the optimum.

The fecundity schedule can affect the optimum in another interesting way. Populations with fecundity schedules of short duration relative to the length of the juvenile phase are likely to have fitness functions with plateaus and no distinct optimum. This occurs when the optimum happens to lie at a time when no individuals are passing through the end of the sensitive period (cf. Figure 15.5), i.e., between generations. The possibility of this occurring is reduced by an emergence distribution with a high variance, which spreads out the age distribution and, consequently, increases the ranges of times when individuals are passing through the end of the sensitive period. This is the only case where the age structure influences the optimum, in this case whether an optimum occurs at all. Age structure otherwise never affects the position of the optimum (Taylor 1986a).

End of the Season

Variability in the end of the season, t_f, affects the position of the optimum directly (Eq. 3). We have investigated how this influences the evolution of the distribution of switching times using a simple model (Taylor and Spalding in preparation). In this model, values of t_f are selected at random from a gaussian distribution having a mean, μ_f, and standard deviation, σ_f. When females reproduce, they produce an offspring distribution, i.e., a gaussian distribution of offspring switching times, having a mean equal to the mother's switching time and a standard deviation characteristic for the population (Roughgarden 1979). This mode of inheritance can be interpreted as simulating a sexual population in which the heritability is 1 and there is perfect positive assortative mating. We chose this scheme to obtain maximal response to selection and because of the large number of seasons simulated (4800 years total). An expectation from this simulation was that the median response of the population would occur at an earlier switching time in an environment with the end of the season varying around some mean, μ_f, than in a constant environment with this same end of season. This results from an asymmetry in the fitness function in which fitness declines from the optimum faster for late than for early switching times, i.e., late switching times suffer lower fitness than do early. We tested this expectation by considering two levels of variability: a moderately high level with σ_f = 40 du and an extremely high level with σ_f = 80 du. Only when σ_f = 80 du did the median switching time occur significantly earlier than that corresponding to the optimal switching time in a constant environment with the same μ_f. This result contrasts with that of Hairston and Munns (1984) in which conservative switching times occurred when σ_f was approximately 40 du. The principle difference between the simulations that would effect such a difference is the mode of inheritance. In their model, each switching time was represented by a clone. This would exaggerate the effect of large reductions of fitness for later switching times in years having unusually early catastrophes because a switching time does not receive contributions from earlier switching times as the catastrophe subsequently occurs at later times. Thus, the median switching time, after being reduced by a particularly early end of season, moves only very slowly to higher values. In contrast, our simulation exaggerates the ability of the population to move to higher switching times under the same circumstances. The evolutionary responses of real populations should lie between these two extremes, but it is possible that variability in the time of the catastrophe will have little effect on the median response of a population.

In conclusion, under a variety of circumstances one can expect the median switching time to correspond to the optimum associated with the mean end of season over a series of years. For the optimum is not influenced by age structure and is often little affected by changes in survivorship or fecundity. The age of first reproduction, α, and the lag time, σ, between the end of the sensitive period and the diapause age are often constants on a physiological time scale. Finally, the median may exhibit little response to variability in the end of the season.

Adult Involvement as a Phylogenetic Constraint on the Timing of Diapause

The most common situation is that assumed so far in which a juvenile diapause age is preceded by a sensitive period in the same individual (Case 1 in Figure 15.6). As we have seen for this case, one can determine the optimal switching time and describe the shape of the fitness function. Often, however, the adult stage is involved as the diapause stage, as part of the sensitive period or in determining the diapause condition of offspring, and these alterations in diapause physiology change the optimum and the shape of the fitness function. Here I shall review the occurrences of the various ways adults are involved in diapause without providing details about the theoretical results. My intention is to summarize the observed relationships and to point out that formulas for the optimum and shape of the fitness function are available for all cases (Taylor and Spalding in preparation). In reviewing the literature on diapause induction, it became clear that there are four additional developmental pathways involved in diapause induction.

In Case 2 (Figure 15.6) there is an adult diapause stage and in Case 3 the adult participates in inducing diapause in the next generation. These two cases are subdivided further according to whether the sensitive period is determinate or indeterminate. When the sensitive period is *determinate*, it ends at a particular age and, subsequently, in Case 2a the adult either diapauses or reproduces and in Case 3a the adult produces offspring that all diapause or all complete de-

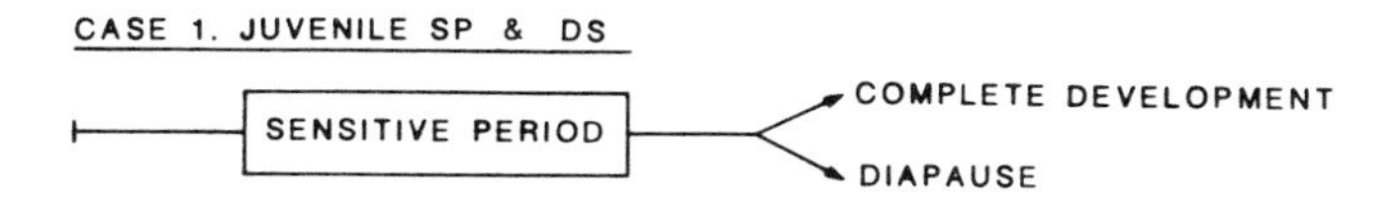

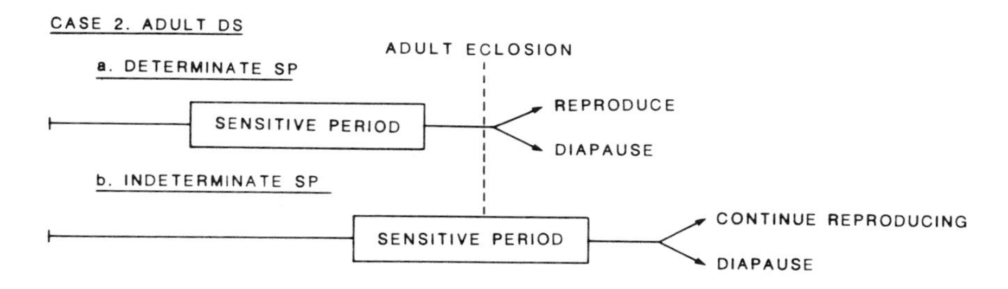

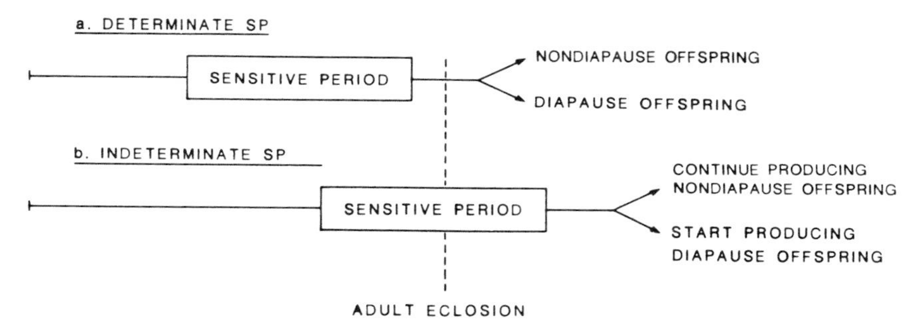

Figure 15.6. Adult involvement in diapause. A schematization of the various ways in which adults are involved in diapause. These are contrasted with Case 1 in which the sensitive period (SP) and the diapause stage (DS) occur during the juvenile period.

velopment. When the sensitive period is *indeterminate,* it can extend indefinitely into the adult stage. Thus, in Case 2b, adults can reproduce before entering diapause and in Case 3b adults can produce nondiapause offspring before producing diapause offspring.

Adult diapause (Case 2) has been observed in 10 orders of insects (Saunders 1982, Appendix Table 4) and in a mite (Hoy 1975). A determinate sensitive period (Case 2a) must occur when the sensitive period is preadult as in some species in six orders of insects (Saunders 1982). The sensitive period occurs entirely in the adult or extends into the adult stage in some species of insects (Saunders 1982), plus a mite (Hoy 1975). The appropriate experiment to differentiate between the two types of sensitive period, when the sensitive period includes the adult, is to rear the organism in diapause averting conditions until reproduction begins and then to transfer it to diapause inducing conditions. In all instances when this has been done, reproduction ceased and diapause ensued, indicating an indeterminate sensitive period. Thus, as a rule, an adult sensitive period associated with an adult diapause stage is indeterminate and falls into Case 2b.

The usual situation for Case 3 of an adult sensitive period followed by an egg or embryonic diapause stage in the next generation has been observed in six insect orders, as well as in a mite (Lees 1953), *Daphnia,* and a copepod (Hairston and Munns 1984). Occasionally the diapause stage in this instance is larval. In every case where the appropriate experiment has been carried out, the sensitive period, if it includes the adult, has been shown to be indeterminate. Thus, again the rule is that an adult sensitive period is indeterminate making these examples of Case 3b. Occasionally, in four insect orders, a juvenile sensitive period in one generation precedes a diapause stage in the next generation and these examples belong to Case 3a.

In summary, all four developmental pathways involving the adult stage, except perhaps Case 3a, commonly occur among the arthropods. In addition it is apparently evolutionarily difficult to make the transition among Cases 1, 2, and 3 (Taylor and Spalding in preparation). The data are not available to assess the transition between deterministic and indeterministic sensitive periods. Thus, these alternative physiological mechanisms constitute phylogenetic constraints on diapause induction. It is, therefore, important to recognize that the evolution of the timing of diapause induction will differ in each case, because the fitness functions differ, and to investigate these differences.

Outstanding Problems

What Is Hibernal Diapause Induction an Adaptation For?

The theory developed so far assumes that a catastrophe at time t_f greatly reduces survivorship for all individuals not in diapause. The obvious candidate for a sudden catastrophe is a hard frost. There is evidence, but actually surprisingly little, that insects in diapause are more resistant to low temperatures than those not (e.g., Hanec and Beck 1960; Miller 1978; Lee 1980). Certainly leaves of plants are damaged by low temperatures (Rosenberg et al. 1983) and this could lead to starvation of the actively feeding stages of phytophagous insects that are not killed by similar, low temperatures.

It is noteworthy that, if the catastrophe selecting for diapause induction is a hard frost, the end of the season may be quite predictable. To study variability in the timing of the first hard frost, Spalding and Taylor (in preparation) analyzed temperature data from over 100 weather stations along two latitudinal gradients in the United States, one from Georgia to Maine on the East Coast and a second from Texas to North Dakota in the Great Plains. Variability between years in the date and day length corresponding to the optimal switching times was computed as follows: The date of the first hard frost of -4 or -8°C was determined for each year of the series of years available for each station. Then each year, starting at this date, we computed backward, using max-min temperature data, 125 du and recorded the date when this occurred. This date approximates the optimal switching time (Eq. 3) on a physiological time scale. The day length on this date provides the optimal day length to use in diapause induction. By using three physiological time scales, we determined the effect of development rate on variability in the optimum. This was done by using three values of R_m, the maximal rate of development associated with the optimal temperature (Taylor 1981). An $R_m = 4\%$ per day means that at the optimal temperature the insect completes development to adulthood in 25 days compared to 8⅓ days for an R_m 12% per day. The results for the East Coast (Figure 15.7) show that the date of the optimum is about equally variable on all scales. Each point represents the interquartile range, or central 50% about the median, for all the years at

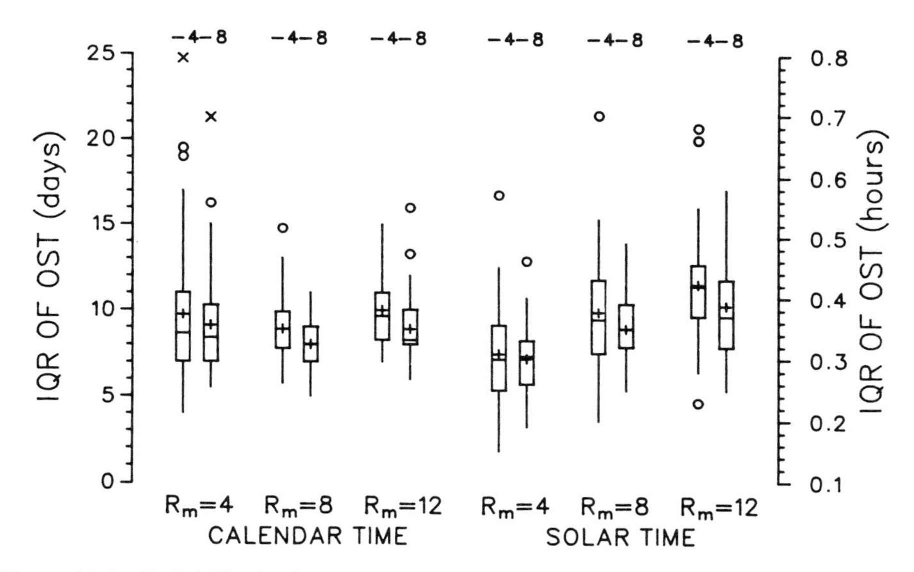

Figure 15.7. Variability in the occurrence of hard frosts on the East Coast. The results for all of the locations on the East Coast are summarized by box plots in which the limits of the box represent the interquartile range (IQR), or central 50%, centered around the median indicated by the line crossing the box. The mean is indicated by the cross in the box. Beyond the limits of the box, the tails extend to the most extreme values within 1.5 IQRs of the end of the box. Mild outliers, denoted by circles, lie between 1.5 and 3 IQRs and extreme outliers, denoted by x's, lie more than 3 IQRs from the end of the box. Each point in the distribution summarized by a box plot here represents the IQR of a distribution for one location as follows. The six box plots on the left give the results for the analysis of variability in the dates, or days of the year, corresponding to the optimal switching times (OST). The six box plots on the right give the results for the day lengths corresponding to these dates. Two criteria for the end of the season were used, −4 and −8°C. Three maximal rates of development were used, $R_m = 4$, 8, and 12% per day. See the text for further explanation.

one station. Surprisingly, the interquartile range for most locations is less than 10 days. When translated into day lengths, these interquartile ranges remain low. Those for slower developing insects tend to be lowest because these optimal switching times correspond to dates well before the autumnal equinox, when day length changes more slowly, compared to those for faster developers, which approach the equinox. In the majority of instances, the variability of day lengths at the optimal switching times is low, being less than 0.4 hr, or 24 min. Thus, if the catastrophe selecting for diapause induction is a hard frost, there may be remarkably little variation among years in the optimal switching time.

Clearly there are other types of catastrophe and the patterns may differ depending on how the insect makes its living, e.g., whether it is an herbivore, detritivore, predator, or parasitoid. Two major possibilities that would complicate the theory can be illustrated by considering phytophagous insects. (1) The probability of survivorship for nondiapause individuals may decline precipitously as above, but there may be great variability among individuals as to when the catastrophe occurs. This pattern could happen when nondiapause individuals have widely differing lower lethal temperatures, which may be brought about by differing rates of acclimation as well as differing ultimate freezing points. It could also happen if individuals are in very different microclimates or if various host plants freeze at different times. In this case, the catastrophe would no longer occur at one time; there would be a distribution of catastrophe times each year. (2) Alternatively, the probability of survivorship for nondiapausing individuals may decline gradually as might occur on a senescing host plant. This is a difficult case to deal with because fecundity may be affected in addition to survivorship and because survivorship may be age-dependent. For example, those ages nearest the diapause age may have higher survivorship because they suffer poor host quality for less time.

It will be important to quantify these complicating processes in natural populations and to determine their implications for the evolution of diapause induction.

Heritability of Switching Times

Estimation of the heritability of switching times would *ideally* proceed by collecting data on switching times of individual insects in a standard analysis of variance design (Falconer 1981). To implement the standard techniques, e.g., full- or half-sib design, data at the family level (sire × dam cross) would have to be actual switching times to obtain a mean and standard deviation for each family. A switching time, however, is observed only as a threshold variable, i.e., each individual either does or does not enter diapause. We are overcoming this problem (Schrader and Taylor in preparation) by using probit or logit assay techniques (Finney 1978) to estimate the family distribution of switching times. In an experiment to estimate heritability, samples from each cross are exposed to naturally declining day lengths at successively later times by using a solar timer. The resulting pattern of increasing percentage diapause over time for samples experiencing shorter day lengths is fitted by a cumulative probability distribution using probit or logit analysis and this provides an estimate of the mean and standard deviation of switching times at the family level. The results are analyzed by modified ANOVA procedures to obtain an estimate of the heritability. In principle, this new method could be used to estimate the heritability of switching times under natural conditions. It may also be applied to any threshold character for which assays can be run at the family level, e.g., susceptibility to freezing or pesticides.

More Realistic Evolutionary Models

As discussed above, two attempts have been made to understand the evolution of the distribution of switching times, but in one case each switching time was treated as a clone (Hairston and Munns 1984), greatly restricting the population response to variability in the time of catastrophe, and in the other, the mode of inheritance may have permitted too rapid a response to this variability (Taylor and Spalding in preparation). Since the response of a population to variable, optimizing selection is likely to be considerably influenced by the details of the mode of inheritance, simulations should be conducted using other, more realistic models of quantitative inheritance (Lande 1976, 1977; Slatkin 1979; Felsenstein 1981). These will delimit the range of feasible responses and, when compared to the results of realistic selection experiments, should indicate which models will most accurately predict the responses of natural populations to yearly variability in the time of the catastrophe.

More About Adult Involvement

The various ways in which adults are involved in diapause suggest several intriguing problems. At the physiological level, for example, virtually nothing is known about how an indeterminate, adult sensitive period ends. A juvenile sensitive period is considered to end at a specific developmental stage and a method has been proposed to estimate this stage (Taylor 1985). An indeterminate sensitive period, in contrast, continues indefinitely until the onset of diapause-inducing conditions. The main question that has yet to be addressed is how, in this case, are environmental signals processed by the adult. For example, is information integrated from the earliest age of the sensitive period; or is there a moving window of time over which information is processed; and, in either case, is the most recently acquired data more important than that acquired earlier? Once the end of the sensitive period has been determined, it will be of interest to know how long a female takes to enter diapause or to begin producing diapausing offspring. As the theory points out (Taylor and Spalding in preparation), these processes are central to understanding the evolution of the timing of diapause.

Further questions of evolutionary interest involve the very existence of the two kinds of sensitive period, deterministic and indeterministic. Is there a fitness advantage to having one or the other in the context of a particular life history and is there sufficient genetic variability to make both feasible evolutionary alternatives, or has the dichotomy become fixed in divergent phylogenies? Simulations comparing populations having either the determinate or indeterminate sensitive periods, but otherwise having similar life history parameters, show miniscule differences in fitness at the optimal switching time (Taylor and Spalding in preparation). If these constitute adaptive differences, either extremely small differences in fitness have favored one alternative or more is going on than was accounted for in the simulations. Comparisons of closely related species having differing types of sensitive periods, for example among predaceous coccinellid beetles (Taylor 1984), might shed light on the evolutionary significance of this dichotomy.

The Role of the Physiological Mechanism

The simplicity of the physiological mechanism for diapause induction, as proposed by Saunders (1981, 1982, Beck 1980), offers a unique possibility for studying the influence of the underlying mechanism of a trait on its evolution. In Saunders' scheme, the insect counts the light-dark cycles (days) experienced in its sensitive period and, if these exceed its required day number, it enters diapause. The observation that the duration of the sensitive period is mainly a function of temperature, whereas the required day number is temperature-compensated (Saunders 1971), and that the required day number is a function of day length, whereas the duration of the sensitive period usually is not, explains the characteristic responses of diapause induction to temperature and day length. A test of the utility of this model in understanding the timing of diapause induction in nature is to see if, after estimating the duration of the sensitive period and the required day number, one can predict the course of diapause induction in samples of insects exposed to changing day length and temperature (Taylor in preparation).

Given an understanding of the physiological mechanism, it is possible to study how it may alter the course of the evolution of the timing of diapause induction in changing environments. As suggested in the chapter (by Taylor and Spalding, in this volume) reviewing geographical patterns in diapause induction, it will be interesting to use the mechanism to understand how marked differences in the variance of the distribution of switching times are maintained.

Testing the Theory

The theory suggests that a population experiencing a catastrophe time that is not too variable (although it certainly does not have to be constant) and that does not suffer survivorship below the critical level should exhibit a median switching time corresponding to the mean optimum over a series of years. Furthermore, if the date of the catastrophe is preserved in the historical record, for example the date of a hard frost, and if development is almost entirely a function of temperature, the historical sequence of optima can be reconstructed from local temperature records (corrected for microclimates experienced by the insects [Fulton and Haynes 1977]). The study by Hairston and Munns (1984) is instructive in illustrating the potential difficulties involved in carrying out this apparently straightforward program. Over a series of years, they observed diapause induction in the copepod, *Diaptomus sanguineus,* for which the catastrophe is a sudden surge in fish predation on gravid females as the water warms in the spring. Their test was hampered by difficulties in estimating the catastrophe date, variability in this date, development time from newborn egg to adult, the age of first egg laying, and replacement time. Thus, it is essential to choose a population where these parameters can be estimated with precision and accuracy. The problems are compounded when the catastrophe date is highly variable, as it may have been for Hairston and Munns, because, as already mentioned, the mode of inheritance may then greatly influence the population response. Attempts to test the theory are valuable, because they add to our knowledge of the processes involved, e.g., variability in the diapause response both within and between years as related to variability in the important parameters of catastrophe time, development rates, survivorship, and fecundity.

Concluding Remarks

In this chapter, I have attempted to review ongoing research into the evolution of the timing of diapause induction. It should support the contention that, at least at a theoretical level, much is known, although even here much remains to be done. Since only one study exists that attempts to apply this theory to a real population (Hairston and Munns 1984), clearly many opportunities remain for further observation and experimentation. An additional point, which I believe is amply supported by this chapter, involves the remarkable depth of knowledge that is required to understand the evolution of even a simple trait, as the timing of diapause induction would surely appear to be (Figure 15.2). This knowledge extends from the minute details of the physiological mechanism of the trait (and I did not consider the biological clock, which underlies the required day number [Saunders 1982]) and its genetic basis to the characteristics of the population being studied and the environment in which it resides. An essential ingredient in understanding any process as complicated as the evolution of a biological trait is a comprehensive theory. I hope that the beginnings of the theory outlined here for the timing of diapause induction will provide some guidance for theoretical and empirical studies yet to be undertaken.

Acknowledgments I appreciate the comments of Ollar Fuller and Ted Nusbaum on an earlier draft of this chapter. Yevanne Ramsey drew Figures 15.1 through 15.4. This research was funded by the following NSF grants: DEB-8104698 and DEB-8208998.

References

Beck, S.D.: Insect Photoperiodism. 2nd edit. New York, Academic Press, 1980.

Charlesworth, B.: Evolution in Age-Structured Populations. Cambridge, Cambridge University Press, 1980.

Denlinger, D.L., Bradfield, J.Y.IV.: Duration of pupal diapause in the tobacco hornworm is determined by number of short days received by the larva. J. Exp. Biol. 19, 331–337 (1981).

Falconer, D.S.: Introduction to Quantitative Genetics. 2nd edit. NewYork, Longman, 1981.

Felsenstein, J.: Continuous-genotype models and assortative mating. Theor. Pop. Biol. 19, 341–357 (1981).

Finney, J.D.: Statistical Method in Biological Assay. 3rd edit. London, Charles Griffing and Company, 1978.

Fulton, W.C., Haynes, D.L.: Use of regression equations to increase the usefulness of historical temperature data in on-line pest management. Environ. Entomol. 6, 393–399 (1977).

Hairston, N.G., Munns, W.R., Jr.: The timing of copepod diapause as an evolutionarily stable strategy. Am. Natur. 123, 733–751 (1984).

Hanec, W., Beck, S.D.: Cold hardiness in the European corn borer, *Pyrausta nubilalis*. J. Insect Physiol. 5, 169–180 (1960).

Henrich, V.C., Denlinger, D.L.: Selection for late pupariation affects diapause incidence and duration in the flesh fly, *Sarcophaga bullata*. Physiol. Entomol. 7, 407–411 (1982).

Hoy, M.A.: Diapause in the mite *Metaseiulus occidentalis:* stages sensitive to photoperiodic induction. J. Insect Physiol. 21, 745–751 (1975).

Lande, R.: Natural selection and random genetic drift in phenotypic evolution. Evolution 30, 314–334 (1976).

Lande, R.: The influence of the mating system on the maintenance of genetic variability in polygenic characters. Genetics 86, 485–498 (1977).

Lande, R.: A quantitative genetics theory of life history evolution. Ecology 63, 607–615 (1982).

Lee, R.E.: Physiological adaptations of coccinellidae to supranivean and subnivean hibernacula. J. Insect Physoil. 26, 138–148 (1980).

Lees, A.D.: Environmental factors controlling the evocation and termination of diapause in the fruit tree red spider mite *Metatetranychus ulmi* Koch (Acarina, Tetranychidae). Ann. Appl. Biol. 40, 449–486 (1953).

Miller, L.K.: Physical and chemical changes associated with seasonal alterations of freezing tolerance in the adult northern tenebrionid, *Upis ceramboides*. J. Insect Physiol. 24, 791–796 (1978).

Rosenberg, N.J., Blad, B.L., Verma, S.B.: Microclimate, The Biological Environment. 2nd edit. New York, John Wiley and Sons, 1983.

Roughgarden, J.: Theory of Population Genetics and Evolutionary Ecology: An Introduction. New York: Macmillan, 1979.

Saunders, D.S.: The temperature-compensated photoperiodic clock "programming" development and pupal diapause in the flesh-fly, *Sarcophaga argyrostoma*. J .Insect Physiol. 17, 801–812 (1971).

Saunders, D.S.: Insect photoperiodism—the clock and the counter, a review. Physiol. Entomol. 6, 99–116 (1981).

Saunders, D.S.: Insect Clocks. 2nd edit. New York, Pergamon Press, 1982.

Slatkin, M.: Frequency- and density-dependent selection on a quantitative character. Genetics 93, 755–771 (1979).

Taylor, F.: Optimal switching to diapause in relation to the onset of winter. Theor. Pop. Biol. 18, 125–133 (1980).

Taylor, F.: Ecology and evolution of physiological time in insects. Am. Natur. 117, 1–23 (1981).

Taylor, F.: Sensitivity of physiological time in arthropods to variation of its parameters. Environ. Entomol. 11, 573–577 (1982).

Taylor, F.: Mexican bean beetles mate successfully in diapause. Int. J. Invert. Reprod. Dev. 6, 297–302 (1984).

Taylor, F.: Estimating the ends of the sensitive period for diapause induction in arthropods. J. Theor. Biol. (117, 319–336) (1985).

Taylor, F.: The fitness functions associated with diapause induction in arthropods. I. The effects of age structure. Theor. Pop. Biol. (In press) (1986a).

Taylor, F.: The fitness functions associated with diapause induction in arthropods. II. The effects of fecundity and survivorship on the optimum. Theor. Pop. Biol. (In press) (1986b).

Part IV
Concluding Remarks

Chapter 16

Pervasive Themes in Insect Life Cycle Strategies

WILLIAM E. BRADSHAW

Two prominent themes recur as organizing constructs for studies on life history evolution. The first envisions life history variation as adaptations to the relative intensity and/or timing of mortality incurred by various stages of the life cycle (Istock 1967; Murphy 1968; Emlen 1970; Schaffer 1974a,b; Livdahl 1979). The second envisions life histories as adaptations responding to the degree of density dependence experienced by populations (MacArthur 1962; MacArthur and Wilson 1967; Pianka 1970). The theoretical and conceptual offspring of these themes are enormous as any review will reveal (e.g., Wilbur et al. 1974; Giesel 1976; Stearns 1976, 1977; Gould 1977). Direct experiments designed specifically to discriminate between or test these theories have been slower to emerge. Reasonable attempts in the laboratory have been undertaken with bacteria (Luckinbill 1978, 1984), protozoa (Luckinbill 1979), *Drosophila* (Giesel and Zettler 1980; Mueller and Ayala 1981; Barclay and Gregory 1981), and copepods (Bergmans 1984).

Tests of theory among natural populations may be harder to establish, but they exist nonetheless. Bradshaw and Holzapfel (this volume) describe a system in which density-dependent selection operates relatively free of the complications of interspecific competition, predation, or variation in microhabitat over geographical distances. Shapiro (this volume) sought to establish at what systematic levels r and K selection might be operating. His examination of pierid butterflies from temperate to tropical and high-elevation populations and species revealed a striking departure from the general pierid pattern in the form of high Andean *Phulia*. Shapiro concluded that the life history tactics of *Phulia* were not surprising in the context of their habitat but what was surprising was the failure of r and K selection theory to explain arctic bumblebees, aphids, and *Phulia*. *Phulia* and the combined responses to laboratory selection cited above suggest that the "simple model of r and K selection based on the effects of density alone seems inadequate to explain variation in life history features of complex organisms" (Barclay and Gregory 1981). Results do conform more generally to the predictions of bet hedging but without a high degree of accuracy.

Even when some adaptations are consistent with predictions, the specific traits involved appear highly idiosyncratic. Having discarded r and K selection and having found a myriad of idiosyncracies surrounding bet hedging, what are the pervasive themes of life cycle evolution?

In even the most thoroughly rejected forms of r and K selection theory, there are elements of actual life histories that drew our attention in the first place. The lure of r and K selection was and is, I believe, the realization that most life cycles do arrange themselves in suites of apparently co-adapted characters that we call strategies. Similarly, many strategies share patterns in common and, along with Dingle (this volume), I would like to call these patterns "syndromes." Syndrome implies only that if we observe the presence of certain traits, we should be able to predict the presence of others as well without being upset if they do not conform to constants in the logistic equation.

We would be making much better progress if we were, at this point, identifying common syndromes of life cycle strategies rather than each of us comparing our own data to theories we all know are not absolute. We already know a number of life cycle syndromes of which, under the weight of new data, some are crumbling, some are being confirmed, and some are still emerging. Since my emphasis is on conceptual progress, I shall concentrate on the latter two categories.

Adaptation Is Alive and Well

One school of thought, stemming from the work of Eldredge and Gould (1972, Gould 1977), perceives evolution to proceed in temporally sporadic spurts separated by longer periods of relative stasis. When this concept is extended from morphological to life historical traits, major changes in life history tactics (tactic: a set of co-adapted traits, designed by natural selection, to solve particular ecological problems; Stearns 1976) are not likely to occur during the lifetime of a species. A variety of empirical studies on insects, however, have shown that life history traits do vary within species, usually over geographical ranges (Walter and Hacker 1974; Stearns 1976; Giesel 1976; Hegmann and Dingle 1982; Dingle et al. 1982; Giesel et al. 1982; Dingle and Baldwin 1983; Allan 1984; Taylor and Spalding, this volume) or in contrasting subhabitats (Spielman 1957; Crovello and Hacker 1972; Stearns 1976; Hairston and Munns 1984). Fisher's "fundamental theorem of natural selection" (1958) can be interpreted to imply that characters such as life history traits that have a large impact on fitness (fitness characters) should have little or no underlying genetic variability as the result of adaptive depletion of such variability. Perhaps this interpretation would have some validity if directional selection were the only form of selection operating. However, selection in varying directions (Haldane and Jayakar 1963; Levins and MacArthur 1966; Powell 1971; Roughgarden 1972; Bradshaw 1973; Istock 1981) or negative genetic correlations between traits (Williams 1957; Lande 1980, 1982a,b; Rose and Charlesworth 1981a,b; Falconer 1981; Rose 1982, 1983) may serve to maintain genetic variability despite apparently strong selective forces. Substantial additive genetic variance persists among fitness characters (Jinks and Broadhurst 1963; Perrins and Jones 1974; Dingle et al. 1977; Istock 1981; Lynch 1984), and even plastic fitness characters may respond rapidly to selection (Stearns 1983; Via 1984).

Still, proponents of evolutionary stasis argue that life history traits are constrained by phylogeny and ontogeny to tight functional and/or genetic units so that more complete understanding of life history evolution is to be gained only by considering the whole organism, not individual traits (Gould and Lewontin 1979; Williamson 1981; Tuomi et al. 1983). I would not argue that evolution, development, and genetic correlations impose constraints to adaptation. However, some traits appear relatively free from genetic correlations (e.g., Dingle et al. 1982; Hegmann and Dingle 1982) and even among those that are genetically correlated with other traits, the degree of genetic correlation may vary from one population to another (Berven and Gill 1983). Thus, the focus of selection may be an independent trait, a complex of genetically correlated traits, and/or the degree of genetic correlation itself.

In a study that provides a model of how we might pursue life cycle strategies, Masaki (this volume) first correlated ovipositor length with body size and then sought causes for deviations from this regression. Ground crickets (*Pteronemobius*) superficially appear constrained by ovipositor length: body size allometry (Masaki, Figure 2.3A, this volume); but, on closer inspection, individual species deviate from the genus mean relationship and these deviations are closely associated with habitat (Masaki, Figure 2.3B, this volume). Within one habitat type, deviations from allometry among *P. taprobanensis* and *P. mikado* are correlated with overwintering stage, latitude, and voltinism (Masaki, Figures 2.5 and 2.6, this volume). Longer ovipositors are then expected in drier habitats among univoltine crickets that overwinter as eggs at northern latitudes; shorter ovipositors are expected for crickets living in the contrasting habitats. Having recognized that there is an allometric correlation between ovipositor length and body size, Masaki did not stop at this relationship but probed further to find that species and populations are able to escape allometric constraints and adapt to their specific habitats.

We should beware of clutching onto convenient, current theory or dogma to explain the latest deviation from our expectations and let our data lead us to new patterns. Several times during the discussion after individual papers or at the end of this symposium, we heard phylo- or ontogenetic constraints invoked to "explain" why an organism cannot do something. I have known for a long time that, whereas a number of insect orders include species that may interpolate instars under poor food conditions, mosquitoes are constrained to only four. I also knew that there was no maternal care among mosquitoes, but, as Lounibos and Machado-Allison have shown (this volume), *Trichoprosopron digitatum* is unaware of its phylogenetic constraints and broods its eggs. I am eagerly awaiting the discovery of a fifth instar mosquito.

I am not advocating the complete dismissal of prior theory. We should continue to measure individual life cycle strategies against known theories and syndromes. But, when results are not congruent with theory, we should not reject either the theory as being wrong or the organisms as being too idiosyncratic. Rather, we should focus on these deviations from expected to see if they fall into a pattern, which identifies a new theory or syndrome.

The potentials for this approach are clear in Brown's chapter (this volume), which has taken perhaps the broadest overview of life history evolution in this symposium. She has asked to what extent the stage of vegetation succession selects for specific herbivore life cycles. From her data (Brown, Table 7.1, this volume) she has been able to test previous theories of Margalef (1968) and Odum (1969) as well as some predictions of MacArthur and Wilson (1967). More importantly, she has established convincingly that insect life cycles show pronounced changes along a successional gradient and that none of these changes are absolute. The latter point means that inhabiting each seral stage are insects whose life histories apparently contradict the general pattern. Rather than despairing these exceptions, I would like to know more about them to see if their life cycles reveal a secondary, underlying pattern. Are the exceptional egg diapausers also exceptional in their other life cycle characters or do they conform to the broader, overlying pattern of life cycles along a successional gradient?

For many years, but especially since Danilevskii's (1965) treatment of photoperiodism and seasonal development, life historians have been aware that photoperiodically mediated diapause represents one of the most consistent ecogeographical relationships: critical photoperiod tends to increase with latitude or altitude. I agree with Taylor and Spalding (this volume) that further "studies directed solely at examining latitudinal trends in the median day length or date of diapause induction response are redundant. It remains of great interest, however, to understand the evolutionary mechanisms maintaining the median response at even a single location." Their plot of latitudinal trends (their Figure 5.2) clearly illustrates considerable variation about the central tendency. Pursuing the theme of this section, I would like to ask what factors are associated with deviations from the overall trend?

The assertion I wish to make is that darkness of habitat can affect the interpretation of day length and, consequently, the timing of diapause. To illustrate this point, I shall use the mosquito species shown in Taylor and Spalding's Figure 5.2 and refer to their species designations in parentheses. Two of the mosquitoes, *Wyeomyia smithii* (4) and *Aedes atropalpus* (V), live in pitcher plants and rock holes, respectively, whereas *Toxorynchites rutilus* (3) and *Aedes sierrensis* (W) live in tree holes. Consequently, diapause in *T. rutilus* and *A. sierrensis* is determined under darker conditions than in *W. smithii* or *A. atropalpus*. At a given latitude (Taylor and Spalding, Figure 5.2) the median day length is lower and the day of the year on which the median day length occurs is later for *T. rutilus* and *A. sierrensis* than for *W. smithii* and *A. atropalpus*. These differences are based on each species' perceiving a day that begins at the onset of civil twilight in the dawn and ends at the termination of civil twilight in the dusk. The effective day length is shorter in tree holes in pitcher plants or rock holes, however, and tree-hole mosquitoes have compensated for this difference with shorter median day lengths than among the more exposed pitcher plant and rock hole species. Evidence consistent with this interpretation comes from *T. rutilus*. Fourth instars of *T. rutilus* captured 26 August at 40.3°N (Bradshaw and Holzapfel 1975) were all in diapause. This date is considerably earlier than 20 September estimated by Taylor and Spalding based on Bradshaw and Holzapfel's (1975) laboratory determination of mediam photoperiod but using as a photoperiodic day sunrise to sunset plus two civil twilights. By contrast, if, due to their dark habitat, the photoperiodic day of *T. rutilus* is based on sunrise to sunset only, the day of median photoperiod advances to 13 August, a date far more consistent with field observations. Thus, taking photic habitat into account can reduce the deviations from the central tendencies in Taylor and Spalding's (Figure 5.2) relationships. Certainly other factors such as generation time and the relationship of the sensitive period to the diapause stage will prove to be important considerations as well.

The important point here is that these considerations would not have been forthcoming without the variation shown in Taylor and Spalding's chapter. The overall trend, the positive correlation of critical day length or date with latitude is clear and familiar but, like ovipositor length in Masaki's crickets, new interesting relationships will be forthcoming by examining deviations from this trend.

Dingle (this volume) has pointed out that among lygaeid bugs of the genus *Oncopeltus,* subgenus *Oncopeltus* produces the more general lygaeid pattern of large clutches of eggs every few days whereas members of the subgenus *Erithrischius* produce small clutches every day. Among the former but not the latter, clutch size is positively correlated with body size. One can then view the dependency of clutch size on body size as a phylogenetic constraint imposed by the timing of oviposition. By contrast, further examination of members of the subgenus *Erithrischius* may reveal that they are the lygaeid equivalent of Shapiro's (this volume) *Phulia* and illustrate that adaptation is flourishing in the genus *Oncopeltus*.

Life Cycle Delays

In a pivotal paper, Lewontin (1965) showed that the intrinsic rate of natural increase (r) was more strongly influenced by a small change in age at first reproduction or in generation time than by an equivalent change in fecundity. Consequently, selection among colonizing species (and subsequently, among r-selected species) "will have long since shortened development time, but will not have acted as efficiently on fecundity." The direct implication is that whenever intrinsic rate of natural increase is a reasonable index of fitness, organisms should minimize both time to first reproduction and mean generation time. Yet, virtually every insect considered in this volume delays its development and/or reproduction beyond physiologically possible times.

Insect life cycle delays may take place over the span of hours, days, weeks, months, or even years. Some of these delays are extrinsically mediated by the direct effects of the environment; others, although environmentally cued or set, are the result of endogenous mechanisms. It is the latter type of life cycle delay that proves to be so intriguing because, whether or not we can identify ultimate causality, we must be left with the conclusion that endogenous delays are genetically programmed, adaptive responses to environmental variability.

Many developmental and reproductive events such as hatching, molting, emergence, mating, and oviposition occur at specific times of the day or night; these events usually persist with circadian regularity under constant conditions (Saunders 1982). Eggs having completed embryogenesis or pupae having completed adult development will delay hatching or molting until the time of an internally programmed, externally phased temporal "gate." Although hundreds of examples exist illustrating circadian rhythms among insects, the adaptive significance of only a relative few is known or reasonably inferred.

The adaptive significance of longer life cycle delays is not always so elusive. One of the finest examples of variation in endogenous orchestration of life cycle delays is presented by marine midges of the genus *Clunio* (Neumann, this volume). Marine *Clunio,* although living in an energy-rich environment, are characteristically slower developing than their freshwater counterparts and have very short-lived adults. To coordinate their reproduction with infrequent or variable exposure of their tidal or subtidal habitat, *Clunio* exhibit endogenous circadian emergence and circasemilunar pupation rhythms. Thus, even though individuals may be fully mature, pupation or eclosion may be delayed up to 2 weeks or 24 hr, respectively, if developmental maturity does not coincide with the proper phase of the endogenous oscillator. Phasing of the circasemilunar pupation rhythm is cued by temperature, moonlight, and tidal flux. In the south where the habitat is exposed only during a few days each semilunar cycle, moonlight is the primary cue; in the north where the habitat is exposed part of each day, tidal or temperature flux is the stronger cue. Tidal flux predominates on exposed coastlines and thermal flux in protected fjords. The interaction of endogenous circadian and circasemilunar rhythms with locally reliable zeitgebers thus effects coincidence of oviposition with the optimal habitat exposure between an insect with a complex life cycle and a temporally heterogeneous but highly periodic environment.

By far one of the most pervasive features of temperate insects is that they enter some form of seasonal developmental or reproductive arrest. Diapause generally effects a life cycle delay of several months so that development or reproduction ceases from the end of one growing season through aestival or hibenal harshness until the beginning of the next growing season. Both Dingle (this volume) and Sauer (this volume) point out that insects have an array of mechanisms available to them to arrive at the critical developmental or reproductive interval at the environmentally proper time. This array of mechanisms is wonderfully illustrated by various chapters in this volume. Wardhaugh (this volume) shows how photoperiod interacts with temperature and humidity to elicit an embryonic diapause syndrome characterized by changes in maternal oviposition behavior, egg-pod morphology, and embryonic physiology. Neumann (this volume) shows how hibernal diapause in *Clunio* is cued by photoperiod, coordinated with, and perhaps mediated through, endocrine mechanisms similar to those of the circasemilunar pupation rhythm. It is now clear that more than a few insects may diapause at more than one stage. The plague locust, *Chortoicetes terminifera* (Wardhaugh, this volume), may diapause as an embryo or as a nymph; under the influence of short days, eggs that did not diapause have a greater propensity to enter nymphal diapause. The flesh fly, *Calliphora vicina* (Vinogradova, this volume), may diapause twice, as a larva or as an adult. All populations are capable of adult, reproductive diapause but individuals among more northern populations may diapause as a larva as well. The burnet moth, *Zygaena trifolii* (Wipking and Neumann, this volume), may diapause in a variety of successive larval stages. Southern populations are generally bivoltine with an aestival diapause of variable duration. Among northern populatons, larvae enter hibernal diapause but may do so up to three times in successive larval stages spanning 3 years or more. Some individuals fail to terminate diapause after the first winter, molt, and develop during the third summer, and diapause again the following winter. The entire range of life cycles from univoltine to 3 years or more "was evident even within the progeny of a single female." Wipking and Neumann propose that females may therefore be "spreading the risk" (den Boer 1968) over several growing seasons as some summers are so inclement that they greatly inhibit reproductive success. Similarly, entry into diapause during successive stages of locusts (Wardhaugh) or flesh flies (Vinogradova) may provide a "fail-safe" mechanism against misinterpretation of environmental cues or against the lack of correlation between environmental cues and weather during a specific year (Lounibos and Bradshaw 1975; Bradshaw and Holzapfel 1977; Holzapfel and Bradshaw 1981).

Zygaena trifolii (Wipking and Neumann, this volume) provides an example of apparently inherited life cycle delays that extend beyond a year. Diapause in *Z. trifolii* is entirely different from diapause in the emperor dragonfly, *Anax imperator* (Corbet 1956), which is capable of completing development within one summer but when food is scarce, may enter diapause twice in the same generation during successive winters. Diapause in *A. imperator* reflects the direct effects of the immediate trophic environment on development whereas diapause in *Z. trifolii* can be preprogrammed by endogenous physiological mechanisms, independently of the environment.

Even longer life cycle delays are possible. Cicadas as a group exhibit longer life cycles and slower growth rates than other xylem feeding homopterans (Karban, this volume). In an extreme case, generation time may extend to 17 years. Karban has proposed that this great increase in preadult development relates to increased fitness accruing to individuals of larger size: larger males are more attractive to females than smaller males; larger females produce more eggs than smaller females. Cicadas emerge with great regularity at 17-year intervals. Regardless of the selective forces maintaining this synchrony, it persists over a range of local subhabitats and among cicadas of variable weight. Consequently, some endogenous counter or timer appears to be regulating this periodic emergence. As with *Clunio* and *Zygaena trifolii*, the endogenous nature of periodic development in cicadas argues that, even if we cannot identify ultimate causality, the periodicity is a genetically programmed response to the environment experienced by the insect during its evolutionary history.

My assertion is that periodic life cycle delays, which are independent of or compensated for the effects of the immediate environment, are characteristic of insect populations in general. When our attention focuses on these delays themselves, we are not likely to belittle their import. When our attention focuses on other life history traits, especially demographic traits, we need to integrate periodic delays into our considerations. Dingle's studies (this volume and literature cited therein) with lygaeid bugs provide a good example of this approach. Traits such as age of first reproduction, mean generation time, clutch size, and clutch frequency are then viewed as part of life cycle syndromes, that is, traits potentially interacting with diapause and migration. A pervasive theme is that insect life cycles are punctuated by exogenously cued but endogenously programmed delays; consequently, these delays must be integrated into any theory of life cycle evolution in insects.

Homeostasis, Polymorphism, and Environmental Uncertainty

Taylor (this volume) considers the onset of winter as a variable catastrophe in response to which insects adaptively vary sensitive periods and switching times mediating diapause. A fundamental consideration "is that the catastrophe be predicted by some event so that it can be prepared for in advance" (Taylor, this volume). Principal events used by insects to predict the onset of winter include photoperiod, temperature, and moisture. These cues exhibit a strong correlation with changing seasons, which of these cues provides the highest predicting power varies with geography, both within and between species. No single cue is perfectly correlated with the optimal time to enter diapause and, as we have seen above, insects use combinations of cues to induce diapause or enter diapause in sequential stages to provide a "fail-safe" mechanism against environmental variation. Even a combination of cues may not always provide very reliable indications of impending catastrophe. Yet, insects do develop, diapause, and persist in highly variable environments with few reliable cues. In this section, I shall propose that there is a consistent relationship between environmental predictability and the form of insect variability.

As early as 1953, Levene showed theoretically that multiple subhabitats could provide opportunity for the persistence of a variety of genotypes in a population. Levene's multiple-niche polymorphism has resurfaced over the years in many guises but serves as an organizing concept for all of them. Roff (this volume) proposes that genetic control for wing length should evolve from single- to multi-locus control so that organisms could present a continuously variable array of phenotypes to the environment. Dingle (this volume) reports that in California as opposed to Iowa, some winters are sufficiently mild that they do not kill nymphs of the milkweed bug, *Oncopeltus fasciatus*. Correspondingly, there is extreme variability in diapause response among California as compared with Iowa strains of *O. fasciatus* in the laboratory. Thus, there has been a maintenance of genetic variability among California *O. fasciatus* due to the uncertainty of California winters. The important point here is that it is the uncertainty of winter conditions, not their harshness, that selects for variability.

Sauer et al. (this volume) point out that whereas day length is the same every year, the end of the favorable season varies about some mean. The mean end of season at any locality may select for a population level critical day length and/or required day number for the initiation of diapause but the standard deviation in the end of season maintains significant genetic variability for these traits. At the same time, polygenic control provides a means for "genotypes that were completely eliminated by selection in one generation (to) be reconstituted by recombination the following generations" (Sauer et al., this volume). Thus, both genetic variability and the degree of polygenic control underlying life cycle traits should increase with environmental uncertainty (Sauer et al., Roff, this volume).

A lack of genetic variability does not necessarily imply a lack of phenotypic variability. When even very harsh conditions are predictable in advance from environmental cues, individual organisms may make physiological or developmental adjustments for this harshness before it impinges upon them. Thus, when environmental conditions are highly correlated, we may expect individual insects to display elaborate homeostases so that the appropriate phenotypes are deployed at the appropriate times.

The evolution of homeostasis provides a possible explanation for the maintenance of genetic polymorphisms within populations. Suppose (Figure 16.1) that there are two alleles at the α locus and that when both alleles occur in a population, they are maintained at frequencies greater than expected from mutation alone by heterosis, multiple subhabitats, or other factors (Figure 16.1, middle). This population is then polymorphic at the α locus. By virtue of its very stability, genetic polymorphism could persist sufficiently long for a variety

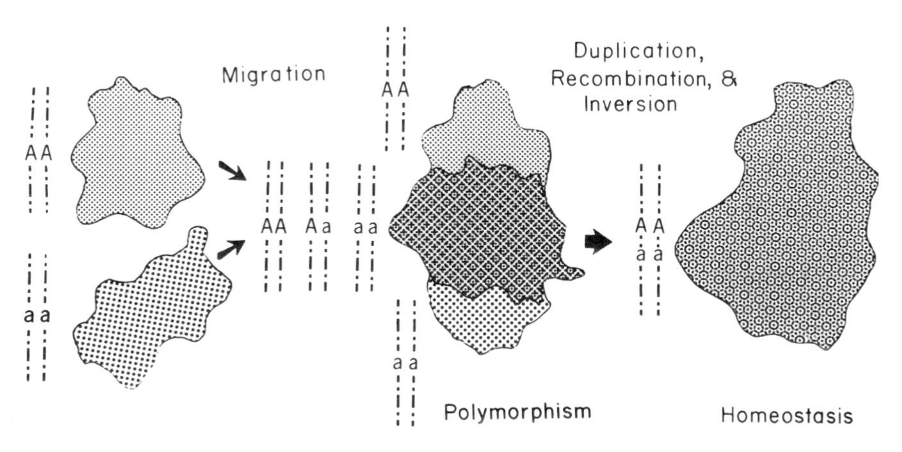

Figure 16.1. Evolution of homeostasis from genetic polymorphism.

of chromosomal events to take place. If these events included duplication, recombination, and inversion, the former alleles could come to occupy separate, homozygous loci protected from recombination (Figure 16.1, right). With a common operator for the two loci, each individual would be effectively heterozygous. But, if each locus was controlled by a separate operator that could respond independently to environmental factors, then each individual in the derived population would be able to assume the phenotypic repertoire of the entire ancestral population without the cost of homozygosity. An individual of the derived, homeostatic, and apparently homozygous population would, on the average, realize greater fitness than an individual of the ancestral polymorphic population. Homeostatic mechanisms should, therefore, replace stable polymorphism.

Why then should not all organisms accumulate progressively elaborate homeostases (Slobodkin 1968)? The answer, I believe, lies in the observation that homeostatic mechanisms enable organisms to overcome specific exigencies only if they can deploy the appropriate phenotype in advance of the exigency. Otherwise, a homeostatic individual has no advantage over an individual with a fixed phenotype. Whether or not polymorphism persists or evolves into a homeostatic mechanism depends on the relative timing of three events (Figure 16.2): (1) the time or age at which organisms must switch between alternate phenotypes, (2) the time of the catastrophe, period of environmental harshness, or other selective force, and (3) the timing of an environmental cue associated with the selective force. If the environmental cue is available at the time of or before the point of developmental decision and both events precede the selective force,then a homeostatic individual will realize greater average fitness than an individual with a fixed phenotype (Figure 16.2A). If the environmental cue occurs after the time of developmental decision (Figure 16.2B), a homeostatic

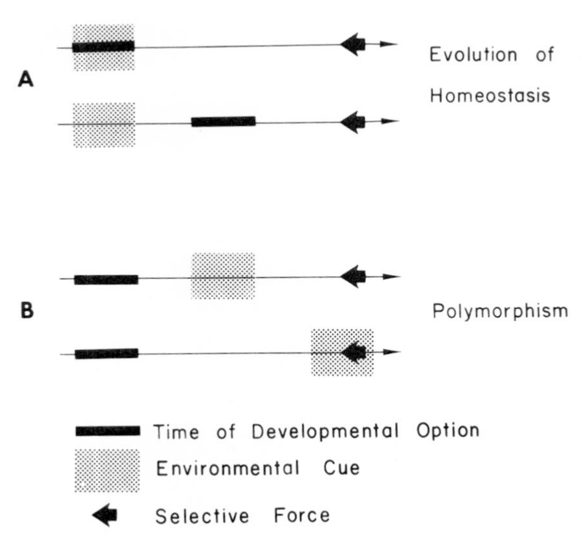

Figure 16.2. Maintenance of genetic polymorphism.

organism will have no advantage over an individual with a fixed phenotype, regardless of whether the environmental cue precedes or coincides with the selective force and genetic polymorphism will persist in the population. Thus, genetic polymorphisms should abound among loci involved in polygenic systems when environmental predictability is low; varying degrees of homeostases coded for by a few loci or tightly linked, co-adapted gene complexes should predominate when environmental predictability is high.

One has to be careful to identify the focus of selection. In a homeostatic population, it is the ability of the individual to deploy the appropriate phenotype in response to the correlated cue. In a genetically polymorphic population, it may be the degree of genetic variance or degree of gene–gene interaction. Finally, the focus of selection may be the range of phenotypes that an individual can produce among its offspring. Dingle (this volume) has shown how maternal effects in aphids can produce fine tuning to the environment, but there are also more elusive mechanisms. Diapause, as discussed above, may occur in successive stages of the life cycle. Roff (this volume) argues that it may be the pattern of wing length variation that is selected for rather than an optimal wing length alone. Bonner (1965) described what he termed "range variation" whereby the range of variation of a phenotype can be genetically determined but the position of an individual within that range is not. Bonner's range variation encompasses more recent manifestations such as "spreading the risk" (den Boer 1968) or "bet hedging" (Stearns 1976) without the implied ultimate causality of the latter terms. Such a mechanism may be producing the pattern of diapause among *Zygaena trifolii* (Wipking and Neumann, this volume) where individual females apparently produce progeny that, among them, show the entire range of variation from univoltine, single diapause generations to generations spanning more than 3 years, accompanied by up to three periods of diapause. Regardless of the mechanism by which range variation is effected, it is probably a far more widespread adaptive response to unpredictable environments than is presently appreciated.

Conclusions

The above considerations, well illustrated by Chapters in this volume, lead me to propose the following pervasive themes of insect life cycle strategies.

1. Life cycle delays are characteristic of most, if not all insect populations. Present knowledge concerning their genetic control indicates that they are underlaid by high heritabilities and low genetic correlations with demographic traits (Dingle, this volume). Consequently, despite apparent ontogenetic or phylogenetic constraints, life cycle delays have an enormous capacity to respond adaptively to environmental exigencies or opportunities.
2. In well adapted insects, phenotypic adjustment to predictable environmental events will involve little or no genetic change in the population;

phenotypic change in response to unpredictably environmental events will involve greater or predominantly genetic change in the population (Slobodkin 1968).

3. In moderately variable environments, developmental fine tuning may be achieved by facultative life cycle delays in successive stages and/or multiple interacting cues (Vinogradova, Wardhaugh, Wipking and Neumann, Neumann, Dingle, this volume).
4. The number of loci involved in life cycle delays will increase with environmental uncertainty (Roff, Sauer et al., this volume).
5. Phenotypic plasticity surrounding life cycle delays will increase with environmental predictability. Genetic variability and heritability of life cycle delays will increase with environmental uncertainty. Phenotypic plasticity and genetic variance-covariance relationships modulating life cycle events are therefore subject to selection and adaptive in nature.
6. When environments become highly unpredictable, individuals should deploy among their offspring the full array of potentially surviving life cycle phenotypes, i.e., by spreading the risk, hedging their bets, or practicing some other tactic of range variation.

Acknowledgments I would especially like to thank Christina Holzapfel for many years of enjoyable discussions concerning insect development, evolution, and life cycle strategies. I have also profited greatly from informal discussions over the years with Hugh Dingle, Fritz Taylor, and L. Phillip Lounibos. I am grateful to all the participants of this symposium for their enthusiastic input in the form of provocative papers and excited discussion in Hamburg and for their prompt submission of the interesting manuscripts that I have been again stimulated by reading. Particularly, I would like to extend my appreciation to Hugh Dingle for helping organize the symposium and to Fritz Taylor and Rick Karban for taking on the large responsibility of editing this volume. Finally, I thank the National Science Foundation for continual financial support, most recently through grant BSR-8114953.

References

Allan, J.D.: Life history variation in a freshwater copepod: evidence from population crosses. Evolution 38, 280–291 (1984).

Barclay, H.J., Gregory, P.T.: An experimental test of models predicting life history characteristics. Am. Natur. 117, 944–961 (1981).

Bergmans, M.: Life history adaptation to demographic regime in laboratory-cultured *Tisbe furcata* (Copepoda, Harpacticoida). Evolution 38, 292–299 (1984).

Berven, K.A., Gill, D.E.: Interpreting geographic variation in life history traits. Am. Zool. 23, 85–97 (1983).

den Boer, P.J.: Spreading the risk and stabililization of animal numbers. Acta Biotheor. 18, 165–194 (1968).

Bonner, J.T.: Size and Cycle. An Essay on the Structure of Biology. Princeton: Princeton University Press, 1965.

Bradshaw, W.E.: Homeostasis and polymorphism in vernal development of *Chaoborus americanus*. Ecology 54, 1247–1259 (1973).

Bradshaw, W.E., Holzapfel, C.M.: Biology of tree-hole mosquitoes: photoperiodic control of development in northern *Toxorhynchites rutilus* (Coq.). Can. J. Zool. 53, 713–719 (1975).

Bradshaw, W.E., Holzapfel, C.M.: Interaction between photoperiod, temperature, and chilling in dormant larvae of the tree-hole mosquito, *Toxorhynchites rutilus* Coq. Biol. Bull. 152, 147–158 (1977).

Corbet, P.S.: Environmental factors influencing the induction and termination of diapause in the emperor dragonfly, *Anax imperator*. J. Exp. Biol. 33, 1–14 (1956).

Crovello, T.J., Hacker, C.S.: Evolutionary strategies in life table characteristics among feral and urban strains of *Aedes aegypti* (L.). Evolution 26, 185–196 (1972).

Danilevskii, A.S.: Photoperiodism and Seasonal Development of Insects. Edinburgh and London: Oliver and Boyd, 1965.

Dingle, H., Baldwin, J.D.: Geographic variation in life histories: a comparison of tropical and temperate milkweek bugs (*Oncopeltus*). In: Diapause and Life Cycle Strategies in Insects. Brown, V.K., Hodek, I. (eds.). The Hague: Junk, 1983, pp. 143–166.

Dingle, H., Blau, W.S., Brown, C.K., Hegmann, J.P.: Population crosses and genetic structure of milkweed bug life histories. In: Evolution and Genetics of Life Histories. Dingle, H., Hegmann, J.P. (eds.). New York: Springer-Verlag, 1982, pp. 209–229.

Dingle, H., Brown, C.K., and Hegmann, J.P.: The nature of genetic variance influencing photoperiodic diapause in a migrant insect, *Oncopeltus fasciatus*. Am. Natur. 111, 1047–1059 (1977).

Eldredge, N., and Gould, S.J.: Punctuated equilibrium: an alternative to phyletic gradualism. In: Models in Paleobiology. Schopf, T.J.M. (ed.). San Francisco: Freeman, Cooper, and Co., 1972, pp. 82–115.

Emlen, J.M.: Age-specificity and ecological theory. Ecology 51, 588–601 (1970).

Falconer, D.S.: Introduction to Quantitative Genetics. London and New York: Longman, 1981.

Fisher, R.A.: The Genetical Theory of Natural Selection. New York: Dover, 1958.

Giesel, J.T.: Reproductive strategies as adaptations to life in temporally heterogeneous environments. Annu. Rev. Ecol. Syst. 7, 57–79 (1976).

Giesel, J.T., Murphy, P., Manlove, M.: An investigation of the effects of temperature on the genetic organization of life history indices in three populations of *Drosophila melanogaster*. In: Evolution and Genetics of Life Histories. Dingle, H., Hegmann, J.P. (eds.). New York, Springer-Verlag, 1982, pp. 189–207.

Giesel, J.T., Zettler, E.E.: Genetic correlations of life historical parameters and certain fitness indices in *Drosophila melanogaster:* r_m, r_s, diet breadth. Oecologia 47, 299–302 (1980).

Gould, S.J.: Ontogeny and Phylogeny. Cambridge: The Balknap Press of Harvard University Press, 1977.

Gould, S.J., and Lewontin, R.C.: The spandrels of San Marco and the Panglossian paradigm: a critique of the adaptationist programme. Proc. Roy. Soc. Lon. B 205, 581–598 (1979).

Hairston, N.G., Munns, W.R.: The timing of copepod diapause as an evolutionary stable strategy. Am. Natur. 123, 733–751 (1984).

Haldane, J.B.S., Jayakar, S.D.: Polymorphism due to selection of varying direction. J. Genet. 58, 237–242 (1963).

Hegmann, J.P., Dingle, H.: Phenotypic and genetic covariance structure in milkweed bug life history traits. In: Evolution and Genetics of Life Histories. Dingle, H., Hegmann, J.P. (eds.). New York: Springer-Verlag, 1982, pp. 177–185.

Holzapfel, C.M., Bradshaw, W.E.: Geography of larval dormancy in the tree-hole mosquito, *Aedes triseriatus* (Say). Can. J. Zool. 59, 1014–1021 (1981).

Istock, C.A.: The evolution of complex life cycle phenomena: an ecological perspective. Evolution 21, 592–605 (1967).

Istock, C. A.: Natural selection and life history variation: theory plus lessons from a mosquito. In: Insect Life History Patterns. Denno, R., Dingle, H. (eds.). New York: Springer-Verlag, 1981, pp. 113–127.

Jinks, J.L., Broadhurst, P.L.: Diallele analysis of litter size and body weight in rats. Heredity 18, 319–336 (1963).

Lande, R.: Genetic variance and phenotypic evolution during allopatric speciation. Am. Natur. 116, 463–479 (1980).

Lande, R.: A quantitative genetic theory of life history evolution. Ecology 63, 607–615 (1982a).

Lande, R.: Elements of a quantitative genetic model of life history evolution. In: Evolution and Genetics of Life Histories. Dingle, H., Hegmann, J.P. (eds.). New York: Springer-Verlag, 1982b, pp. 21–29.

Levene, H.: Genetic equilibrium when more than one ecological niche is available. Am. Natur. 87, 331–333 (1953).

Levins, R., MacArthur, R.: The maintenance of genetic polymorphism in a spatially heterogenous environment: variations on a theme by Howard Levene. Am. Natur. 100, 585–589 (1966).

Lewontin, R.C.: Selection for colonizing ability. In: The Genetics of Colonizing Species. Baker, H.G., Stebbins, G.L. (eds.). New York: Academic Press, 1965, pp. 77–94.

Livdahl, T.P.: Environmental uncertainty and selection for life style delays in opportunistic species. Am. Natur. 113, 835–842 (1979).

Lounibos, L.P., and Bradshaw, W.E.: A second diapause in *Wyeomyia smithii:* seasonal influence and maintenance by photoperiod. Can. J. Zool. 53, 215–221 (1975).

Luckenbill, L.S.: r- and K-selection in experimental populations of *Escherichia coli*. Science 202, 1201–1203 (1978).

Luckenbill, L.S.: Selection and the r/K continuum in experimental populations of protozoa. Am. Natur. 113, 427–437 (1979).

Luckenbill, L.S.: An experimental analysis of a life history theory. Ecology 65, 1170–1184 (1984).

Lynch, M.: The limits to life history evolution in *Daphnia*. Evolution 38, 465–482 (1984).

MacArthur, R.H.: Some generalized theorems of natural selection. Proc. Natl. Acad. Sci. USA 48, 1893–1897 (1962).

MacArthur, R.H., and Wilson, E.O.: The Theory of Island Biogeography. Princteon: Princeton University Press, 1967.

Margalef, R.: Perspectives in Ecolgical Theory. Chicago: University of Chicago Press, 1968.

Mueller, L.D., Ayala, F.J.: Trade-off between r-selection and K-selection in *Drosophila* populations. Proc. Natl. Acad. Sci. USA 78, 1303–1305 (1981).

Murphy, G.I.: Patterns in life history and the environment. Am. Natur. 102, 391–403 (1968).

Odum, E.P.: The strategy of ecosystem development. Science 164, 262–270 (1969).

Perrins, C.M., Jones, P.J.: The inheritance of clutch size in the great tit (*Parus major* L.). Condor 76, 225–229 (1974).

Pianka, E.R.: On r- and K-selection. Am. Natur. 104, 592–597 (1970).

Powell, J.R.: Genetic polymorphism in varied environments. Science 174, 1035–1036 (1971).

Rose, M.R.: Antagonistic pleiotropy, dominance, and genetic variation. Heredity 48, 63–78 (1982).

Rose, M.R.: Theories of life history evolution. Am. Zool. 23, 15–23 (1983).
Rose, M.R., Charlesworth, B.: Genetics of life history in *Drosophila melanogaster*. I. Sib analysis of adult females. Genetics 97, 173–186 (1981a).
Rose, M.R., Charlesworth, B.: Genetics of life history in *Drosophila melanogaster*. II. Exploratory selection experiments. Genetics 97, 187–196 (1981b).
Roughgarden, J.: The evolution of niche width. Am. Natur. 106, 683–718 (1972).
Saunders, D.S.: Insect Clocks. Oxford: Pergamon Press, 1982.
Schaffer, W.M.: Selection for optimal life histories: the effects of age structure. Ecology 55, 291–303 (1974a).
Schaffer, W.M.: Optimal reproductive effort in fluctuating environments. Am. Natur. 108, 783–790 (1974b).
Slobodkin, L.B.: Toward a predictive theory of evolution. In: Population Biology and Evolution. Lewontin, R.C. (ed.). Syracuse, Syracuse University Press, 1968, pp. 187–205.
Spielman, A.: The inheritance of autogeny in the *Culex pipiens* complex of mosquitoes. Am. J. Hyg. 65, 404–425 (1957).
Stearns, S.C.: Life history tactics: a review of the ideas. Q. Rev. Biol. 51, 3–47 (1976).
Stearns, S.C.: The evolution of life history traits: a critique of the theory and a review of the data. Annu. Rev. Ecol. Syst. 8, 145–171 (1977).
Stearns, S.C.: The evolution of life history traits in mosquitofish since their introduction to Hawaii in 1905: Rates of evolution, heritabilities, and developmental plasticities. Am. Zool. 23, 65–75 (1983).
Tuomi, J., Hakala, T., Haukioja, E.: Alternative concepts of reproductive effort, costs of reproduction, and selection in life history evolution. Am. Zool. 23, 25–34 (1983).
Via, S.: The quantitative genetics of polyphagy in an insect herbivore. II. Genetic correlations in larval performance within and among host plants. Evolution 38, 896–905 (1984).
Walter, N.M., Hacker, C.S.: Variation in life table characteristics among three geographic strains of *Culex pipiens quinquifasciatus*. J. Med. Entomol. 11, 541–550 (1974).
Wilbur, H.M., Tinkle, D.W., and Collins, J.P.: Environmental certainty, trophic level, and resource availability in life history evolution. Am. Natur. 108, 805–817 (1974).
Williams, G.C.: Pleiotropy, natural selection, and the evolution of senescence. Evolution 11, 398–411 (1957).
Williamson, P.G.: Morphological stasis and developmental constraint: real problems for neo-Darwinism. Nature 294, 214–215 (1981).

Index

D

N

O

Q

R

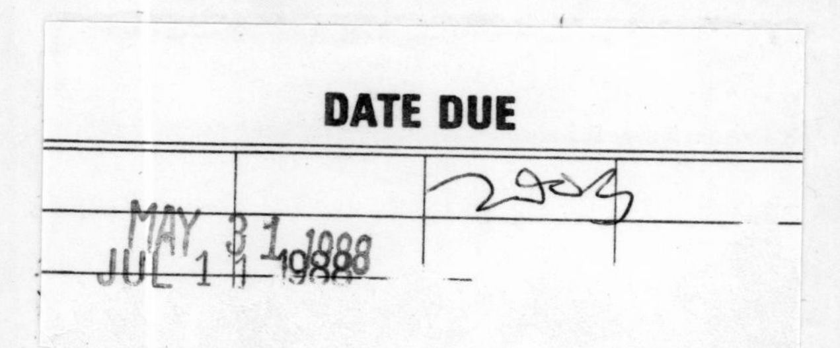
DATE DUE

MAY 31 1988
JUL 11 1988
2003